Lecture Notes in Computer Science 2865

Edited by G. Goos, J. Hartmanis, and J. van Leeuwen

Springer
Berlin
Heidelberg
New York
Hong Kong
London
Milan
Paris
Tokyo

Samuel Pierre   Michel Barbeau
Evangelos Kranakis (Eds.)

Ad-Hoc, Mobile, and Wireless Networks

Second International Conference, ADHOC-NOW 2003
Montreal, Canada, October 8-10, 2003
Proceedings

 Springer

Series Editors

Gerhard Goos, Karlsruhe University, Germany
Juris Hartmanis, Cornell University, NY, USA
Jan van Leeuwen, Utrecht University, The Netherlands

Volume Editors

Samuel Pierre
Ecole Polytechnique de Montreal
Department of Computer Engineering
P.O. Box 6079, Station Centre-Ville, Montreal, Canada, H3C 3A7
E-mail: samuel.pierre@polymtl.ca

Michel Barbeau
Evangelos Kranakis
Carleton University, School of Computer Science
5376 Herzberg Laboratories, 1125 Colonel by Drive
Ottawa, Canada, K1S 5B6
E-mail: {barbeau,kranakis}@scs.carleton.ca

Cataloging-in-Publication Data applied for

A catalog record for this book is available from the Library of Congress.

Bibliographic information published by Die Deutsche Bibliothek
Die Deutsche Bibliothek lists this publication in the Deutsche Nationalbibliografie;
detailed bibliographic data is available in the Internet at <http://dnb.ddb.de>.

CR Subject Classification (1998): C.2, D.4.4, H.4.3, H.5.3, K.4.3

ISSN 0302-9743
ISBN 3-540-20260-9 Springer-Verlag Berlin Heidelberg New York

Springer-Verlag Berlin Heidelberg New York
a member of BertelsmannSpringer Science+Business Media GmbH

http://www.springer.de

© Springer-Verlag Berlin Heidelberg 2003
Printed in Germany

Typesetting: Camera-ready by author, data conversion by Olgun Computergrafik
Printed on acid-free paper   SPIN: 10963562   06/3142   5 4 3 2 1 0

Preface

Ad Hoc Networks are wireless, self-organizing systems formed by co-operating nodes, within communication range of each other which form temporary networks. Their topology is dynamic, decentralized, and ever-changing, and the nodes may move around arbitrarily. The last few years have witnessed a wealth of research ideas on Ad Hoc networks which are moving rapidly into implemented standards.

Mobile computing, particularly wireless-enabled mobile computing, covers a large area of applications in mobile computing environments, networking, communication devices and systems. This conference exposes experimental as well as theoretical research in ad hoc, mobile and wireless networks. The range of topics covered includes management of power consumption, architectures and protocols, quality of service, and security. The aim of the conference was to provide a unique opportunity for researchers and students in industry and academia to participate at an annual forum and share their research results and experiences.

This conference followed the first successful conference (held at the Fields Institute in Toronto during September 20–21 of last year), and was held at the Holiday Inn, Midtown in Montreal during October 8–10, 2003. It was co-sponsored by the *Mobile Computing and Networking Research Laboratory* (LARIM) of the École Polytechnique de Montréal, the *School of Computer Science* (SCS) of Carleton University, MITACS (Mathematics of Information Technology and Complex Systems), and the *Association for Computing Machinery* (ACM).

Forty-two papers were submitted, of which 23 regular and 4 short papers were selected for presentation. All papers were reviewed for technical merit by the program committee. We would like to thank the invited speakers Adrian Perrig (Carnegie Mellon University, USA) and Violet R. Syrotiuk (Arizona State University, USA) for their presentations. Many thanks also go to Khaled Laouamri for helping with the conference logistics, as well as all the following people for their helpful contribution as paper reviewers: Gustavo Alonso, Ronald Beaubrun, Paul Boone, Steven Chamberland, Ali Chamam, Soumaya Cherkaoui, Roch Glitho, Norm Hutchinson, Jeannette Janssen, Mike Just, Danny Krizanc, Thoma Kunz, Peter Marbach, Fabien Nimbona, Paolo Penna, Alejandro Quintero, S.S. Ravi, Daniel Rossier, Sunil Shende, Ivan Stojmenovic, Tao Wan, and Yufei Wu. Special thanks to Amir Ghavam, Jeyanthi Hall and Zheyin Li for publicity, and Mark Vigder for Web site contributions. Finally, we would like to thank all the members of the organizing committee, as well as Raymond Lévesque and Sébastien Lévesque from BCU.

Michel Barbeau
Evangelos Kranakis
Samuel Pierre

Organizing Committee

Conference Co-chairs

Michel Barbeau
Carleton University

Evangelos Kranakis
Carleton University

Samuel Pierre
École Polytechnique de Montréal

Publicity and Tutorials Chair

Alejandro Quintero
École Polytechnique de Montréal

Local Arrangements Chair

Sabine Kébreau
École Polytechnique de Montréal

Program Committee

G. Alonso, ETHZ, Switzerland
M. Barbeau, Carleton University, Canada
R. Beaubrun, Université Laval, Canada
S. Chamberland, École Polytechnique, Canada
S. Cherkaoui, U. de Sherbrooke, Canada
R.H. Glitho, Ericsson Research, Canada
J. Janssen, Dalhousie University, Canada
M. Just, Treasury Board, Canada
N.C. Hutchinson, UBC, Canada
E. Kranakis, Carleton University, Canada
D. Krizanc, Wesleyan University, USA
T. Kunz, Carleton University, Canada
R. Liscano, Mitel Networks, Canada
P. Marbach, U. of Toronto, Canada
L. Narayanan, Concordia U., Canada
I. Nikolaidis , U. of Alberta, Canada
H. Mouftha, Ottawa U., Canada
P. Penna, University of Rome, Italy
S. Pierre, École Polytechnique, Canada
A. Quintero, École Polytechnique, Canada
S. Ravi, SUNY Albany, USA
D. Rossier, Swisscom, Switzerland
S. Shende, Rutgers University, USA
I. Stojmenovic, U. of Ottawa, Canada
S. Tohmé, ENST, France

Keynote Speakers

Adrian Perrig, Canergie Mellon U., USA
Violet R. Syrotiuk, Arizona State U., USA

Tutorials

Ivan Stojmenovic, University of Ottawa, Canada
Ramiro Liscano and Amir Ghavam, University of Ottawa, Canada
Michel Barbeau, Carleton University, Canada

Table of Contents

Space-Time Routing in Ad Hoc Networks*

Henri Dubois-Ferrière[1], Matthias Grossglauser[1], and Martin Vetterli[1,2]

[1] School of Computer and Communication Sciences, EPFL, Lausanne, Switzerland
{Henri.Dubois-Ferriere,Matthias.Grossglauser,Martin.Vetterli}@epfl.ch
[2] Department of EECS, University of California Berkeley, USA

Abstract. We introduce Space-Time Routing (STR), a new approach
to routing in mobile ad hoc networks. In STR, the age of routing state is
considered jointly with the distance to the destination. We give a general
description of STR, which can accommodate various temporal (age) and
spatial (distance) metrics. Our formulation of STR describes a family of
routing algorithms, parameterized by a choice of node clock scheme, a
neighbor-distance function and a binding spatio-temporal metric which
allows the algorithm to compare potential routes taking into account
both their age and their distance to the destination. We discuss possible
instantiations of a Space-Time Routing protocol. In particular, we review
FRESH (FResher Encounter SearcH), a routing algorithm using temporal information only, and GREP (Generalized Route Establishment
Protocol), a routing protocol which uses jointly spatial and temporal information about routes. We discuss a third STR algorithm using only
physical notions of space and time, and finally show that STR provides
loop-free routes.

1 Introduction

An ad hoc network is a communication medium where users or nodes also provide the infrastructure for communication. That is, nodes play both the role of
terminals (i.e. source and destination of messages) and of relays. Thus, a message traverses an ad hoc network by being relayed from node to node, until it
reaches its destination. When, in addition, nodes are moving, this becomes a
challenging task, since the topology of the network is in constant flux. How to
find a destination, how to route to that destination, and how to insure robust
communication in the face of constant topology change are major challenges in
mobile ad hoc networks.

Routing in ad hoc networks is a well studied topic, with a number of proposed
protocols like AODV [1] and DSR [2], as well as simulation studies. A common
point of existing algorithms is that their computations involve almost exclusively
distance (or *spatial*) types of information. This approach can be traced all the
way back to the classic Dijkstra, Bellman-Ford, and Floyd-Warshall algorithms,
which are driven by quantities measuring *distances*[1] between nodes.

---

* The work presented in this paper was supported (in part) by the National Com-
petence Center in Research on Mobile Information and Communication Systems
(NCCR-MICS), a center supported by the Swiss National Science Foundation under
grant number 5005-67322

[1] equivalently, *transmission costs*.

S. Pierre, M. Barbeau, and E. Kranakis (Eds.): ADHOC-NOW 2003, LNCS 2865, pp. 1–11, 2003.

However these spatial routing algorithms were designed with an assumption of static or near-static topologies, where nodes do not move and links change at a slow rate (if at all). In previous work [3], we considered the situation where all nodes are constantly moving, making therefore topology change the norm rather than the exception. In such a scenario, we showed that a routing algorithm that was driven *exclusively by temporal metrics* could significantly outperform spatial approaches. Specifically, we introduced an algorithm named FRESH (FResher Encounter SearcH). Using a simple flood-based search primitive, FRESH advances toward the destination by searching iteratively for a node which has encountered the destination *more recently* than the current node.

FRESH took the extreme approach of using only temporal information in order to demonstrate the value of such information for routing in highly mobile ad hoc networks. However it is clear that spatial information can still be useful, and that ignoring spatial state that exists in the network is highly suboptimal. Now, given that temporal information can increase routing efficiency, and that spatial information remains useful, the question is: Are spatial and temporal approaches incompatible and distinct, or can we design routing algorithms which incorporate seamlessly both aspects?

The purpose of this paper is to answer the above question by introducing a unifying view of routing in highly mobile networks using jointly *both temporal and spatial* information. We call such an approach *Space-Time Routing* (STR).

The central intuition underlying STR is the following. When the rate of topology change increases, the average time during which spatial information remains exact is reduced. For example, a routing entry saying that the destination is reachable from node S in 8 hops through neighbor N becomes inexact if N moves, or if intermediate nodes move such that the number of hops is different than 8. However, even if the routing entry is not perfectly accurate anymore, it can still be helpful. In other words: *aged, inexact routing state is valuable, and incorporating temporal information about the age of routes allows the algorithm to make full use of all available information, including partially outdated routes.* This can be contrasted with spatial-only approaches which are predicated on routing state being exact (since they have no way of 'weighing' the accuracy of aged state). As a result, when spatial algorithms (conceived for mostly-static graphs) are transposed to mobile ad hoc routing protocols, the protocols must be very aggressive in timing out state in order to avoid as far as possible having outdated routes which the protocols are not equipped to handle. For example the default route timeout in AODV [1] is 3 seconds.

Just as spatial routing algorithms can use different distance metrics, STR is amenable to various spatial, temporal, and joint spatio-temporal metrics. Specifically, a particular STR algorithm is defined by the choice of

- physical or logical notion of time,
- a spatial neighbor-distance function $\triangle$, and
- a binding spatio-temporal (S-T) metric f.

Therefore we provide a general formulation of STR which is independent of the specific metric choices. The neighbor-distance metric can be logical (e.g.,

number of hops) or physical (e.g., euclidean distance, energy cost). The binding
S-T metric is used to compare two route entries to a destination of different
distance and age and to decide which is closest in the joint spatio-temporal
space.

The rest of the paper is organized as follows. In Section 2, we give a general
formulation of STR and discuss some properties. In Section 3, we give examples
of two specific STR algorithms: FRESH, GREP, and outline a third algorithm
using physical notions of space and time. In Section 4, we discuss some properties
of STR, including loop-freedom. Section 5 concludes the paper.

2   Space-Time Routing

2.1   Notation and Assumptions

We note $V = \{1 \ldots n\}$ the set of nodes in the network, and E the set of edges
$(i, j) \in E$ for $i, j \in V$. Associated with the set of edges is a distance function[2]
$\triangle : E \to \mathbf{R}$. For example if distance is counted as the number of hops, we
would have $\triangle(i, j) = 1$. We assume that any node can obtain the distance to
its neighbors (trivially in the case of hop-count distance, or for example using a
signal-strength based estimation in the case of euclidean distances). Each node
maintains its own clock, which is used to stamp every packet with the clock time
of the node which originates it. Simple examples of a node clock are a *physical*
(oscillator-based) clock giving a continuous reading, for example in seconds, or
a *logical* clock providing a discrete ordering of routing events relative to that
source. Whichever temporal representation is used, STR does not require any
form of inter-node clock synchronization.

Then STR requires a *binding spatio-temporal metric*, which is a function
$f : \mathbf{R}^2 \to \mathbf{R}$, taking as input a (spatial) distance value and a (temporal) clock
value and returning a scalar representing the "norm" of this pair in the spatio-
temporal space. The binding S-T metric must satisfy the following conditions:

$$argminf(s, t) = (0, 0) \tag{1}$$

For fixed d, f is an increasing function of t

$$sgn(f(d, t_1) - f(d, t_2)) = sgn(t_1 - t_2) \tag{2}$$

For fixed t, f is an increasing function of d

$$sgn(f(d_1, t) - f(d_2, t)) = sgn(d_1 - d_2) \tag{3}$$

Routing Table Entries. Each node maintains a distance-vector routing table
containing one entry for each destination node. In addition to the *next hop* and
distance fields which are used in spatial routing algorithms, STR routing entries
also include the *age* of the entry.

---

[2] Note that given node mobility, E and $\triangle$ vary over time. For simplicity of notation
we drop the time index since we only refer to the values of E and $\triangle$ "at the present
time".

Table 1. Routing table entries.

n_D^N	Next hop to node D in N's routing table.
d_D^N	Distance to node D in N's routing table.
t_D^N	Source Time of the routing entry to D in N's routing table.

Table 1 summarizes the notation used to describe routing state at each node. We drop the superscript and use the notation n_D, d_D, t_D when the context allows doing this unambguously. We use the convention that when a node has no entry for D, $n_D = null$, $d_D = \infty$, and $t_D = \infty$.

Packet Types. Beside regular data packets, STR uses *route request* (RREQ) and *route reply* (RREP) packets. A node sends a RREQ packet when it has no route to the destination, or if the next hop along the route is broken. It sends a RREP packet in reply to a route request when it has a route fresh enough and short enough to satisfy that request. Note that STR does not use any route error packets: since link breaks are always repaired locally, there is no need to inform the source and upstream nodes when this occurs.

Apart from the usual source and destination addresses of a packet, we introduce the following STR-specific fields: The *source time* of a packet ($p.st$) is the clock time at the packet's source node when it originated the packet. Each packet is stamped with the clock time of the source that originates it. This field is present in all packets.

The *source distance* of a packet ($p.sd$) is the distance this packet has traversed since leaving the source. It is updated at each hop to reflect the new distance from the source. This field is present in all packets.

The *destination distance* ($p.dd$) and *destination time* ($p.dt$) of a packet are present only in RREQ and RREP packets. In the case of a RREQ packet, they come from the routing entry that the source of the RREQ has to the requested destination. In the case of a RREP packet, they represent the distance to the requested destination from the replying node.

2.2   STR Algorithm

DATA Processing. We first describe originating and forwarding of data packets. A node S originating a data packet initializes the source distance field to 0 and initializes the source time field to the present value of its clock. A node N receiving from neighbor M a packet originated by node S first increments the source distance field of the packet to reflect the distance that the packet has now traversed: $p.sd \leftarrow p.sd + \triangle(N, M)$.

Then, if the packet has come over a shorter (in the spatio-temporal metric space) route than the route it currently has, N updates its routing entry to S. Formally, if $f(p.sd, p.st) < f(d_S, t_S)$, then N updates its routing entry for S as:

$$d_S \leftarrow p.sd \qquad t_S \leftarrow p.st \qquad n_S \leftarrow M.$$

If the destination D of the packet is N itself, then no further processing is needed. If the destination is another node, N forwards the packet to its next hop n_D.

If $n_D = null$, or if forwarding fails, N buffers the packet and initiates a route request procedure.

RREQ Processing. A node N initiating a route request procedure for destination D sets the source distance and source time fields as for a DATA packet. The destination distance and destination time fields on the packet are set respectively with the values d_D and t_D from N's routing table (with a suitable encoding when $d_D = \infty$ and $t_D = \infty$).

A node N receiving from neighbor M a RREQ packet originated by S increments the source distance of the packet and (possibly) updates its routing entry for S according to the same procedure as for a DATA packet. N then verifies if the spatio-temporal distance of its routing entry to D is smaller than the sum of S's spatio-temporal distance to D and the distance traveled by the packet, and originates a RREP to M if this is true.

Formally, if $f(d_D, t_D) < f(p.dd, p.dt) + f(p.sd, 0)$, then N initiates a RREP packet to S. Otherwise N re-broadcasts the RREQ packet (RREQ floods will be scoped with a time-to-live (TTL) field; we omit the details).

RREP Processing. A node R originating a RREP packet sets the source distance and source time fields as for a DATA packet. The destination distance and destination time fields are set respectively with the values d_D and t_D from N's routing table.

We consider now a node N receiving from neighbor M a RREP packet originated by R, in response to a route request for a route to node O (if $R = O$ then the route reply was initiated by the destination itself, otherwise we say that R sent a route reply *on-behalf-of* O. N first increments the source distance of the packet and (possibly) updates its routing entry for R following the same procedure as for a DATA packet. Then, N updates its routing entry to O, if this will result in a shorter route (in the spatio-temporal metric space).

Formally, if $f(d_O, t_O) < f(p.dd, p.dt) + f(p.sd, 0)$, then N updates its routing table as:

$$d_O \leftarrow p.sd + p.dd \qquad t_O \leftarrow p.st \qquad n_O \leftarrow M.$$

Then, if N has updated its routing entry to O, it forwards the RREP packet toward the originator of the route request (as determined by the destination field in the RREP packet). Otherwise N silently discards the RREP packet. When the RREP arrives at the node which originated the route request, this node can now forward its buffered DATA packets along the newly established route.

2.3   Discussion

Protocol-Specific Optimizations. The above exposition of STR is voluntarily simple and does not include possible optimizations that would be present in a full, practical protocol. As a first example of optimizations that might be present in a full protocol specification, a node receiving (or overhearing) any packet from a neighbor can update its routing entry to have a current, one-hop route to that neighbor. A second example pertains to route requests which in practice will be scoped using a TTL mechanism, and will likely proceed according to an expanding ring search. Expanding ring searches are used in many ad hoc routing protocols; the specifics of this procedure are omitted here. We refer to [4] as an example of a complete, practical STR protocol formulation.

A final example relates to proactive operation of STR. The formulation given here is *purely reactive*, meaning that a route is only computed when it is required to send packets. However STR can also accommodate proactive, or hybrid proactive/reactive operation, whereby nodes proactively inform other nodes of some or all of routing entries. This is done using *route advertisement* packets, and a route update decision mechanism similar to that used when receiving a regular packet: if an advertised route is shorter in the S-T space than the one in the receiving node's routing table, then it overrides the existing route. We refer to [4] for a simple example of route advertisement operation in a STR protocol. Many schemes for controlling the proactive dissemination of routing information are possible. For example, [5] explore schemes to adjust the relative amount of reactive and proactive overhead. With STR another possibility would be to define a threshold value ω such that a node N proactively disseminates only the routing entries which satisfy $f(d_D, t_D) < \omega$.

On Explicit versus Implicit Use of f. We remark that in this exposition of STR, the actual value of the spatio-temporal distance metric need not be explicitly computed. Specifically, the STR algorithm only uses the S-T metric to compare two values, in order to decide if $f(d_1, t_1)$ is smaller or greater than $f(d_2, t_2)$. This has two consequences. The first is that two S-T metrics will result in identical STR protocols, if they induce the same ordering. For example all functions in the one-parameter family $f_k(d, t) = k(d + t)$ will result in the same ordering for any choice of $k > 0$. The second consequence is that the function f need not be explicitly defined. For example, in the case of the GREP protocol (see Sect. 3), the 'natural' presentation does not define f explicitly, though of course a function f resulting in the equivalent ordering can be defined.

3   Three Instances of STR

We now give three instances of specific STR algorithms which provide a sample of the wide range of protocols that fit under the STR umbrella.

3.1   FRESH: FResher Encounter SearcH

FRESH [3] is a simple route discovery algorithm using exclusively temporal information. Nodes keep a record of their most recent encounter times with other nodes. Instead of searching for the destination, the source node searches for any intermediate node that encountered the destination *more recently than did the source node itself*. The intermediate node then searches for a node that encountered the destination yet more recently, and the procedure iterates until the destination is reached. Therefore, FRESH replaces the single network-wide search of current proposals with a succession of smaller searches, resulting in a cheaper route discovery.

The formulation originally employed in [3] was a direct algorithmic transposition of the above description, using at each iteration of an underlying *search primitive* which roughly corresponded to the flooding and reverse-path setup phase of STR.

We now show that FRESH can be expressed as a STR algorithm. First, FRESH uses physical time, so packets are stamped with the clock time of the node which originates them. Second, the spatial distance is measured in hops: $\triangle_{FRESH}(i,j) = 1$. Finally, the binding S-T metric ignores all spatial information and only compares one-hop encounter times:

$$f_{FRESH}(d,t) = \begin{cases} \infty & \text{if } d > 1; \\ t & \text{if } d \leq 1. \end{cases}$$

We note that the function f_{FRESH} alone, when inserted into the STR description of Sect. 2 does not result in the exact FRESH algorithm [3]. This would require distinguishing in STR two binding spatio-temporal metrics, one of which (corresponding to this f_{FRESH}) to be used in deciding whether a node's route is suitable to answer a route request, the other to be used in deciding whether to update the reverse-path entry to the source of an incoming packet.

3.2 GREP: Generalized Route Establishment Protocol

GREP [4] is a complete, practical routing protocol which demonstrated that a protocol incorporating both spatial and temporal metrics was not only feasible but also highly efficient compared to spatial-only approaches. Though the original proposal for GREP predates the general formulation of STR given in this paper, we now show how GREP can be defined as an instance of STR.

First, GREP uses logical clocks, similar to Lamport's clocks [6]. Each node maintains its own integer-valued clock, and increments it each time it transmits a packet. Packets are stamped with the logical clock time of the node which originates them, and are therefore similar to *sequence numbers* as used in many routing protocols [1]. As in FRESH, neighbor distances are measured in hops: $\triangle_{GREP}(i,j) = 1$.

In the original proposal for GREP, the binding S-T metric was not explicitly computed. Rather, the ordering between two (s,t) pairs was obtained as:

$(s_1,t_1) < (s_2,t_2)$ if $t_1 < t_2$ or $((t_1 = t_2)$ and $s_1 < s_2)$.

We believe that the above formulation is the most expressive for GREP since it captures the notion that GREP advances (in spatial mode) along a route segment of particular age, then switches to temporal mode if the next hop along this route is broken, and looks for any route of younger age. However, as with any STR algorithm, it is possible to explicitly define the binding S-T metric of GREP:

$$f_{GREP}(d,t) = d + k * t$$

where k is a suitably large constant (for example $k = |V|$).

3.3   STR with Physical Space and Time

Our third example of STR is based on a physical representation of both spatial and temporal distances, and illustrates how a priori knowledge of the mobility process may be exploited in designing the binding metric f.

We note X_n the euclidean position of node n. Now $\triangle$ measures euclidean distance between neighbors: $\triangle(i,j) = \|X_i - X_j\|$. Node clocks measure physical time as for FRESH.

The binding metric in this case has the form

$$f(d,t) = d + cvt^\alpha.$$

Note that the unit of v here is [m/s]. One possible choice would be to set v to the average velocity of nodes, in order to have the S-T metric reflect a quantity related to the expected present distance to a particular node. A suitable choice for the parameter α would depend on the mobility process assumed. For example, in a waypoint model, nodes traverse a distance which is proportional to time elapsed, which would indicate the choice $\alpha = 1$. Or with a random walk, one may choose $\alpha = 1/2$, since the time taken to traverse a distance d is $O(d^2)$.

4   Analysis and Properties

Loop Freedom. In this section we show that STR is free of routing loops. We distinguish between *packet loops* and *route loops*. A packet loop happens when a unicast packet traverses the same node twice. A route loop happens when a unicast packet traverses the same node twice, and the routing state pertaining to the packet's destination at that node does not change between both traversals. A route loop is potentially infinite (unless some mechanism is used to kill packets which have traversed more than some number of hops). In other words a packet gets "stuck" in a route loop but not in a packet loop, since the routing state has changed when it traverses the same node for the second time.

The route loop-free operation of STR comes from a simple observation: *At each hop, a packet advances to a node which is closer to the destination in the spatio-temporal metric space..* This can be stated equivalently in terms of node routing tables:

Lemma 1. *If $n_D^N = M$ then $f(d_D^M, t_D^M) < f(d_D^N, t_D^N)$.*

The proof of this lemma follows from the fact that at each step in the protocol operation, (see Sect. 2) a new routing entry can only override an existing one if it offers a shorter $f(d,t)$ S-T distance to the destination.

We can now show that STR is route loop-free:

Theorem 1. *A packet routed by STR can never enter a route loop.*

Proof. Let us assume that a packet p is in a route loop, that is that it traverses node N twice with t_D^N, n_D^N, and d_D^N having identical values at both times. This is immediately contradicted by Lemma 1 since at each hop the packet advances in the spatio-temporal metric space.

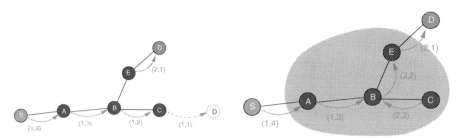

Fig. 1. On the left side: A network with a route from S to D having sequence number 1. D has moved, breaking the last hop. A packet sent by S to D is buffered at C while C sends a route request. On the right side: C's route request is answered by D, resulting in a new route with sequence number 2. The packet for D buffered at C can now be forwarded back through B, resulting in a packet loop ($S - A - B - C - B - E - D$). A packet loop only occurs once; subsequent packets from S to D will be routed by B directly to E.

On Packet Loops. We have shown above that STR routes are loop-free. Our analysis has made the distinction between *packet loops* and *route loops*. This distinction is usually not made in the analysis of routing protocols because they often prove loop freedom by showing that a packet will not traverse the same node twice; therefore both packet and route loops are excluded.

STR, on the other hand, excludes route loops but does not exclude packet loops. Therefore it offers a weaker guarantee than protocols such as [1] [2] which establish routes on an end-to-end basis and require a route to be converged before sending packets. This weakened guarantee can been seen as a consequence of STR's distributed hop-by-hop operation which uses only local repairs without involving the end-points of a route. On the other hand, relaxing protocol guarantees to allow packet loops allows an increase in efficiency which make this particularly worthwhile in highly mobile networks.

We discuss a small example of a packet loop showing that even when a packet loop does occur, subsequent packets will shortcut the loop and therefore packet loops cannot happen on back-to-back packets. In Fig.1, there is a route from S to D, which might have been established by a packet sent earlier from D to S with sequence number 1. D has since moved and therefore the last hop of this route is broken.

This example shows an instance of a packet loop since the data packet traverses node B twice. Note that this is not a *route loop* since B's routing entry for destination D has changed between the first and second traversals, and therefore the packet does not get "stuck" in a loop between B and C. Note also that subsequent packets for D will now be forwarded by B to E; *each instance of a packet loop can only occur once*.

Exploiting Outdated Routing State. Most ad hoc routing protocols attach some notion of *useful lifetime* to their routing state. Typically each route (or individual routing entry) is expired when it remains unused past a certain timeout (3 seconds in AODV for example).

In GREP *routing state never times out*: a routing entry can only be deleted when a newer entry overrides it. This is because a past route, which was often acquired at a high flooding cost, can still carry noisy, but useful information about the present topology.

We consider two simple scenarios to illustrate this. The first scenario is straightforward and concerns a short-term timescale. Consider a route which has been left unused for some time. In this time, one of the nodes in the route has moved. Clearly, timing out the whole route at this point would impose a costly re-discovery if the route is needed again; if we keep all routing entries only a small, local repair is necessary.

In the second scenario we consider a long-term timescale, on the order of the time required for nodes to traverse the whole area that they inhabit. One intuition might be that routing state has no value at this timescale, since the current topology has no relationship with the topology at the time when the route was established. However, this timescale is precisely the one considered in FRESH [3] where we have shown that one-hop routing entries, however old, can be used to constrain new route discoveries and significantly decrease the flooding overhead.

Hop-by-Hop Routing. Routing protocols typically view a route as a consistent end-to-end structure. In this model a route must be set up and converged from source to destination before data can flow across it. Clearly this is well-suited (and has been proven) to wired networks, where topology changes are rare events. For more dynamic networks, where change is the norm rather than exception, the brittle nature of this model can degrade performance. For example, a single link break can bring down an entire route, even when most of the route remains valid. As networks grow larger and routes get longer, the probability of a link break along a route increases, and the amount of time when a route is available correspondingly goes down. This reduces the overall throughput available to an application.

GREP does away with the notion of end-to-end routes and views routes as distributed structures which continuously adapt to change rather than be entirely torn down and rebuild from scratch at each topology change. In this hop-by-hop routing approach a source does not have to stop sending when a link changes along the route to the destination; in fact it is in most cases not even aware that a local repair happened further along the route.

We should also note that the exclusive use of local repairs has one drawback, namely this may result in suboptimal routes that are longer than the shortest possible path. Though our simulation results show that this is effect is not severe enough to damage GREP's performance, we believe that a worthwhile extension to GREP will allow a node to progressively 'shorten' a suboptimal route as a session goes on.

5 Conclusions

In this paper we have introduced a new approach to routing in mobile ad hoc networks, which we call *space-time routing* (STR). This approach uses both spatial and temporal distance information to determine which routing entries can be used to advance a packet towards its destination; it can be contrasted with existing protocols, which are grounded in the classic Dijkstra or Bellman-Ford algorithms and use only spatial information. We have given a general formulation and discussed possible instances of STR, including FRESH [3] and GREP [4], and discussed STR properties and loop-freedom.

We believe that STR offers many opportunities for future research. One direction of future investigation consists in exploring other STR instances, such as those outlined in Sect. 3.3. Another concerns the design of schemes to allow progressive optimization of routes computed by STR, since STR may provide routes of suboptimal length. Finally, we will consider STR in context where topology change occurs as a result of dynamics other than node mobility. For example, nodes in a sensor network usually have static positions, but topology may be dynamic as a result of duty cycling.

References

1. Perkins, C.E., Belding-Royer, E.M., Das, S.R.: Ad hoc on demand distance vector (aodv) routing. IETF, Internet-Draft (2003)
2. Johnson, D.B., Maltz, D.A., Hu, Y.C., Jetcheva, J.G.: The dynamic source routing protocol for mobile ad hoc networks (dsr). IETF, Internet-Draft (2003)
3. Dubois-Ferriere, H., Grossglauser, M., Vetterli, M.: Age matters: Efficient route discovery in mobile ad hoc networks using last encounter ages. In: Proceedings of The Fourth ACM International Symposium on Mobile Ad Hoc Networking and Computing, Annapolis, MD (2003)
4. Dubois-Ferriere, H., Grossglauser, M., Vetterli, M.: Generalized route establishment protocol (grep): Proof of loop-free operation. In: EPFL Technical Report IC/2003/40. (2003)
5. Boppana, R.V., Konduru, S.: An adaptive distance vector routing algorithm for mobile, ad hoc networks. In: Proceedings of the Twentieth Annual Joint Conference of the IEEE Computer and communications Societies. (2001) 1753–1762
6. Lamport, L.: Time, clocks and the ordering of events in a distributed system. In: Communications of the ACM. (1978)

SAFAR: An Adaptive Bandwidth-Efficient Routing Protocol for Mobile Ad Hoc Networks

Jigar Doshi and Prahlad Kilambi

Sri Venkateswara College of Engineering
University of Madras
Pennalur, Sriperumbudur 602 105
jigar@doshi.com, prahlad@acm.org

Abstract. A mobile ad hoc network suffers from the same cost constraints as most wireless networks. In particular bandwidth constraints of wireless links are severe. We present a scalable adaptive fitness-based routing protocol, SAFAR, for mobile ad hoc networks in which we try to optimize the usage of this bandwidth at every stage. The protocol is hybrid, i.e. it makes use of both proactive and reactive procedures for routing in an attempt to reduce route acquisition latency. Using a fitness function, a node decides how many other nodes can be proactively maintained by it. Each node tries to know the best(most fit) nodes in its neighborhood. Hence, high bandwidth nodes are well known. Most of the traffic is routed through these nodes and hence performance is optimized. We present simulation results to substantiate the protocols performance. We also extend this protocol to show how it can be used for power aware routing.

Keywords: MANETs, Hybrid, Adaptive, Fitness.

1   Introduction

Mobile Ad hoc Networks (MANETs) are multi-hop wireless infrastructure less networks. All nodes are capable of movement and can be connected dynamically in an arbitrary manner. Nodes in these networks function as routers that discover and maintain routes to other nodes in the network. Applications of ad-hoc networks have been widely studied and they find extensive use in emergency services. Mobile ad hoc networking can support robust and efficient operation in mobile wireless networks by incorporating routing functionality into mobile nodes. The topology in such networks is dynamic and sometimes rapidly changing. Therefore a protocol for such a network has to be robust as well as efficient.

Mobile networks have many interesting characteristics, which differ from traditional wired networks as described in detail in [1]. In particular Wireless links will have significantly lower capacity than their hardwired counterparts. The throughput of wireless communications has to take into account the effects of multiple access, fading, noise, and interference conditions, etc. Therefore, transmission rate is bound to suffer [1]. One effect of the relatively low to moderate link capacities is that congestion is typically the norm rather than the exception.

S. Pierre, M. Barbeau, and E. Kranakis (Eds.): ADHOC-NOW 2003, LNCS 2865, pp. 12–24, 2003.

Existing routing protocols can be classified into two - proactive routing protocols and reactive routing protocols. Proactive routing protocols in general have not been favored in ad hoc networks because of the volume of routing information exchange (overhead) in a volatile environment. Proactive protocols such as [2] are limited in their application by this overhead. Reactive protocols on the other hand, aim to solve this problem by discovering routes as and when necessary in an on-demand fashion. However these suffer from high route acquisition latencies Data packets have to wait while a route to the destination is found. Reactive protocols such as [3,4], and [5] also cause excessive network traffic when the number of routes required is more. A few hybrid protocols like [6] which, try to combine the best of both worlds, have also been proposed. But all of them are homogenous in their view of an ad hoc network. Nodes of varying bandwidth and power characterize ad hoc networks [1] and these parameters themselves vary over time. Bandwidth, hence, becomes a prime factor of optimization for an ad hoc routing protocol. They do not take into account this diversity in bandwidth and power capacity.

We present a hybrid protocol, Scalable Adaptive Fitness-based Ad hoc Routing (SAFAR), which maintains a restricted active view of the surroundings and uses route discovery for nodes, which are not in the active routing neighborhood. SAFAR is essentially a hybrid protocol, which routes based on the concept of 'node fitness'. Each node is assigned a fitness value, which can be its bandwidth, power or a cost metric (like the weighted average of the bandwidth and power). The protocol then uses the node's fitness to decide its role in routing and the extent of its proactive nature. Thus the protocol can dynamically adjust to network characteristics. As the node's fitness changes so does its role in routing.

2   Related Work

There have been two main approaches to hybrid routing. One approach is to use node election to elect a landmark for a zone. This approach is used in [7]. The landmarks are then proactively maintained. The other approach involves forming overlapping zones like those used in [6] with each node maintaining a proactive zone thus distributing the work of the landmark leader. The disadvantage of the first approach is that the election of the landmark may consume resources and may have to be repeated as topology changes, thus significantly increasing overhead. In the case of [6], its performance depends on the selection of the zone radius, which cannot be done dynamically but instead is set by some administrative means. Unlike [7] our protocol does not use leaders. And unlike [6] our protocol does not use a statically set zone radius. The extent of proactive routing is wholly dynamic.

Many power efficient routing schemes have been proposed as in [8,9]. However these schemes mainly optimize transmission power by using longer routes. Such protocols are again not dynamic and cannot be adapted to topologies where performance and low latency are the overriding concerns.

[10] introduces an adaptive hybrid protocol. Our protocol is similar to [10] in that we dynamically adjust the proactive region of the protocol. However

our protocol differs from [10] in various ways. In [10] the proactive region is uniform depending on the zone radius. [10] adjusts the zone radius dynamically. In our protocol the zone is not uniform. Each node selectively adds only the best (most fit) nodes in its neighborhood to its proactive region. In [10], the adjustment of the zone is based on an approximation cost model. [10] adjusts the proactive region in order to make a node more accessible. We change the proactive region in order to reduce route acquisition latencies. Unlike [10], we use the concept of FITNESS(a Genetic Algorithm-based technique) to determine the node's participation in proactive routing. This yields a more realistic proactive region as it takes into account the changing environment of a node.

This paper is organized as follows. In section 3, we give an overview of the proposed protocol, bringing out its salient features. Section 4 contains the complete functional description of the protocol. Section 5 contains simulation results where we investigate performance of our protocol and the issues of scalability and route acquisition latencies. In section 6 we investigate adapting the protocol for power oriented routing.

3   Overview of SAFAR

Our protocol uses a fitness-based routing table buildup scheme. The fitness of a node is based on node characteristics like power or bandwidth value and can change over time. We query nodes, which have a "good" idea about its surroundings. This minimizes the overhead to maintain the routes when compared to proactive protocols and other hybrid protocols. In comparison to purely reactive schemes, our protocol reduces route acquisition latency. Existing protocols are not adaptive in terms of overhead. The disadvantage of a non-adaptive approach is that the diversity of the nodes is completely ignored. A node with low bandwidth (or power - in battery time left) cannot afford the overhead involved in having actively maintained routes. But a node with good bandwidth, which can afford this, helps boost its own performance as well as that of its neighborhood. This is an example of an adaptive "overhead" scheme.

For routing we use a two-stage approach with a limited discovery scheme in the first stage. There is a very high probability of finding a route in the first stage itself. Thus the overhead of a reactive route discovery stage is avoided. If the route is not found in the first stage then we use a route discovery procedure similar to the on demand protocols. In our protocol, data is routed be-tween two mobile nodes. We do not explore multicast operation of SAFAR. Each mobile node maintains a routing table. The routing table contains the node's address, next-hop address, bandwidth (in kbps) and power(in battery minutes remaining). Ancillary information like neighbor and update lists, required by the protocol, need to be maintained separately or as part of the routing table itself. Each node also maintains a node heap which is a max heap maintained in terms of a cost value (power or bandwidth). This heap is used during the table buildup procedure explained in 4.2. Power optimizations will be explained in section 6.

4   Protocol Details

The protocol has four main subdivisions. In section 4.2, we explain the neighbor discovery mechanism, which leads to the table buildup procedure(Section 4.3). In section 4.4, we describe how routing is carried out and the route discovery procedure. Section 4.5 describes how routes already acquired are maintained.

Assumptions:
- Every node has information about its own bandwidth in terms of link capacity from the link layer.
- Every node has information about its own power at any point of time. This information is available in terms of battery time remaining.
- All links are bi-directional

4.1   Fitness Function

Our protocol uses the concept of fitness function applied extensively in [11]. We use this because each node exists in a diverse environment (nodes have differing bandwidth and power attributes). Thus, it draws a parallel to genes of differing fitness value existing in the population of a Genetic Algorithm(GA). GAs use fitness functions to determine which members from the population are to be selected.

The fitness function, is given by, $F = \frac{\delta}{2\Delta}$ where, δ is node's cost metric and Δ is average cost metric of surrounding network environment. The choice of δ ,can vary in a volatile ad hoc environment. In most cases where neither bandwidth nor power is an overbearing concern, the most versatile choice would be the weighted average of bandwidth and power, $\delta = \frac{\alpha B + \beta P}{\alpha + \beta}$ where, $\alpha, \beta > 0$, B is the node's own bandwidth(in kbps) and P is the node's remaining power (in battery time left). But in certain cases, when bandwidth(or power) alone may require optimum usage,β (or α) could be set to 0. For the remainder of the paper, we describe the protocol behavior in terms of optimizing bandwidth alone ($\alpha = 1$ and $\beta = 0$). The choice of Δ is obvious and is given by $\Delta = \frac{\delta}{n}$ where, n = no. of nodes in the immediate vicinity of the current node. A more effective choice for Δ might be the average of averages from different environments. This may be given by, $\Delta = \sum_{k=1}^{N} \frac{\Delta_k}{N}$ where, Δ_k is the average of node cost metric δ for environment k, and N is the number of environments considered.

4.2   Neighbor Discovery

Initially when a node (say A with bandwidth 1500 kbps) joins a network (Fig. 1), it sends a hello packet to its neighbors by means of a broadcast as it does not know their addresses. The hello packet carries A's address and cost metric(bandwidth) value.

A's neighbors (B, C, D, E, F), on receiving this hello packet, update their routing tables, adding A as a neighbor. They then reply, with a hello reply packet that reports their address and bandwidth value.They also have the option of preventing A from actively maintaining information on them using a blocking bit(explained subsequently). For example in Fig. 1, node F sets the blocking bit and hence prevents A from querying it although it has high fitness.

Node A, now adds all the neighbors who responded into its neighbor list and those with blocking bit not set to the node heap. Now the node begins the table buildup procedure. Node A, for the remainder of its lifetime, keeps polling its neighborhood with hello messages to stay informed of its neighbors.

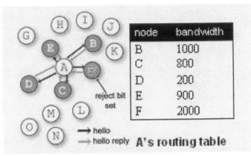

Use of the Blocking bit: This bit may be set if a node finds the ratio of routing information to data information exceeds some threshold. This safeguards the proactive routing over-

Fig. 1. Node A's immediate neighborhood on its entry into the network

head of the protocol and hence a node with high fitness is prevented from being overloaded (Node F in Fig. 1). The blocking bit is also set when the environment around that node is volatile.This forces nodes around it to switch over to reactive routing, which performs better in such scenarios. In a stable environment, nodes will be encouraged to use active information. Thus the protocol adapts to changing environment characteristics.

4.3   Routing Table Buildup

APPLICATION OF FITNESS FUNCTION

We extend the concept of fitness function, discussed earlier, to MANETs. Each node uses the fitness function to find its role in the environment and how many nodes it can query. If it has higher fitness, it has to assume a role of facilitator and allow less fit nodes to communicate. This function F(fitness) is now used to select the number of nodes to be queried (M) from subsequent hop nodes using the formula,

$$M = \texttt{min}(\texttt{round}(F \times \exp^{-i^2} \times n), n1) \tag{1}$$

where, i is the iteration number (0,1,...,n), n is the sum of number of hop (i+1) nodes to be queried and number of nodes not selected up to i iterations and n1 is the number of nodes currently in heap. In Fig.2, Node A selects B, C and E based on this fitness function as they have sufficiently high bandwidth. It then starts the next iteration of the table buildup by querying these selected nodes. This is explained in the next paragraph(Querying Fit nodes). The exponential fall-off in the order of 2 is used because it becomes that much more expensive to maintain nodes proactively with increasing hop radius. This is multiplied by n, as n is dynamic and varies from hop to hop.Thus the fitness is effectively 'scaled' [11].

Choice of fitness function: The ratio F(fitness), remains fairly constant throughout the table buildup procedure. Hence, it makes for a stable fitness function. The node is likely to query half the nodes if its bandwidth matches the average.

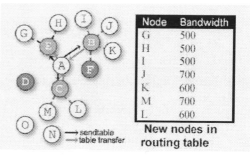

Node	Bandwidth
G	500
H	500
I	500
J	700
K	600
M	700
L	600

New nodes in routing table

Fig. 2. Node A, relaying a send-table message to nodes B,C and E

QUERYING FIT NODES

This involves querying the selected nodes once the fitness function has been applied. In Fig. 2, Node A queries nodes B and C with a send table message. This message is a request to these nodes to return their neighbor lists back to the source node A. B and C are added to the proactive list(the proactive field in the routing table is set to true). As illustrated in Fig. 2, nodes B, C and E, on receiving a send table message from node A, generate a transfer table packet which reports address and bandwidth information of its neighbors (I,J,K,L,M,G,H respectively) back to A. It also adds A to its update list (i.e. the update field in the routing table is set to true). This list is explained in section 4.4. Node A, on receiving a table transfer message, updates its routing table and its heap with the newly found nodes (I, J, K, L, M, G and H).These nodes will be considered in the next iteration of fitness function application along with the rejected nodes from the previous query (D). The node A calculates average bandwidth again using new data in its routing table and recalculates number of nodes to query. As shown in Fig. 3, nodes J and M are selected.

It has to be noted that the M nodes can be chosen from any hop on the basis of maximum bandwidth value. Thus the radius of the active neighborhood is not maintained uniformly. The procedure of querying fit nodes is applied again on J and M (Fig. 3). At this stage the number of nodes to query becomes zero and hence the table build-up procedure ends.

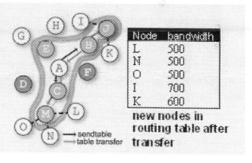

Node	bandwidth
L	500
N	500
O	500
I	700
K	600

new nodes in routing table after transfer

Fig. 3. Formation of the proactive zone

4.4   Routing

When a node requires a route to the destination (say X), it starts a route discovery procedure, which follows a two-stage mechanism as given below.

TYPE-1 MESSAGING

The node, in need of the route, sends a type-1 message to everyone in its proactive list. If anyone of these queried nodes has a path to the destination,it responds with a Type-1 reply packet. This packet also reports the maximum cost(lowest bandwidth) link along the route it has to the destination (it would be able to calculate this from its own routing table).

The first Type-1 reply for the destination X, triggers an entry being added for the destination in the source node's routing table with the responding node's address set as the next hop. For subsequent Type-1 replies for the same destination, the response with a higher bandwidth is chosen and the routing table is updated appropriately. An intermediate node, which receives a Type-1 reply, adds an entry for destination X to its routing table before forwarding the packet to the source A. In the event of no response from any node(within a timeout period), route discovery proceeds to the next stage, otherwise, it ends here.

Type-2 messaging (This is an on demand route discovery, following the same pattern as [5] or [6]).
When the members of the proactive list do not respond positively, a Type-2 request is created. The Type-2 request packet contains a request-ID, a bandwidth field and a node list. The source adds its address to this list and broadcasts the message. This prompts a route discovery among the nodes that receive this message. Every intermediate node G(say), that receives this message checks the source-destination-request-ID tuple to see if it has already handled this packet. If it has, it discards the packet, otherwise, it adds this tuple to its list. This also prevents loop formation.
• If G does not have knowledge of X, it adds its own address to the node list of the Type-2 message, compares its bandwidth to the bandwidth field of the packet and sets the bandwidth field of the packet to the lower of the two values. This helps to maintain the minimum bandwidth link along the path so that the source node can choose the maximum least bandwidth reply in case of multiple responses for destination X. This is because the minimum bandwidth along a path is its bottleneck. Node G, now forwards the packet to its neighbors.
• If G has knowledge of X, it simply reverses the node list of the Type-2 request and adds it to the Type-2 reply message that it generates. It also copies the value of the bandwidth field from Type-2-request to the reply. It then forwards this message to the first node on the node list. The source just adds an entry to its routing table choosing the least cost route if it has multiple routes.

In case there are multiple replies for the same destination, the source chooses the response with the highest value in the bandwidth field as explained earlier. The source then transmits the data packets based on this entry. In case there is a change in the network configuration over this period and the route no longer exists,the node that is not able to find the destination(say Z) in its routing table, upon receipt of the data packet, returns a "route error" message to the source. This is likely to occur if the destination leaves the routing table of Z before the arrival of the data packet but after the type-2 reply has been dispatched.

Advantage of the two-stage approach: The probability of finding the destination node using a Type-1 message is high as the members of the proactive list have a high bandwidth. Hence, they are likely to be well known to others. Therefore, in the average case, the route is likely to be found at the first stage itself. Even if it is not found at this stage, the overhead is negligible compared to the gain in performance. It has to be noted that although the node exchanges neighbor information when it is queried during the table buildup procedure, it does not exchange the complete routing table. This is why a Type-1 message is required. Exchanging the complete table would cause problems during update because of chained updates. Also, Type-2 relies on source routing in its reply. Hence, the routing overhead is reduced.

4.5   Route Maintenance

Proactive region maintenance

If a node A leaves the neighborhood of node X, it is removed from both the neighbor list and routing heap and all entries with A as the next hop are removed. If a node B sends X a send table packet, then X adds B to its update list before sending A its neighbor list. In case of any change in the immediate topology of X, i.e. a neighbor being added or removed, an update message will be sent to all nodes in the update list(including B). Thus, the proactive routes are maintained.

In a mobile scenario, where the topology is bound to change rapidly, nodes will be continuously added and deleted from the proactive zone. This presents a challenge because we take the trouble to dynamically buildup our "proactive zone". Hence we describe a zone replacement strategy, as follows: The node X keeps monitoring the status of its routing table and the number of nodes proactively maintained. Assume, initially the number of nodes in the proactive list is n. Over time, due to addition and deletion of nodes, n may fall below a certain threshold. In such a case, X deletes nodes from the heap and applies the table buildup procedure again to rebuild its proactive list from scratch. This threshold value is calculated as follows: The node X keeps querying neighbors with hello messages and also updates its lists when it gets an update packet. When the total number of proactively maintained nodes drops below 40% of the initial value, the node may begin the table buildup again. The threshold can be tweaked according the environment in concern. In a highly mobile environment, the threshold may be decreased because repeatedly applying the table buildup procedure will be expensive while in an environment which is expected to be stable the threshold value can be increased. The whole buildup procedure is repeated instead of incremental updates because as all the nodes are mobile, it is assumed that the node will find itself in a totally "alien" environment after some time.

Handling Route Errors

If a node forwarding a data packet finds out that there is no route to destination then it tries a local route repair. This differs from traditional reactive schemes because we resort to the Type 1 discovery mechanism. If the Type-1 discovery does not yield a route to the destination, we return a route-error message to the source. In the very unlikely case, that we do not have a route to the source itself, we drop the packet.

5   Simulation

We simulate SAFAR under varying conditions of a mobile ad hoc network and show that it achieves its objectives of being highly adaptive. We also evince how SAFAR routes data packets expeditiously whilst remaining bandwidth efficient. The simulator is multi threaded and has been developed in C++. The simulator is, essentially, real time in nature and uses preprocessing of mobility information to relieve some burden off its operation.

The number of nodes in our simulation has been fixed at 50 to replicate the scenarios in [12]. The initial position of nodes is chosen from a uniform random distribution over an area of [670mx670m]. The simulation itself runs for a period of 900 seconds. Every cycle in the simulation corresponds to 1/10 second. The node movement follows the random way point model of mobility.The nodes are assumed to be moving with constant velocity. They move from one location to another by setting an initial velocity, which is maintained throughout their movement. Once they reach this location the mobile host pauses for a random amount of time before moving to another destination. Within the next few time cycles, a change of direction occurs.

Each node is assumed to be equipped with a transceiver whose transmission range (or radius) T_R can be varied. It is assumed that all transmissions within this radius are reliable and have a specified channel error rate. Using the transmission radius we calculate the adjacency lists of each node during any point of the simulation and store it in files, which are accessed by the real time simulator at every clock cycle. Thus in each clock cycle a node knows which nodes are adjacent to it without any expensive calculations.

A unique number known as node-id identifies each node. Every node differs from the other in terms of its bandwidth. The nodes are randomly assigned bandwidth values from a distribution that runs very close to the values present in actual mobile networks. Other than this, all nodes are homogenous in their functional characteristics, i.e. an individual thread handles every node. Each thread has data structures private to it, which simulates the node's internal data structures. All nodes also access shared global structures at every time instant to simulate the channel access and contention characteristics of the MAC layer. In particular, the channel may be busy, in which case, a node cannot send a message and has to wait till it becomes free. The lower layers(MAC and physical) have been simulated with a packet drop rate of around 5%. Data packets are handled differently from other routing packets. Their size is randomly generated between 64 bytes and 1024 bytes. Each node has sufficient buffer capacity to handle data packets of varying sizes.

Assumptions:
• The link layer can report varying bandwidth conditions in the environment.
• The hello and hello reply packets used are not part of the routing overhead. This is because these packets are generated, transmitted and received by the link layer. Any change to the bandwidth information is made through a shared structure, which is accessible at a higher layer.

We measure the performance of the SAFAR protocol against the following performance metrics:

Size of the Proactive Zone

Since the proactive zone is adaptive, the size of the proactive zone varies according to the node's bandwidth and also its neighborhood. We show change in average size of the proactive zone of the node with varying standard deviations of bandwidth under fairly constant average bandwidth.

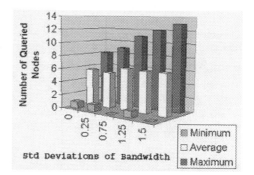

Fig. 4. Number of nodes in proactive zone vs Standard deviation of Bandwidth

The graph (Fig. 4), shows the variations in number of nodes queried by a node for proactive maintenance during a simulation run to the standard deviations of bandwidth. Each node in the simulation run is given a different bandwidth value, with the standard deviation of all 50 nodes being shown in the graph. The average bandwidth of nodes remains constant. As shown in Fig. 4, it has been found that the maximum number of nodes queried shows a sharp increase with increase in standard deviation of bandwidth. This is to be expected since with a larger standard deviation, there will be nodes whose bandwidth is far greater than its surroundings.

Thus the factor F(fitness) increases and it queries more nodes. The average number of nodes remains fairly constant(as expected), since the average bandwidth is held constant. However, the difference between the average and maximum value shows a steady increase. This shows that the fitness function allows adaptive table buildup.

Type-1 Success

This is a measure of the percentage of successful routes discovered in stage 1 of routing. It is measured against bandwidth of the node that initiates this node discovery.

For analyzing the Type-1 success ratio, nodes of different bandwidth values are analyzed as they try to route packets to random destinations. The percentages of such requests that are successful are logged. As seen in Fig. 5, the Type-1 success ratio increases with the bandwidth as expected. The graph also shows that a very high percentage of destinations are found with a Type-1

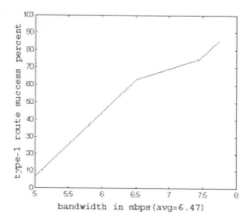

Fig. 5. Percentage of Type-1 queries successful vs Node Bandwidth

query itself thus reducing the need for an expensive Type-2 query mechanism and hence decreasing route acquisition latency.

Packet Delivery Ratio

This is defined as the ratio of the number of data packets delivered successfully to the total data packets sent. We have considered the busy transmission channel condition, which might affect this ratio. The ratio thus, gives us a measure of the reliability of a route and is measured against a varying transmission radius.

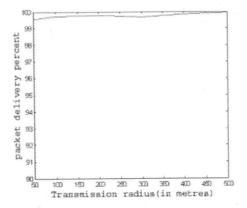

The percentage of data packets routed successfully shows that SA-FAR finds routes quickly and accurately. The percentage of packets routed successfully increases with transmission range. This is because

Fig. 6. Packet Delivery Ratio vs Transmission Radius

there is more probability of loss in a route with many hops, as the movement information might not have been registered.

Route Acquisition Latency

It is defined as the delay between generating a route query and receiving the corresponding reply. During our simulation, we assume that the processing delay of messages is negligible but take into consideration the delay due to queuing of messages in the transmission buffer.

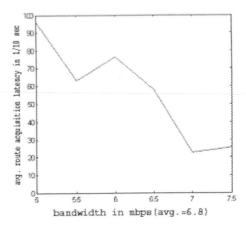

The route latency graph (Fig. 7) shows, how the latency of discovering routes decreases with increasing bandwidth. This substantiates the advantages of SAFAR. By being highly adaptive, it allows a high bandwidth node to use its capacity effectively. It knows more destinations and hence can route faster and more effectively. Even if it does not have a

Fig. 7. Route Acquisition Latency vs. Bandwidth

route in the routing table, it can query using a Type-1 packet, which is sent to more nodes. Thus, a high bandwidth node need not resort to Type-2 querying which slows the protocol down and increases latency. Hence, the average route acquisition latency drops with increasing bandwidth.

6   Optimization of Power

As mentioned earlier, our protocol can switch over from the bandwidth to the power domain during its operation. In this section, we show that bandwidth and power are interchangeable.

To make SAFAR power centric, the bandwidth field of a node is replaced with a power field(in battery time remaining). Using this, it again selects nodes which will be 'alive' for a longer time.

Table build-up phase: This is in direct correspondence with a high bandwidth node, since a node with higher battery power would be able to sustain more data and routing traffic. This would also provide relief to nodes with low power, as they would not have to handle requests other than their own traffic. A node with low power would not be concerned with route efficiency and latency. Since it has low power, it would have few nodes in the proactive list and hence would not spend too much time with Type-1 querying. Instead, it proceeds directly to Type-2 querying which is more likely to yield a route, and also consumes less power in that node, than a combined Type-1 Type-2 mechanism, as it has to transmit more packets which corresponds to more transmission power. Again it has to be noted that this is achieved completely dynamically. In case of multiple responses in the power domain, the path having the maximum least battery time is chosen.

7   Conclusion

In this paper, we have presented and evaluated a bandwidth adaptive hybrid protocol that is well suited for operation in mobile ad hoc networks. The protocol uses a table build-up procedure whose little overhead is well used in the dynamic maintenance of neighboring nodes on the basis of bandwidth fitness. This also leads to reducing the route acquisition latency and hence reduces routing overhead. It was also seen that the protocol adapts to varying high traffic conditions, wherein a node is given the option of reducing its traffic by preventing others from using it. An improvement to the protocol will be in using a hybrid cost factor, which includes both power and bandwidth instead of one of them purely, making it a better estimate of the network's performance constraint.

References

1. J Macker, S Corson: "Mobile Ad hoc Networking (MANET): Routing Protocol Performance Issues and Evaluation Considerations", *Internet draft, January '99.*
2. Charles E. Perkins and Pravin Bhagwat: "Highly dynamic Destination-Sequenced Distance Vector routing (DSDV) for mobile computers",pages 234-244, In *Proceedings of the SIGCOMM '94 Conference on Communications Architectures, Protocols and Applications, August '94.*
3. Charles E. Perkins and Elizabeth M. Royer: "Ad Hoc On Demand Distance Vector (AODV) algorithm", In *Proceedings of WMCSA'99, New Orleans, LA, February '99.*

4. David B. Johnson and David A. Maltz: "Dynamic source routing in Ad hoc wireless networks", In *Mobile Computing, edited by Tomasz Imielinski and Hank Korth, chapter 5, pages 153-181, Kluwer Academic Publishers, '96.*
5. Vincent D. Parka and M. Scott Corsonba: "A Highly Adaptive Distributed Routing Algorithm for Mobile Wireless Networks-Temporally-Ordered Routing Algorithm(TORA)", In *Proceedings of Infocom '97.*
6. Zygmunt J. Haas, Marc R. Pearlman: "Zone Routing Protocol", *Internet draft,<draft-zone-routing-protocol-00.txt>, MARCH '03.*
7. Mario Gerla, Xiaoyan Hong, Guangyu Pei: "Landmark Routing for Large Scale Wireless Ad Hoc Networks with Group Mobility", In *Proceedings of Mobihoc 2000, Boston, MA, November '00.*
8. Suresh Singh, Mike Woo and C S Raghavendra: "Power Aware Routing in Mobile Adhoc Networks", pages 181-190, In *Proceedings of MobiCom'98.*
9. Javier Gomez, Andrew T. Campbell, Mahmoud Naghshineh and Chatschik Bisdikian: "Conserving Transmission Power in Wireless ad hoc Networks", In *Proceedings of IEEE 9th International Conference on Network Protocols, Riverside, California, November '01.*
10. Venugopalan Ramasubramaniam, Zygmunt J.Haas and Emin Gün Sirer: SHARP: A Hybrid Adaptive Routing Protocol for Mobile Ad Hoc Networks, pages 303-314, In *Proceedings of Mobihoc '03.*
11. David E Goldberg: Genetic Algorithms in Search, Optimization and Machine Learning, pages 76-80, *Pearson Education - 1999.*
12. David B. Johnson, Josh Broch, David A. Maltz, Yih-Chun Hu, and Jorjeta Jetcheva: "A performance comparison of Multi-Hop Wireless Ad Hoc Network Routing Protocols", In *Proceedings of MobiCom '98.*

Evaluation of the AODV and DSR Routing Protocols Using the MERIT Tool

Priya Narayan and Violet R. Syrotiuk*

Computer Science & Engineering, Arizona State University, Tempe, AZ 85287-5406

Abstract. Selecting the most appropriate routing protocol for a given set of conditions in a mobile ad hoc network (MANET) remains difficult. While the quantitative performance metrics are helpful in the selection, the problem is that the metrics of the protocols under consideration be generated by the same simulator in order to be comparable. The MERIT framework proposes to compare a routing protocol to a theoretical optimum rather than to a competing protocol. The goal is to achieve an implementation independent, scalable comparison methodology for protocols. In this paper we evaluate the DSR and AODV routing protocols in the MERIT tool. Our results agree with performance studies comparing these two protocols, validating the MERIT methodology.

1 Introduction

Routing is a fundamental problem in *mobile ad hoc networks* (MANETs). What contributes to the difficulty of the problem is that a MANET is a collection of mobile wireless nodes with no supporting infrastructure. As a result, routing information must be computed in a distributed manner that is responsive to the continuously changing topology induced by node mobility.

Over the past few decades, a considerable number of routing protocols have been proposed for MANETs (also called packet radio networks or multi-hop networks in earlier work). More recently, more than a dozen protocols have been documented in the form of Internet-Drafts in the Internet Engineering Task Force (IETF) MANET working group [13]. Given the large choice of protocols, each with its own strengths, it remains a difficult problem to select the one most appropriate for a given set of network conditions.

Routing protocols for MANETs are traditionally evaluated by simulation since few testbeds exist. There is general consensus on the quantitative metrics to use for comparing protocols, namely end-to-end throughput and delay, route acquisition time, percentage of out-of-order delivery, and the efficiency of the protocol in terms of control traffic overhead [4]. It has been shown that the results from different simulation platforms cannot be directly compared. Specifically, factors such as the implementation of the physical layer not only affect the absolute performance of a routing protocol, it can change the relative ranking among protocols for the same scenario because its impact on the different protocols is not uniform [18].

---

* This work was supported in part by NSF grant ANI-0220001.

S. Pierre, M. Barbeau, and E. Kranakis (Eds.): ADHOC-NOW 2003, LNCS 2865, pp. 25–36, 2003.

The MERIT framework [6,7] takes a new approach to routing protocol assessment for MANETs. In MERIT a protocol is compared to a theoretical optimum rather than to a competing protocol. In particular, the measure proposed is the MERIT ratio, the mean ratio of the cost of the route actually used by the protocol to the cost of the optimal mobile path under the same network history. Since we take a ratio, we believe that some of the dependencies on the simulator cancel out, yielding an implementation independent measure. The MERIT spectrum is the ratio taken as a function of some network parameter, such as the node velocity. MERIT is a scalable comparison methodology, since the MERIT spectra for a protocol are computed once and then can be compared to spectra of other protocols directly, rather than requiring that protocols of interest be ported to the same simulator.

The definitions and theoretical foundations of the MERIT framework were laid out in [6]. In this paper, we focus on the implementation of MERIT in the ns-2 network simulator [14] and perform an evaluation of two well established MANET routing protocols. Specifically, we present MERIT spectra for the Dynamic Source Routing (DSR) [11] and the Ad hoc On Demand Distance Vector (AODV) [16] routing protocols for several network parameters.

The remainder of this paper is organized as follows. In Section 2 we overview the MERIT framework as well as the DSR and AODV routing protocols. Section 3 details the MERIT tool, the implementation of the MERIT framework, describing the generation of a mobile graph in ns-2. We also describe how we extract the actual routes for each of DSR and AODV and use them to compute a MERIT ratio for a given run. We then use the MERIT tool to produce spectra for various parameters. These results are presented and discussed in Section 4. Conclusions and ongoing work are given in Section 5.

2   Overview

2.1   Overview of the MERIT Framework

Since a MANET is a mobile network, MERIT models the history of network topology changes over some time scale T by a sequence of graphs. A *mobile graph* $\mathcal{G} = G_1 G_2 \ldots G_T$ is defined as any sequence $G_i, i = 1, \ldots, T$, of graphs where the vertices of G_i correspond to the nodes in the network, and its edges correspond to communication links between nodes. Similar models with the goal to capture dynamics in graphs include [1,8,12].

Given a mobile graph, a *mobile path* between a source-destination pair is defined as a path sequence $\mathcal{P} = P_1 P_2 \ldots P_T$ where P_i is a path in the corresponding G_i between the same source-destination pair.

It is assumed that a *cost model*, expressed as a weight function on edges, underlies each graph. The *weight function* $w_i(u, v)$ for graph G_i is a function of vertex pairs (u, v) such that

$$w_i(u, v) = \begin{cases} m(u, v) & \text{if edge } (u, v) \text{ exists in } G_i \\ \infty & \text{if edge } (u, v) \text{ does not exist in } G_i \end{cases} \tag{1}$$

where $m(u, v)$ is the value of the link metric on the edge (u, v).

The weight of a path P_i in G_i is denoted by $w_i(P_i)$. For additive path metrics, it is the sum of the link weights along the edges of the path. The weight of a mobile path includes the individual weights of each path in the sequence, and a transition cost c_{trans} that represents the overhead incurred by the protocol to respond to a topology change. Thus, the weight of a mobile path $\mathcal{P}$ in $\mathcal{G}$, denoted by $w(\mathcal{P})$, is defined as

$$w(\mathcal{P}) = \sum_{i=1}^{T} w_i(P_i) + \sum_{i=1}^{T-1} c_{trans}(P_i, P_{i+1}). \tag{2}$$

Given these definitions and assuming a given cost model, the *shortest mobile path (SMP) problem* is defined within the framework as follows.

Problem 1. Given a mobile graph $\mathcal{G} = G_1 \ldots G_T$ and a specified source-destination pair (s, t), find a mobile path $\mathcal{P} = P_1 \ldots P_T$ from s to t, such that the weight

$$w(\mathcal{P}) = \sum_{i=1}^{T} w_i(P_i) + \sum_{i=1}^{T-1} c_{trans}(P_i, P_{i+1})$$

of the mobile path is minimum.

In [6], a simple 2-valued transition cost function was considered initially because the SMP problem is tractable in this case.

MERIT Assessment Measures. Two assessment measures are proposed within the MERIT framework: the MERIT ratio and the MERIT spectrum.

For a given mobile graph $\mathcal{G}$ and s-t pair, let $\mathcal{P}_{real}$ be the actual mobile path generated by the MANET routing protocol $\mathcal{R}$. The weight $w(\mathcal{P}_{real})$ of this mobile path is computed directly from the routing state trace for the s-t path in each G_i. The paths generated in turn directly fix the transition costs. Similarly, let $\mathcal{P}_{ideal}$ be the shortest mobile path for $\mathcal{G}$. Both the path $\mathcal{P}_{ideal}$ and its weight $w(\mathcal{P}_{ideal})$ are computed by the SHORTEST MOBILE PATH algorithm [6] run on this instance $\mathcal{G}$ of the mobile graph for source-destination pair s-t.

The MERIT ratio $= \mathbf{E}\left(\frac{w(\mathcal{P}_{real})}{w(\mathcal{P}_{ideal})}\right)$ is the expected value of the cost ratio of the actual mobile path to the shortest mobile path. The ratio represents how far the routes in protocol $\mathcal{R}$ deviate in cost from the theoretical optimum. We compute the mean of the ratio of a large enough sample size of randomly drawn s-t sessions in order to obtain a 95% confidence interval. The distribution of the sample is implicit from the simulation parameters.

The value of the MERIT ratio becomes meaningful when we understand how it changes as a function of a some parameter. The MERIT ratio expressed as the function of some independent parameter defines the MERIT spectrum of the protocol $\mathcal{R}$. Some examples of such parameters include the node velocity, the average actual path length, the average node density, transmit power levels and the related energy consumption, and even the transition cost.

2.2   An Overview of the DSR and AODV Routing Protocols

Dynamic Source Routing. The Dynamic Source Routing (DSR) protocol [11] is an on-demand routing protocol that allows nodes to dynamically discover a source route to any destination in the network. In source routing, the source is responsible for computing the route that a packet should take.

When a node wishes to communicate with another node, it employs route discovery to flood a control packet, called a route request (RREQ), through the network, in search of a route to the destination. A RREQ packet with a target of t is broadcast in the network. It is forwarded hop-by-hop from the node initiating the route discovery. When the RREQ reaches the destination t or a node that is aware of a route to t, forwarding stops. A route reply (RREP) packet is sent back to the source s on the reverse path, including a full source route to the destination t. This source route is included in the header of each data packet sent to t and enables stateless forwarding. Data sent to t by an application is buffered at s until a route reply with a route to t is received, at which point, these packets are forwarded to t along the acquired route.

The route maintenance mechanism monitors the status of source routes in use, detects link failures and repairs routes with broken links. When route maintenance indicates that a source route is broken, s can attempt to use any other route it happens to know to t, or the source s can invoke route discovery again to find a new route.

Ad Hoc on Demand Distance Vector. The Ad hoc On Demand Distance Vector (AODV) [16] routing protocol is also an on-demand protocol. Similar to traditional distance vector protocols, AODV maintains routing tables with one entry per destination.

AODV builds routes using RREQ and RREP control packets similar to the route discovery mechanism in DSR. A node receiving the RREQ packet may send a RREP if it is either the destination or if it has a route to the destination with corresponding sequence number greater than or equal to that contained in the RREQ. Otherwise, it rebroadcasts the RREP. As a RREP propagates back to the source, nodes set up forward path entries to the destination in their route tables. Once the source node receives the RREP it may begin to forward data packets to the destination.

There are several differences in the route discovery mechanisms of DSR and AODV. The source routing mechanism used in DSR enables s to learn routes towards each intermediate node on the route to t. Additionally, each intermediate node on the path from s to t may learn routes to every other node on the route. In DSR, the destination node replies to all RREQs sent towards it during a single request cycle. Thus, in DSR, the source learns many alternate routes to the destination in contrast to AODV where the destination replies only to the first arriving request.

2.3   Earlier Comparisons of DSR and AODV

Several attempts have been made to compare DSR and AODV with respect to path optimality among other measures. Broch et al. [2] use an internal mechanism of ns-2 to measure the path optimality for the routing protocols. Essentially an omniscient observer of the simulation stores the total number of mobile nodes and a table of shortest number of hops required to reach from one node to another at a particular instant of time. This information is used to compute the difference between the length of the actual path used between the source and destination at a particular instant of time in the instantaneous snapshot of the network graph. The approach in MERIT is different from [2] in that it considers the entire history of the network topology for the duration of the data flow and the cost of updating the routes to the destination.

The studies in [2] show that DSR uses routes that are very close to optimal whereas AODV finds a greater number of routes which are further apart from the optimal in the instantaneous topology.

In [5], the authors compare the protocols DSR and AODV using several quantitative metrics. The general conclusion is that for application oriented metrics such as delay and throughput, DSR outperforms AODV in less stressful conditions, i.e., for a smaller number of nodes and/or lower mobility.

3   The MERIT Tool

Figure 1 shows a high level block diagram of the MERIT tool. The tool takes, as input, a mobile graph as well as actual route traces for the routing protocol under consideration. Once there is sufficient confidence in the ratio for one parameter, runs for other parameters may begin. From a sequence of MERIT ratios for a given parameter, a spectrum is generated.

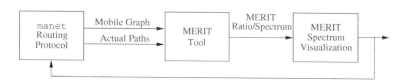

Fig. 1. The MERIT tool.

3.1   Simulation Environment

We generate the input for the MERIT tool using an extended version of ns 2 [14] because it has the ability to simulate MANETs [3] and provides a reference implementation of the DSR and AODV routing protocols.

We create mobility scenarios as in [2], which include 20 nodes in a 1500 × 300 m^2 area in order to force the use of longer routes. In our scenarios, every node has an omni-directional transmission radius of 250 m and moves using the random waypoint model [3] with a maximum velocity v, which varies from $0\,\mathrm{m/s}$ to $10\,\mathrm{m/s}$. The mobility model is also characterized by a pause time.

Our experiments use constant bit rate (CBR) data sources over UDP for data communication. The primary goal of our current experiments is to illustrate the behaviour of routing protocols for simple cases. Therefore, we choose UDP as our transport layer protocol in order to eliminate the influence of congestion and flow control mechanisms on the performance of the routing protocol.

3.2   Traffic and Mobility Patterns

For a typical ns-2 wireless simulation, a connection pattern file designates the configuration and behaviour of data connections in the network scenario to be simulated. It specifies the end points between which the data flow takes place, when the data flow over the connection should start and stop, and the type of application data sent by the source. In our experiments, each connection pattern file has one s-t pair and a CBR connection between them to eliminate any possibility of collisions or buffer overflow due to high network load.

For each s-t pair, the data flow starts and ends at a random time. A script to generate connection pattern files is provided with the CMU extensions to ns-2 [3]. We modify the CMU script to impose the restriction that each connection last for at least 100 seconds.

A mobility pattern file defines the motion of the nodes in the network and the changes in the paths between the nodes over time. A program to generate the mobility pattern is also provided [3]. By default, this program assumes a fixed transmission range of 250 m to denote one hop. The initial positions of the nodes are designated at random and the move sequences are generated according to the random waypoint mobility model.

Using an approach similar to Holland and Vaidya [9,10], we generate mobility patterns for the network of 20 nodes moving with a mean speed of $2\,\mathrm{m/s}$ for 1800 seconds and a pause time of $0\,\mathrm{s}$. We call these mobility pattern files base patterns. We generate 5 such base pattern files. These base pattern scripts are used to generate mobility patterns for mean speeds of $4, 6, 8$ and $10\,\mathrm{m/s}$, respectively.

3.3   Extraction of the Mobile Graph

In order to generate the mobile graph for a given scenario, we sample the current position of every mobile node at regular intervals and output the link transitions that occurred within each interval. The sampling interval is such that we obtain 100 samples before the node travels through its transmission radius once.

The initial connectivity matrix translates into the first graph sequence in the mobile graph. By applying link additions and breakages recorded at different time intervals, we obtain the graphs in the graph sequence over the time scale equal to the total simulation time.

3.4   Mobile Path Computation for DSR

To facilitate the computation of the mobile path for our evaluation of the DSR protocol, we require the actual paths used by the protocol to route data packets from the source to a destination. In order to obtain this information, we modify the DSR implementation in ns-2 to trace the primary and secondary cache each time an entry is either added to or deleted from the cache.

For every simulation run, we gather the route cache trace and use this to determine the actual path, which DSR would use to route a packet from source to destination for a particular time sequenced graph in the mobile graph $\mathcal{G}$. This approach is generalized and can be used to gather route information for protocol implementations in other network simulators such OpNet or QualNet [15,17].

The procedure we adopt to select the actual path is consistent with the algorithm used by the ns-2 implementation of DSR to select a route. For example, Fig. 2 shows a three graph subsequence $G_{i-1}G_iG_{i+1}$ in a mobile graph for a session between source-destination pair 4-5. The solid path between nodes 4 and 5 is the actual route computed by DSR while the dashed path (which coincides with the last two hops of the actual route) is the shortest hop path.

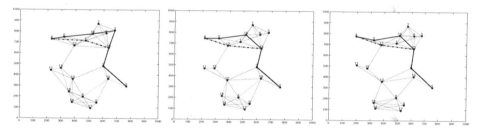

Fig. 2. Subsequence $G_{i-1}G_iG_{i+1}$ of a mobile graph for source-destination 4-5.

3.5   Mobile Path Computation for AODV

Every node that implements the AODV protocol maintains a separate routing table with the next hop information. Each entry in the routing table has an expiration time and a sequence number. To compute the actual mobile path for AODV, we trace the route table for every node in the simulation whenever a new entry is added or updated in the routing table.

One can easily trace the valid path to the destination from the routing table information for each node. The expiration time for each routing table entry is used to judge whether or not the route calculated is stale. We use the same algorithm used by the ns-2 AODV implementation to calculate the actual mobile path for our mobile graph.

3.6   Computation of the MERIT Ratio

Every connection in the connection pattern file has a designated start (t_{start}) and stop time (t_{stop}). Our mobile graph $\mathcal{G}$ consists of a graph sequence for the entire duration of the simulation T. Since DSR and AODV are reactive, the trace is valid only between t_{start} and t_{stop} when data flow actually exists.

Only a subsequence of mobile graph $\mathcal{G}$ corresponding to the time between t_{start} and t_{stop} is used to extract actual paths, run the SMP algorithm, and compute the MERIT ratio. Let us denote this subsequence as $\mathcal{G}' = \mathcal{G}_{t_{start}} \cdots \mathcal{G}_{t_{stop}}$. For every graph in the mobile graph $\mathcal{G}'$, we compute the actual path from the source to the destination.

To compute the MERIT ratio for the mobile graph $\mathcal{G}'$, the SMP algorithm requires the shortest path in each graph of the mobile graph and the shortest path in the intersection of all subsequences of the graph sequence. We implement Dijkstra's shortest path algorithm to find the shortest hop path in any graph, while computing the cost matrix for the SMP algorithm. We use hop count as the metric for our implementation of the SMP because both DSR and AODV use this metric in their path computation.

For this work, we consider the cases where we assign the constant values of $0, 0.5, 1.0, 1.5$ and 2.0 for the update cost. We chose a maximum value of 2.0 for our update cost because we found that the average path length for our simulation experiments did not exceed 3.5 hops.

4   MERIT Spectra for DSR and AODV

We conduct two sets of experiments. In the first set, we measure the MERIT ratio for different scenarios by varying the degree of mobility. In the second set, we keep the mobility rate constant and measure the MERIT ratio by varying the data rate for the CBR flow from the source to the destination.

In all cases, our results are averaged over a sufficiently large number of runs to ensure a small variance (95% confidence interval). From the confidence interval calculations on our experimental results, we find that the MERIT ratio varies within 2.85% of the plotted value.

4.1   Parameters of Interest

The parameters against which we plot MERIT spectra include:

Mean speed: A mean speed of $2\,\text{m/s}$ equates to the speed of a pedestrian and a mean speed of $10\,\text{m/s}$ corresponds to the speed of a moving vehicle.

Data arrival rate: The data arrival rate represents the number of packets sent per second. For our experiments, we use data arrival rates of $2, 4, 6, 8$ and 10 packets/second, with the traffic being sent out at a constant rate and the packet size being constant at 64 bytes.

Average path length: The average path length for a simulation run denotes the average actual path length (hop count) over all the paths for the specified source-destination pair.

Average end-to-end delay: The average end-to-end delay for a run denotes the average end-to-end packet delay (in units of number of seconds) computed from the router trace for the data packets sent by the candidate protocols.

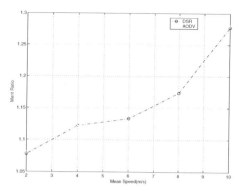

Fig. 3. MERIT ratio versus mean speed.

4.2   MERIT Spectra

In this section, we present the various MERIT spectra generated for the set of experiments where we vary the speed from 2 m/s to 10 m/s.

Figure 3 shows that the MERIT ratio computed for DSR and AODV grows with increasing mobility rate. It is clear that as the speed increases, the topology changes at a faster rate. Since both DSR and AODV are on-demand protocols, a certain delay is experienced before the routes are repaired and the routing caches/tables are updated to reflect the latest changes in topology. The computed values of the MERIT ratio reflect that the paths in the routing table are closer to optimal for scenarios with lower speeds than for scenarios with higher speed, mostly because the topology changes are more frequent and route repairs or route maintenance is done more frequently. Additionally, we observe that the DSR protocol yields MERIT ratios that are closer to the optimal cost than AODV. The maximum and average differences between the computed MERIT ratios of DSR and AODV expressed as a percentage is 1.25 and 0.5136 respectively. These differences are consistent with the observations made in [2] where the authors suggest that DSR's caching is more effective than AODV at lower mobility rates where the cached information goes stale more slowly.

In Fig. 4, we see that the MERIT ratio increases with respect to the actual path length. The path length we plot is the average of the actual path lengths from the route trace. The results in the figure indicate that when the actual path lengths computed by the AODV routing protocol are larger, the MERIT ratio is also higher. AODV computes paths that are longer than the paths computed by DSR primarily because DSR has access to significantly greater amount of route information than AODV because it employs aggressive caching and promiscuous listening. AODV has access to less information because it maintains only one entry per destination in its routing table and relies significantly on a higher frequency of route discovery flooding to keep its routing table up to date.

When the paths between a source-destination pair are longer (more hops), it could result in a higher end-to-end delay. Also, when the speed is higher and the topology changes at a faster rate, there could be a higher end-to-end delay

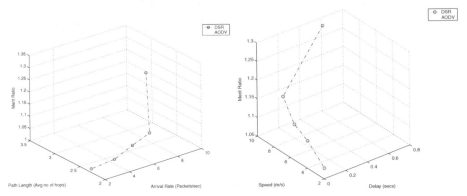

Fig. 4. MERIT ratio versus (a) arrival rate and path length; (b) delay and speed.

because a greater amount of route maintenance and repair will have to be done whenever the topology along the path changes. We conclude that in scenarios where the mobility rate is higher and in scenarios where the path between the source and the destination is longer, the paths are further away from the optimal. Since the average delay is affected by the combination of these factors and other factors like the network load as well, we can expect to see that higher delays would lead to higher MERIT ratios.

The SMP algorithm used to compute the ideal cost in the MERIT framework, involves a two-valued transition cost. The transition cost can be viewed as a trade-off between the importance of the stability of the path and the length of the path. When the cost of updating path information is small, the SMP algorithm places more emphasis on searching for the shortest path without regard to whether or not it changes the path. When the cost is larger, the algorithm gives more importance to searching for stable paths since it is costly to change paths. We observe that MERIT ratios generally increase with an increase in the update cost values because the actual routing protocol implementation makes its routing decisions by using only the instantaneous information recorded in its data structures at any given time. With a higher cost of update, the importance of using the past history to make routing decisions is more critical because stable paths are preferred when the update cost is higher. This is effectively captured in the higher values obtained for the MERIT ratio with an increase in cost.

We see that the MERIT ratio increases with an increase in the data arrival rate. The path lengths for these scenarios are also noticeably shorter than the ones observed for the previous set of experiments. This can be attributed to the fact that both DSR and AODV are on-demand protocols, hence a higher data rate causes the routing tables to be updated more frequently, resulting is smaller path lengths.

Our results clearly indicate that for the scenarios we have considered, DSR provides paths which are closer to the optimal than AODV. In our experiments, we choose simulation scenarios where we have only one *s-t* pair, which is communicating. We use a single CBR source that sends data at a steady rate in all our experiments, therefore we effectively consider scenarios where the network load is not bursty. In [2,5] it is shown that DSR outperforms AODV in scenarios where the offered network load is low. Our results show the same trend.

5 Conclusions and Ongoing Work

The main goal of this work is to evaluate the performance of two MANET routing protocols in the MERIT tool, an implementation of the MERIT framework. Our results show that DSR outperforms AODV in the scenarios we consider. These results agree with the observations made in a similar context in [5] where the protocols were compared in terms of their path optimality and other performance measures. Also, the MERIT spectra which plots the MERIT ratio against various network parameters of interest show that the MERIT ratio intuitively captures the network dynamics and behaves as expected.

In our ongoing work, we are evaluating the performance of routing protocols implemented in different network simulators to verify the implementation independence, and scalability aspects of the MERIT framework. As well, we are working towards defining a transition cost function that will capture more accurately the cost of updating the paths in the routing table.

Acknowledgments

We thank A. Faragó for helpful discussions, J. Boleng for sharing code to determine link state transitions in `ns-2`, and K. Vadde for help with the figures.

References

1. S. Bhadra and A. Ferreira, "Computing Multicast Trees in Dynamic Networks using Evolving Graphs," Institut National de Recherche en Informatique et Automatique (INRIA), Research Report No. 4531, August 2002, revised October 2002.
2. J. Broch, D.A. Maltz, D.B. Johnson, Y.-C. Hu, and J. Jetcheva. "A Performance Comparison of Multi-Hop Wireless Ad Hoc Network Routing Protocols," *Proceedings of the 4th Annual ACM/IEEE International Conference on Mobile Computing and Networking* (Mobicom'98), Dallas, Texas, pp. 85–97, October 1998.
3. Wireless and Mobility Extensions to `ns-2`. Carnegie Mellon University Monarch (Mobile Networking Architectures) Project. http://www.monarch.cs.cmu.edu
4. M.S. Corson and J. Macker. "Mobile Ad hoc Networking (MANET): Routing Protocol Performance Issues and Evaluation Considerations," *Network Working Group, RFC 2501*, January 1999. http://www.ietf.org/rfc/rfc2501.txt
5. S.R. Das, C.E. Perkins, and E.M. Royer. "Performance Comparison of Two On-Demand Routing Protocols for Ad Hoc Networks," *Proceedings of the 19th Annual Joint Conference of the IEEE Computer and Communication Societies* (Infocom 2000), Tel Aviv, Israel, pp. 3–12, March 2000.
6. A. Faragó and V.R. Syrotiuk. "MERIT: A Unified Framework for Routing Protocol Assessment in Mobile Ad Hoc Networks," *Proceedings of the 7th Annual ACM International Conference on Mobile Computing and Networking* (Mobicom'01), Rome, Italy, pp. 53–60, July 2001.

7. A. Faragó and V.R. Syrotiuk. "MERIT: A Scalable Approach for Protocol Assessment," *Mobile Networks & Applications*, Special Issue on Mobile Ad Hoc Networks, A. Campbell, M. Conti and S. Giordano (eds.), 8, pp. 567–577, 2003.
8. P. Haxell, A. Rasala, G. Wilfong, and P. Winkler, "Wide-Sense Nonblocking WDM Cross-Connects," *Proceedings of the Tenth European Symposium on Algorithms* (ESA) 2002, LNCS 2461, pp. 538–550.
9. G. Holland and N. Vaidya. "Analysis of TCP Performance over Mobile Ad Hoc Networks," *Proceedings of the 5th Annual ACM/IEEE International Conference on Mobile Computing and Networking* (Mobicom'99), Seattle, Washington, pp. 219–230, August 1999.
10. G. Holland and N. Vaidya. "Analysis of TCP Performance over Mobile Ad Hoc Networks, Part II: Simulation Details and Results" *Technical Report 99-005*, Department of Computer Science, Texas A&M University, 61 pages, February 1999.
11. D.B. Johnson, D.A. Maltz, Y.-C. Hu, and J.G. Jetcheva. "The Dynamic Source Routing Protocol for Mobile Ad Hoc Networks (DSR)," Internet Draft, November 2001. Work in progress. http://www.ietf.org/ids.by.wg/manet.html
12. E. Köhler, K. Langkau, and M. Skutella, "Time-Expanded Graphs for Flow-Dependent Transit Times," *Proceedings of the Tenth European Symposium on Algorithms* (ESA) 2002, LNCS 2461, pp. 599–611.
13. Mobile Ad-Hoc Networking (MANET) Working Group.
    http://www.ietf.org/html.charters/manet-charter.html
14. The Network Simulator — ns-2. The University of California, Berkeley.
    http://www.isi.edu/nsname/ns/
15. OpNet, by OpNet Technologies. http://www.opnet.com/
16. C.E. Perkins and E.M. Royer. "Ad hoc On-Demand Distance Vector Routing," *Proceedings of the 2nd IEEE Workshop on Mobile Computing Systems and Applications*, New Orleans, LA, February 1999, pp. 90-100.
17. QualNet, by Scalable Network Technologies.
    http://www.scalable-networks.com/
18. M. Takai, J. Martin and R. Bagrodia. "Effects of Wireless Physical Layer Modeling in Mobile Ad Hoc Networks," *Proceedings of the 2001 ACM International Symposium on Mobile Ad Hoc Networking & Computing* (MobiHoc'01), Long Beach, California, pp. 87–94, October 2001.

On-demand Routing in MANETs:
The Impact of a Realistic Physical Layer Model

Liang Qin and Thomas Kunz

Carleton University, Ottawa, Ont., Canada K1S 5B6
{lqin,tkunz}@sce.carleton.ca

Abstract. Most simulations and performance comparisons of mobile ad hoc network routing protocols are based on a simplistic and idealistic physical layer model. In real applications, there are different kinds of noise or interference that impact the signal power. We use a shadowing propagation model in our simulation evaluation of two on-demand routing protocols: AODV and DSR. Because of signal power fluctuation, active routes will break, which causes significant throughput degradation and longer packet delay. In this paper, we analyze the impact of a shadowing model on the performance of these two routing protocols. Then we set a new signal power threshold during the route discovery process so that only those links with strong enough signal power will be chosen; we also reduce some control messages for DSR. The simulation results show significant increases in packet delivery ratio and decreases in packet latency for both protocols.

1 Introduction

A mobile ad hoc network (MANET) is an infrastructure-less network. Because the radio transmission range is limited and mobile nodes are free to move randomly, the routes are subject to frequent failure. Dozens of routing protocols for MANET have been proposed, for example, Destination-Sequenced Distance-Vector Routing (DSDV)[1], Temporally-Ordered Routing Algorithm (TORA) [2], Dynamic Source Routing protocol (DSR)[3], Signal Stability-Based Adaptive Routing (SSA) [4], and Ad-hoc On-Demand Distance Vector Routing (AODV)[5]. Performance evaluations and comparisons between several routing protocols have been published in [6][7]. But these evaluations are based on simulations using a two-ray ground propagation model. In real applications, the path between transmitter and receiver can be line-of-sight, or obstructed by physical obstacles between them, thus the signal strength on the receiver not only depends on the distance, but also on the environment. At the time when this research began, few simulations tried to use different propagation models to evaluate routing protocol performance, and most routing protocols assume that packet loss over a link indicates a link breakage due to node mobility. Takai [8] presents simulation results using free space, Rayleigh and SIRCIM (Simulation of Indoor Radio Channel Impulse Response Models with Impulse Noise) propagation models in a 130m by 130m area with 20 mobile nodes. Goff [9] proposes some ways to verify whether the signal fluctuation is caused by channel fading.

This paper concentrates on the impact of a shadowing propagation model on the performance of two on-demand routing protocols: AODV and DSR. Simulations of

S. Pierre, M. Barbeau, and E. Kranakis (Eds.): ADHOC-NOW 2003, LNCS 2865, pp. 37–48, 2003.
© Springer-Verlag Berlin Heidelberg 2003

these two protocols show that the signal strength fluctuations cause routes to be assumed broken, and in turn the packet delivery ratio significantly decreases and packet delay increases. We apply a new signal power threshold during the route discovery process, so that only the routes with strong signal strength will be chosen. For DSR, we also reduce some control messages to reduce traffic. Our simulation results show that this can significantly increase the packet deliver ratio and deceased packet latency for both protocols.

An overview of free space, two-ray ground and shadowing propagation models are provided in Section 2. Section 3 analyses the impact of a shadowing model on the performance of routing protocols, and presents our proposal. Section 4 contains the simulation results under different mobility patterns and parameters. Section 5 draws our conclusion and directions for future research. A more detailed introduction to DSR and AODV can be found in [3] [5].

2 Propagation Models

The free space propagation model is used for the situation when the transmitter and receiver have a clear line-of-sight path. The received power at receiver P_r is given by the Friis free space equation [10]:

$$P_r(d) = \frac{P_t G_t G_r \lambda^2}{(4\pi)^2 d^2 L} \tag{1}$$

where P_t is the transmitted signal power, G_t and G_r are the antenna gains of the transmitter and the receiver respectively. L ($L \geq 1$) is the system loss, and λ is the wavelength. The two-ray ground reflection model considers both the direct path and a ground reflection path between the transmitter and receiver. At a long distance, this model is more accurate than the free space model. The received signal power at a distance d from the transmitter can be expressed by [10]:

$$P_r(d) = \frac{P_t G_t G_r h_t^2 h_r^2}{d^4} \tag{2}$$

where h_r and h_t are the heights of the transmit and receive antenna respectively.
The free space model and two-ray ground model predict that received power decays as a function of distance, the radio transmission range is a perfect circle. The shadowing model is a statistical model. The mean received power at distance d is computed relative to $P_r(d_0)$, represented by (in dB) [10]:

$$[\frac{P_r(d)}{P_r(d_0)}]_{dB} = -10\beta \log(\frac{d}{d_0}) + X_\sigma \tag{3}$$

Equation (3) is also called log-normal shadowing model. In this paper, we will simply refer to it as shadowing model, and the free space and two-ray ground model as ideal model or ideal environment. The shadowing model consists of two parts. The first one is the path loss model, d_0 is a reference distance, β is called loss exponent. The second part reflects the variation of the received power at certain distance. X_σ is a Gaussian random variable with zero mean and standard deviation σ, which is called shadowing deviation. β and σ are obtained by measurement. For example, β is 2 for free space, 2 to 3 for obstruction inside factories [10]; σ ranges from 4 to 12 for outdoor environments.

3 Mobile Ad Hoc Network with Shadowing Model

When applying a shadowing model in our simulations of DSR and AODV, we notice severe performance degradation compared to the use of a two-ray ground model. We will first discuss the behavior of DSR and AODV with a shadowing model, then present our proposal.

3.1 The Network Simulator (NS2)

All the modifications and simulations in this paper are based on The Network Simulator (NS2). NS2 is a discrete event simulator developed by the University of California at Berkeley and the VINT project [12]. The Monarch research group at Carnegie-Mellon University developed support for simulation of multihop wireless networks in NS2. It provides tools for generating data traffic and mobile node mobility scenario patterns and NS2 comes with implementations of DSR and AODV. In NS2, the Distributed Coordination Function (DCF) of IEEE 802.11 for wireless LANs is used as the MAC layer protocol. The radio model uses characteristics similar to a commercial radio interface, Lucent's WaveLAN, which is modeled as a shared-media radio with nominal bit rate of 2Mb/s and a nominal radio range of 250 meters with omnidirectional antenna. We choose the traffic sources to be constant bit rate (CBR) sources. Transmission rate is 4 packets per second. Each packet is 64 bytes long.

3.2 Impact of Shadowing Model on DSR and AODV

In an ideal environment, for a transmitter and receiver pair, the received signal power only depends on the distance between them. So if the receiver receives a packet with a signal strength is above the reception threshold, it can be sure it is in the transmission range of the transmitter, and also this link will last for a while, depending on the two nodes' mobility pattern. With a shadowing model, the received signal power has a Gaussian distribution fluctuation. When a packet transmitted successfully on a link, it is not guaranteed that the next packet can be delivered even if the time between these two packets is short. Figure 1 shows how the signal strength for packet reception changes for a pair of nodes over the same period of time (i.e., the distance between two nodes is the same), using two propagation models. Notice that these graphs record all the packets sent by both nodes, including all control packets at the MAC layer, and the two nodes are moving apart. All the data are obtained from NS2, the shadowing model sets β=2 and δ=4, corresponding to a free space environment.

From Figure 1(a) we can see that in a shadowing model, the signal strength fluctuates over at least 2 orders of magnitude. This will cause many active links to be considered broken during transmission, and performance will be affected. Table 1 shows the comparison for DSR and AODV with ideal and shadowing models for different β values, which correspond to different natural environments. All the simulations are run in a 1500x300 m area with 50 mobile nodes, 20 sources, pause time 0 second, maximum speed is 20m/s. Simulation time is 500 seconds. Each scenario is run only once to get an initial idea about the performance degradation. As the β value determines average transmission range, comparing the results has to be done with case, we will expound on this issue later, where we will consider node density for fair comparison.

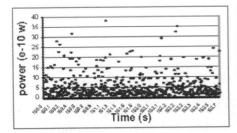 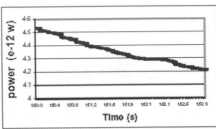

Fig. 1. Signal Power Change over a Wireless Link for (left) Shadowing Model, (right) Ideal Model. They do not use the same power scale.

Table 1. Performance Comparisons of DSR and AODV with Different Propagation Models.

	DSR			AODV		
	Delivery Ratio %	Control Message	Packet Delay (s)	Delivery Ratio	Control Message	Packet Delay (s)
Ideal Model	97.69	12386	0.036	98.36	31617	0.137
β=2.0	98.60	18616	0.131	92.15	25597	0.807
β=2.1	82.36	50725	1.498	83.89	61062	0.988
β=2.2	24.62	114533	2.541	71.08	87871	1.202
β=2.3	18.81	124180	2.779	58.41	127451	1.459
β=2.4	9.98	150840	2.768	43.32	126284	1.752
β=2.5	4.31	204613	3.587	32.95	128958	2.053
β=2.6	4.08	242623	3.623	26.55	120622	2.314

Delivery Ratio: total number of received packets/total number of sent packets.
Control Message: total number of control messages sent.
Packet Delay: average time a packet is in transit from source to the destination.

The impact of the shadowing model on the performance of DSR and AODV is a significantly decreased packet delivery ratio (PDR), increased number of control messages and increased packet latency. The bigger the β value, which corresponds to shorter transmission range, the lower the PDR for both protocols. Also, the average packet latency with a shadowing model can reach several seconds. Comparing the two protocols, AODV has better PDR, and packet delay does not increase as fast as DSR with the β value. Furthermore, we observed the following:

(1) Under an ideal model, most packets are dropped because a target node is out of transmission range. With a shadowing model, more packets are dropped because the Interface Queue is full. The routes found during a Route Discovery process are not valid because of power fluctuation. These invalid routes will trigger a new Route Discovery process, which significantly increases the number of control messages, which have higher priority in the interface queue than data packets. Then data packets have less chance to be sent out. When more new data packets are coming, packets have to be dropped.

(2) Because of power fluctuation, a packet may need multiple retransmission to be forwarded to the next node. Sometimes, the receiver actually received the packet, but the transmitter did not receive an ACK from the receiver. So the transmitter assumes this link is broken. In DSR, the transmitter will salvage this packet if it can find another route to the destination from its route cache. Each salvaging introduces one Route Error message to be sent to the source node. But the previous receiver still transmits this packet based on the source route in the packet, so the same packet may take several paths to the destination. It will consume network resources and increase packet latency.

(3) A packet may take more or less hop counts than the "optimal" number as determined by General Operations Director in NS2, which is based on the ideal environment, with different β values. For example, when $\beta=2.0$, the mean transmission range will be much longer than 250 m, as listed in Table 2. Even if the shortest path needs 3 hops based on the ideal environment, sometimes the source can send the packet to the destination directly. Especially for DSR, there are more paths shorter than the "shortest" paths pre-calculated in NS2 because of shortening mechanism. So for $\beta=2.0$, the PDR of DSR could be higher than the ideal model. On the other hand, the signal power fluctuation can cause a high rate of transmission failures and bigger hop counts than under an ideal model.

(4) In DSR, there exist Route Request messages that are not sent by the source of data packets. It happens when a node wants to send a Route Reply or Error message and finds that the source route is not valid and also it cannot find any route in its route cache. Another source of protocol overhead is due to extra Route Error messages. When Route Error and Route Reply messages are salvaged, a Route Error message is sent to the source node.

3.3   Improving Routing Performance under a Shadowing Model

From the last section, we can see that both routing protocols have low PDR and long packet latency, caused by power fluctuation, in shadowing model. The performance is not acceptable in real applications. In this section, we present some proposals to increase the PDR based on our studies. However, before we discuss the specific proposals and results, we need to address another issue first: how to fairly compare results for different β values, which determine the average transmission range and therefore average network connectivity.

3.3.1   Mean Transmission Range and Required Node Density
Different β values correspond to different average transmission ranges; we can change Equation (3) to:

$$P_r = P_{r0} * 10^{\frac{(-10\beta \log(\frac{d}{d_0})+X_\sigma)}{10}} = \frac{P_t G_t G_r \lambda^2}{(4\pi)^2 d_0^2 L} * (\frac{d_0}{d})^\beta * 10^{0.1X_\sigma} \tag{4}$$

Because the mean of X_σ is zero, if we set P_r to the threshold P_s, we will find a mean transmission range for different β values with Equation (4). According to the parameters used in NS2, the radio frequency is 914MHz, P_t=0.281838 W, L=1, and threshold P_s=3.652x10^{-10} W. The reference distance d_0 is 1 m. The resulting average transmission ranges are shown in Table 2.

Table 2 shows that for a transmission range equivalent to the ideal model, which is 250 m, β=2.385. β values less than these correspond to a longer mean transmission range than ideal model, bigger values correspond to a shorter mean transmission range. To fairly compare the performance, we should conduct simulations under the same or similar conditions. If we still use the same number of nodes in the same area for higher β values, some nodes may be isolated. Bettstetter [13] proposed an algorithm for obtaining the minimum radio transmission range for a homogeneous node density so that every node in the network is connected. But in NS2 simulations, a random waypoint model is used, and this model does not result in a uniform node distribution, so this algorithm cannot be applied. We apply a simple rule for fair comparison: for two different simulations with different transmission range, the total node coverage for a certain area should be equal, i.e:

$$n_1 \frac{\pi r_1^2}{w_1 l_1} = n_2 \frac{\pi r_2^2}{w_2 l_2} \tag{5}$$

where r_1, r_2 are average radio transmission ranges, w_1, w_2 and l_1, l_2 are width and length of the simulation area, and n_1, n_2 are number of mobile nodes respectively. Based on our simulations with an ideal radio model with 50 nodes in a 1500x300 m area with transmission range of 250m, we calculate the required number of nodes for different β values in Table 2.

Table 2. Mean Transmission Range and Required Nodes for Different Beta Values.

β	2.0	2.1	2.2	2.3	2.385	2.4	2.5	2.6	2.7	2.8	2.9	3.0
r $_{(m)}$	725	530	398	307	250	242	194	158	131	110	93	80
#nodes	6	12	20	34	50	54	84	126	183	259	362	489

3.3.2 Improving Packet Delivery Ratio

In the shadowing model, signal power strength fluctuates and existing routes may become unstable if some links of a route are on the edge of the radio transmission range. The results in the previous sections also demonstrate that even a route discovered a very short time ago still might be assumed broken when the Route Reply is going to be sent. So if nodes can have more stable routes that resist to the fluctuation the route will live longer and a source node does not have to find a new route frequently, which in turn will increase the chance to successfully deliver data packets.

We can divide the stable route requirement into two parts. First we try to find some stable routes, second we have to maintain these routes because nodes are moving randomly. In this paper, we solve the first part and will give some suggestion for the second part. To find a stable route in a shadowing environment, it means that in most cases two nodes can communicate successfully though the signal strength is fluctuating. Looking at the shadowing model in Equation (3), X_σ represents the fluctuation, which is a random variable with normal distribution. To achieve stable routes, we want to minimize the probability that a transmitted data packet will be received at below the reception threshold. In a 99% Confidence Interval (CI), the interval width (for normal distributions) is $\pm 2.57583\sigma$. So applying a suitably higher reception threshold during route discovery can guarantee with high probability that the link appears stable. The CI used here is our first try; we can adjust this parameter to adapt to different environments. In the NS2 default setting, $\sigma=4$, so 2.57583x4=10.3. We use 10db as shown in Equation (3), which equals 10 times the base power strength. So we set a new threshold to 10 times the NS2 reception threshold, $10*3.652 \times 10^{-10}$ W=3.652×10^{-9} W. Note this new threshold is only used for the Route Discovery process. When a node receives a Route Request message, it will check whether the power strength is below the new threshold. If it is, the node will drop this packet. We will discuss if this new threshold should also apply to Route Reply messages in the next section.

SSA also uses signal strength to select long-lived routes. A node has to send a link layer beacon to its neighbors at a fixed rate to determine if the links are strong. Our proposal is different, the link selection only happens in Route Discovery process and without the overhead. We also provide a quantification of how to select this higher threshold and suggest further changes, as discussed next.

Section 3.2 showed that there are some extra control messages in DSR under a shadowing model. For example, a Route Discovery process will be triggered if Route Reply or Route Error messages do not have a route to send because the source route is invalid. But if the source route is not valid, this Route Reply message is not valid either. For the Route Error message, the upstream link is also not valid; the upstream node will handle the link failure. So the Route Discovery for these messages is not necessary. We modified the DSR protocol in NS2 so Route Discovery process will not start. Also NS2 salvages Route Reply and Route Error messages if the link is broken. We believe this is not necessary and causes more control messages, so a node will simply drop these kinds of packets.

4   Simulation Results

We start this section with DSR, and will discuss AODV in the second part. First we set $\beta=2.385$ for a mean transmission range of 250m, and run different mobility patterns. Next we change the β values and the total number of mobile nodes to explore the results under different environments.

4.1   Modifications to DSR

In DSR, we not only set a new threshold for the Route Discovery process, we also remove some unnecessary control messages. Figure 2 and Figure 3 show the performance for different combination of schemes. The simulation parameters are 50 nodes, β=2.385, pause time=0 s, max speed is 20 m/s. We run simulations for different modification schemes in DSR for the same mobility pattern and data traffic. The different schemes are as follows:

- with shadowing model and original DSR
- no route request for undeliverable Route Error and Route Reply messages
- no route request and salvage for undeliverable Route Error/Reply messages
- set new threshold only for Route Request messages
- set new threshold for both Route Request and Route Reply messages
- combining all the modifications listed above

The results for different modification schemes show that reducing the number of control messages and setting a new threshold in the Route Discovery process all improve the packet delivery ratio, and the combination of both achieves the best result. For the new threshold, if we also apply it for Route Reply messages, it can improve the PDR further. So in DSR, this new threshold is used for both messages. Because of the route shortening mechanism of DSR, a stable route may become unstable after transmission starts. So we turn the "shortening" option off in DSR and run the simulation again, the PDR increases from 43.77% to 56.70%, the average packet delay is further reduced from 2.06s to 1.168s. The shortening option is therefore turned off in the following DSR simulations.

4.2 Modifications to AODV

We also expand our idea to AODV. First we set the new signal strength threshold for the Route Request message, and then we set it also for the Route Reply message. The packet delivery ratio in the latter case is lower than when imposing the high threshold only for the Route Request message. This is because in AODV the route reply ratio is much lower than DSR. If we block Route Replies from less strong links, some packets may not have any routes and get dropped. So in AODV, we apply the new threshold only for Route Request messages.

Figures 2 and 3 are the simulation results for different mobility patterns using a shadowing model compared to the original DSR and AODV protocols. Here we set β=2.385, 50 nodes and 20 sources. We run seven simulations for each mobility pattern and the average values are shown in the figures. The mobility patterns are represented in pause time-max speed format. So 0-20 represents 0 second pause time and max 20m/s speed, which is the highest mobility rate in our simulations. 500-20 means pause time is 500 seconds, which equals to our simulation time. So all the nodes are stationary during the simulation time. The average hop count is the average number of hops a packet takes from the source to the destination.

Figures 4 and 5 show the simulation results of different β values for AODV and DSR. As presented in Table 2, the required number of nodes are different in a fixed simulation area with β varying, but we still use 20 source nodes to impose the same overall traffic. Pause time is set to 0 second; max speed is 20m/s. We run seven simulations for each β value.

From Figure 2 to Figure 5, we can draw the following conclusions:

Besides low β values (2.0 and 2.1), the packet delivery ratios for both DSR and AODV increase, and total numbers of control messages and packet latency decrease significantly by using a new threshold. The negative results of PDR for low β values are because of sparse node density, and limited Route Reply messages are reduced further by new threshold. When β=2.5, the packet latency under DSR increases. This is because the average PDR in the original DSR is very low, and delivered packets under our modifications are often several hops away. Considering the average transmission range is 80 meters, some packets may need more than 10 hops to reach the destinations. This longer route contributes to the longer latency.

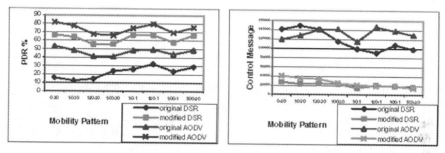

Fig. 2. PDR (left) and Total Control Messages (right) for Different Mobility Patterns

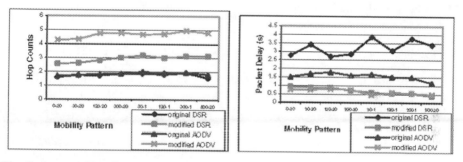

Fig. 3. Average Hop Counts (left) and Average Packet Delay (right) for Different Mobility Patterns

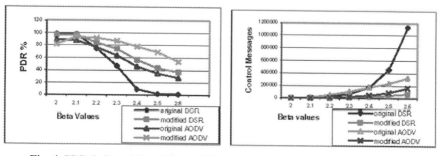

Fig. 4. PDR (left) and Total Control Messages (right) for Different Beta Values

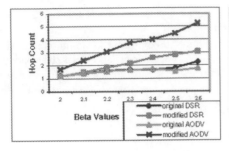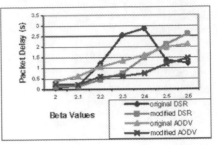

Fig. 5. Average Hop Counts (left) and Average Packet Delay (right) for Different Beta Values

Different mobility patterns do not show an apparent impact on the packet delivery ratio. The shadowing effect is the main cause for the reduced PDR. For DSR, the average improvement for PDR is at least 31 percentage points and 24 percentage points for AODV. For total number of control messages, the new threshold can reduce at least 77% for DSR, and 65% for AODV

For different mobility patterns and different β values, AODV has higher packet delivery ratio than DSR with and without the new threshold. Part of the reason is that in DSR a node can reply to a Route Request from its route cache. Though the new threshold has blocked the weak links, the rest of the route still cannot be verified. Under a shadowing model, the cached routes are more likely out of date.

When the new threshold is applied, the route reply ratio decreases rapidly, especially in DSR, as low as one tenth of original DSR's. This means that a number of weak links have been blocked. Also, DSR has higher route reply ratio than AODV, because in DSR nodes are set in promiscuous mode, so that they can copy the routes from the packets to their route caches.

The average hop count increases faster in AODV than in DSR with the new threshold, but its average packet latency is comparable to DSR. This is because of DSR's salvaging mechanism. A node might try several times to send a packet to the next hop, if it failed, then it will try another route in its route cache to send the packet.

The packet delivery rate decreases with increasing β value, which means shorter transmission range. A packet requires more hops to reach the destination and it has higher possibility to drop because of the power fluctuation.

5   Conclusions and Future Work

With a shadowing model, the signal strength will fluctuate by several order of magnitudes, so the links appear to become unstable. The consequence is decreased packet delivery ratio, increased control messages and longer packet latency. In this paper, we propose a new signal strength threshold applied in the Route Discovery process to select reliable links, then the route can resist to fluctuation caused by shadowing effect. Also unnecessary control messages are eliminated for DSR. After these modifications to DSR and AODV, in most cases the packet delivery ratio significantly increases and the number of control messages and packet latency are considerably reduced.

In the ideal environment, hop count is a valid metric for selecting routes. But with shadowing model, the shortest path means that there is high possibility that two nodes are on the edge of their transmission ranges. If signal power fluctuates, the link will be assumed broken, or more retransmissions are needed to send packets successfully. Based on the wok we did so far, we believe link status may be a better metric for selecting routes in shadowing model. The link status of a node is a historic record with its neighbors. A node may not add the link to the route if the link is weak. This link status will be monitored passively to reduce overhead. A node may keep multiple routes to the same destination, and these routes can be ordered based on their status. Updating the route status is a future research topic, especially for inactive routes. However, salvaging is a mechanism that has to be used carefully. Sometimes the sender may salvage the packet only because of not receiving the ACK. The consequence of aggressively using salvaging is that the same packet is transmitted on multiple routes to the same destination, and multiple Route Error messages are sent to the source, which cause longer packet delay, duplicated packets and wastes network resource.

We are currently working on

(1) Route maintenance: because of node mobility, selected solid routes may become unstable again. A node may monitor the signal strength for packets received/overheard from its neighbors, and inform the source once the link quality drops below a certain threshold. Alternatively, a node could monitor the number of retransmission attempts necessary to forward a packet to the next node and deduce the link quality and its rate of change in that way.

(2) Finding an appropriate threshold for different node density and β values. We can also use different thresholds for Route Requests and Route Replies so that a source can discover at least one route to send packets. This may be done adaptively, where a source node may have to reduce the threshold if it cannot discover a route.

(3) More formally defining equivalent simulation scenarios under different parameters of the physical layer model as well as exploring additional physical layer models.

References

1. C. E. Perkins and P. Bhagwat, Highly Dynamic Destination-Sequenced Distance-Vector Routing (DSDV) for Mobile Computers, Proceedings of the Conference on Communications Architectures, Protocols and Applications, pages 234-244, London, England August, 1994.
2. V. Park and M.S. Corson, A Highly Adaptive Distributed Routing Algorithm for Mobile Wireless Networks, Proceedings of the IEEE Conference on Computer Communications (INFOCOM '97), Kobe, Japan, April 1997.
3. D. B. Johnson and D. A. Maltz, Dynamic Source Routing in Ad Hoc Wireless Networks. In Mobile Computing, Chapter 5, pages 153-181, Kluwer Academic Publishers, 1996.
4. R. Dube et al., Signal Stability-Based Adaptive Routing (SSA) for Ad Hoc Networks. IEEE Personal Communications, February 1997.
5. C. E. Perkins and E. M. Royer, Ad-hoc On-Demand Distance Vector Routing. Proceedings of the 2nd IEEE Workshop on Mobile Computing Systems and Applications, pages 90-100, New Orleans, LA, February 1999.

6. J. Broch et al., A Performance Comparison of Multi-Hop Wireless Ad Hoc Network Routing Protocols. Proceedings of the Fourth Annual ACM/IEEE International Conference on Mobile Computing and Networking, Dallas, TX, October 1998.
7. R. Samir et al., Performance Comparison of Two On-demand Routing Protocols for Ad Hoc Networks. Proceedings of the IEEE Conference on Computer Communications (INFOCOM), pages 30-12, Tel Aviv, Israel, March 2000.
8. M. Takai, R. Bagrodia, K. Tang and M. Gerla, Efficient Wireless Network Simulations with Detailed Propagation Models. Wireless Networks, pages 297-305, Vol. 7 No. 3, May 2001.
9. T. Goff and N. B. Abu-Ghazaleh, Preemptive Routing in Ad Hoc Networks. Proceedings of the Seventh Annual International Conference on Mobile Computing and Networking, pages 43-52, Rome, Italy, July 2001.
10. T. S. Rappaport, Wireless Communications: Principles and Practice. Second edition. 2002, Prentice Hall PTR. ISBN 0-13-042232-0.
11. E. M. Royer and C. -K. Toh, A Review of Current Routing Protocols for Ad-Hoc Mobile Wireless Networks. IEEE Personal Communications Magazine, pages 46-55, April 1999.
12. K. Fall and K. Varahan, editors. NS Notes and Documentation. The VINT Project, UC Berkeley, LBL, USC/ISI, and Xerox PARC, November 1997, see http://www.isi.edu/nsnam/ns/.
13. C. Bettstetter, On the Minimum Node Degree and Connectivity of a Wireless Multihop Network. Proceedings of the 3rd ACM International Symposium on Mobile Ad Hoc Networking and Computing (MobiHoc), pages 80-91, Lausanne, Switzerland, June 2002.

Architecture and Algorithms for Real-Time Mobility Management in Mobile IP Networks

Marcellin Diha[1] and Samuel Pierre[2]

[1] Mobile Computing and Networking Research Laboratory (LARIM),
Department of Computer Engineering,
Ecole Polytechnique of Montreal, P.O. Box 6079,
Station Centre-ville, Montreal, Quebec, Canada, H3C 3A7
marcellin.diha@motorola.com
[2] Mobile Computing and Networking Research Laboratory (LARIM),
Department of Computer Engineering,
Ecole Polytechnique of Montreal, P.O. Box 6079,
Station Centre-ville, Montreal, Quebec, Canada, H3C 3A7
samuel.pierre@polymtl.ca

Abstract. This paper proposes new network architecture and algorithms for real-time mobility management in mobile IP networks. The proposed architecture and algorithms offer a better performance based on the call-to-mobility ratio (CMR) and require less time for the location update and the tunneling compared with the Mobile IP model. These results are very useful and interesting for a real-time context where the factor time is very important.

1 Introduction

The development of cellular networks brought many changes in the telephony area. Voice and mainly data transmission have increased a lot. The traditional telephony made place to new types of services that integrate both voice and data. It leads to the design and implementation of new hybrid networks able to fulfill this need. The mean-term goal is to have cellular networks entirely based on the IP protocol. But several problems must be solved before achieving this goal. Among these problems is the mobility management in IP networks because originally these networks were designed fix. In addition, the real-time aspect that is not supported in the actual IP protocol must also be addressed.

The IETF developed a mobile version of IP protocol able to manage users' mobility in IP networks [4], [6]. Three main algorithms are proposed for mobility management: users' registration to a local router, foreign routers' discovery and location update when users are away from their home network, and finally tunneling and data routing to mobile users. The Mobile IP protocol is implemented in several local area networks and works well. But real-time features are not supported in the protocol.

This paper proposed new network architecture and real-time mobility management algorithms based on the existing ones. The remainder of this paper is organized as follows. In Section 2, background and related work are presented. Section 3 described

S. Pierre, M. Barbeau, and E. Kranakis (Eds.): ADHOC-NOW 2003, LNCS 2865, pp. 49–59, 2003.

the proposed architecture and algorithms as well a performance analysis. Section 4 presents some simulation results and analysis. Finally, the paper concludes with a discussion on open problems faced by real-time mobility management in IP networks.

2   Background and Related Work

There are two major components in mobility management: handover management and location management [1], [9].

Handover management is the way a network uses to maintain connection to a mobile user as it moves and changes its access point to the network. In general, there are two types of handover: intra-cell handover and inter-cell handover [1]. The first type occurs when within a cell a user experiences degradation of signal strength. This leads to a choice of new channels having better signal strength at the same Base Transceiver Station (BTS). The second type occurs when a user moves from one cell to another cell. In this case, the user's connection information is transferred from the old BTS to the new one. In both intra-cell and inter-cell handover, the following procedure is performed. First, the user initiates a handover procedure. Then the network or the mobile (depending on the unit that controls the handover operation) provides necessary information and performs routing operations for the handover. Finally, all subsequent calls to the user are transferred from the old connection to the new one.

Location management is the process a network uses to find the current attachment point of a mobile user for call delivery. The first step of the procedure is the *location registration*. In this phase, the mobile user periodically notifies the network of its new access point. The notifications allow the network to authenticate the user and update its location profile. The second step is the *call delivery*. When a call belonging to a user reached the network, a search for the user's profile is made usually in a local database. Then the call is forwarded to the user based on the information contained in its profile.

Mobility support in the IP protocol has been developed by the IETF leading to the Mobile IP protocol [3], [6], [8]. Currently two versions of   Mobile IP are available, versions 4 (IPv4) and 6 (IPv6). In this paper we focus on IPv4 since it is actually the most implemented one.

A Mobile Node (MN) is a node able to move from one subnet to another without any need of changing its IP address. The MN accesses the Internet via a Home Agent (HA) or a Foreign Agent (FA). The Correspondent Node (CN) is a node establishing a connection with the MN. The HA is a local router on the MN's home network and the FA is a router on the visited network.

The following operations are introduced by the Mobile IP protocol [4], [6].

1. *Discovery*: How an MN finds an agent (HA or FA).
2. *Registration*: How an MN registers with its HA.
3. *Routing and Tunneling*: How an MN receives datagrams when visiting a foreign network [5], [7].

Location management operations include agent discovery, movement detection, forming care-of-address, and location update. Handover operations include routing and tunneling.

Figure 1 illustrates Mobile IP network architecture.

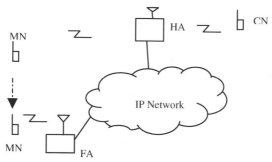

Fig. 1. Mobile IP network architecture

3 Architecture and Algorithms Proposed

This section presents a new Mobile IP network architecture and new algorithms supporting real-time features as well as performance analysis.

3.1 Proposed Architecture

Figure 2 shows the proposed Mobile IP network architecture.

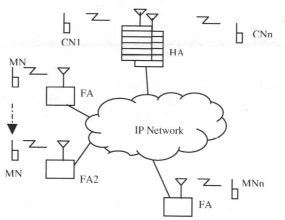

Fig. 2. Proposed architecture of Mobile IP network

The architecture introduces the following main features:
1. Real-time algorithms support.
2. Connection of MNs and CNs to an FA or HA with different arrival rates in the network.

3. All procedures associated with an MN (registration, discovery, tunneling and routing) represent different tasks with a specific priority.
4. Multiprocessor agent (HA or FA). In this paper the emphasis is put on the HA. Also the Home Agent is redundant to allow failure recovery.
5. A main processor dispatching the different tasks arriving on an agent.
6. A set of faster processors is defined to process high-priority tasks.
7. Architecture allowing different speeds for the processors.

3.2 Proposed Algorithms

Based on the architecture described above, a set of new algorithms has been defined for mobility management. These algorithms are derived from Mobile IPv4 algorithms [6]. They introduce the notion of priority management in a real-time context. The new discovery algorithm adds the ranging concept in addition to the lifetime used in Mobile IPv4. Also it allows the MN to initiate Foreign Agents search at startup in stead of waiting advertisements. In addition, this algorithm allows the MN to keep a list of the most recent Foreign Agents that it tries to contact first before initiating any broadcast search.

3.2.1 Tasks Scheduling and Assignment Algorithm

```
-Given i counter of tasks and j counter of processors.
-Given n tasks TSK_1, ...,TSK_i, ..., TSK_n with the priorities p_1, ... p_i, ... p_n.
-Given S a set of faster processors and p_s threshold of a critical task.
-Given u(i) the utilization rate of the task TSK_i and U(j) a vector of u(i) on P_j.
BEGIN
  -i= 1, U(j)= 0, S =1.
  -Sort the n tasks based on u(i) on P_0.
  WHILE i ≤ n DO
     j = min{k|U(k) + u(i) = 1}
     IF ( (p_i > p_s) && (∃ P_R ∈ S | ∑ U(P_R) < 1)) ) THEN  TSK_i → P_R
     ELSE TSK_i → Pj
     i ← i + 1
  END WHILE
END BEGIN
```

Fig. 3. Task scheduling and assignment algorithm

The scheduling part of the algorithm is based on the EDF algorithm [2] while the assignment part is a totally new concept since it based on a multiprocessor architecture. The tasks are sorted based on the deadline and assigned to the processors. If a task is critical (short deadline), it is assigned to a faster processor. If not it is assigned to a normal processor. A task is assigned to a processor only if its current utilization rate is less than 1. This ensures that a processor is not used at its full capacity while others are unused.

3.2.2 Registration Algorithm

The *registration procedure* is a task running on the HA with the *highest priority*. It can preempt any other mobility management task for a given user. For example, during a tunneling procedure, if a registration request is received for the same user, the tunneling process will be delayed until the registration is done. The different stages of the algorithm are described as follow.

1. MN sends a registration request to the HA.
2. HA verifies IF a task other than the registration is in process for the same user.
   IF yes THEN the task is preempted by the registration task.
3. HA sends a response to the MN.
4. IF request accepted THEN registration procedure done ELSE MN retries UNTIL request accepted.

3.2.3 Discovery Algorithm

The discovery algorithm introduces also the notion of priority in a real-time environment and it is based on the lifetime expiration and the ranging. The *discovery procedure* has the *second highest priority*. The different steps are described as follow.

1. IF first time startup THEN MN sends a broadcast advertisement.
2. FAs verify if no higher priority task is being executed for the same MN.
   IF yes THEN delay discovery process UNTIL high-priority task execution is done.
3. FAs send responses back to MN.
4. MN chooses FA with most strong signal strength and records the lifetime, the care-of-address and the FA's IP address.
5. IF lifetime expires or the MN starts going out of range (wick signal strength) THEN send registration request to Foreign Agents in the MN's local database.
   IF no FA responds back THEN broadcast a discovery advertisement message.
6. REPEAT steps 2 through 4 UNTIL registration succeed.
7. IF registration succeeds THEN MN sends new location information to HA for location update.

3.2.4 Routing and Tunneling Algorithm

The new routing and tunneling algorithm also introduces the notion of priority in a real-time environment. *This procedure* has the *lowest priority*. Thus, during a tunneling procedure, if a registration procedure is received for the same user, the location procedure will be suspended until the registration is done. The steps of the algorithm are the following:

1. HA receives data for an MN.
2. HA verifies if a registration request is made for the same user.
   IF yes THEN HA suspends tunneling process until registration is done.
3. IF MN in local network THEN delivered packets using normal IP packets delivery procedure ELSE forward packets to MN via its current FA.

3.3 Performance Analysis

The performance analysis is based on the CMR (Call-to-Mobility Ratio). The CMR is the average number of messages send to a user divided by the average number of networks or subnets visited by the user in a given time stamp. The goal of the CMR analysis is to determine the ratio by which the proposed model reduces the location update and the tunneling times.

$$CMR = \frac{\lambda}{\mu} \tag{1}$$

Where λ is the average number of messages send to a user and μ the average number of subnets or networks visited by the user between two consecutives messages.

We define the following parameters to compare the CMR in the Mobile IP and the proposed models:

U cost for location update procedure execution in the Mobile IP model;

L cost for tunneling procedure execution in Mobile IP model;

u cost for location update procedure execution in the proposed model;

l cost for tunneling procedure execution in the proposed model;

T cost to cross a boundary between two subnets;

U_{MIP} total cost for the location update procedure in Mobile IP model;

L_{MIP} total cost for tunneling procedure execution in Mobile IP model;

C_{MIP} total cost for location update and tunneling procedures execution in Mobile IP model;

U_{PROP} total cost for location update procedure in the proposed model;

L_{PROP} total cost for tunneling procedure execution in the proposed model;

C_{PROP} total cost for location update and tunneling procedures execution in the proposed model.

The total costs are obtained as follows:

$$U_{MIP} = \frac{U}{CMR} \tag{2}$$

$$C_{MIP} = U_{MIP} + L_{MIP} + T = \frac{U}{CMR} + L + T \tag{3}$$

$$U_{PROP} = \frac{u}{CMR} \tag{4}$$

$$C_{PROP} = U_{PROP} + L_{PROP} + T = \frac{u}{CMR} + l + T \tag{5}$$

Our goal is to reduce the location update and the tunneling times by respectively at least 50% and 25%. Then $u = U/2$ and $l = L/4$. We make the following assumptions to simplify the analysis $L = U$, $T = U/4 = u/2$ and $U = 1$, that leads to:

$$\frac{U_{MIP}}{U_{PROP}} = \frac{U}{u} = \frac{1}{u} \tag{6}$$

$$\frac{L_{MIP}}{L_{PROP}} = \frac{L}{l} = \frac{2}{u} \tag{7}$$

$$\frac{C_{MIP}}{C_{PROP}} = \frac{U(4+5CMR)}{4u+2CMR} = \frac{4+5CMR}{4u+2CMR} \tag{8}$$

Figures 4 shows the location update cost for different values of the CMR and $u = 0.2$.

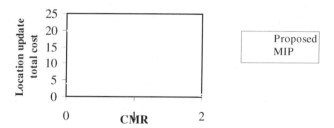

Fig. 4. Comparison of location update total cost (u = 0.2)

Figures 5 illustrates the location update cost for different values of the CMR and $u = 0.5$.

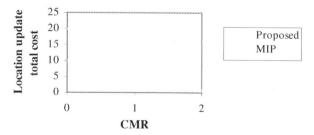

Fig. 5. Comparison of location update total cost (u − 0.5).

Figure 6 shows the location update cost and the ratio C_{MIP}/C_{PROP} for different values of the CMR for $u = 0.2$ and 0.5.

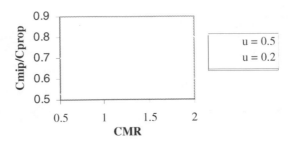

Fig. 6. Comparison of ratio C_{MIP}/C_{PROP}

On Figures 4 and 5 we noticed that the proposed model reduces by 80% the location update time for u = 0.2 and by 50% for u = 0.5 for small values of the CMR. The reason is that in this case the mobile users make an important number of location update requests and the processing on the multiprocessor HA in the proposed architecture is faster compared with the mobile IP model. When the CMR increases, the mobile users stay longer in the same network and, in this case, the location update time decreases and is near 0 for the two models. For the total costs (location update plus tunneling) shown in Figure 6, the reduction is between 67% and 80% for u = 0.2 and between 50% and 60% for u = 0.5.

We can conclude that the proposed model offers a better performance based on the CMR. Indeed, it takes less time for the location update and the tunneling in the proposed model compared with the Mobile IP model. These results are very useful and interesting for a real-time context where the factor time is very important. To validate the analysis, we conducted different simulations in section 4.

4   Computational Results

Figure 7 shows the setup for the different simulations. The network used is an Ethernet based LAN 10/100 Mbps with an 8-Port Ethernet Hub.

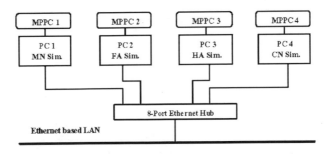

Fig. 7. Simulation Setup

We simulated the implementation of the current mobile IP algorithms as well as the proposed architecture and algorithms in a real-time environment using VxWorks as real-time Operating System running on a MPPC (Motorola™ Power PC). Our simulations focus on the location update average time, the tunneling average time, the number of tasks missing their deadline depending on the number and the speed of the processors on the HA.

4.1 Location Update Average Time

In the Mobile IP model, the location update time is constant and does not depend on the user's arrival rate in the network. The location update average rate value is around 1 sec. In the proposed model, the location update time increase with the user's arrival rate. The maximum value is around 0.8 s. For small value of the arrival rate, few users arrived in the network and only few processors are used for the processing increasing the location update time. But when the arrival rate increase, many users arrived in the network and the processing is faster for the multiprocessor architecture in the proposed model. Overall the location update time is reduces by 20 % to 80% in the proposed model which is above the targeted objective of 50%.

Figure 8 shows the location update average time for different arrival time in the networks.

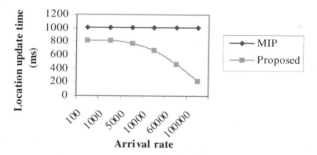

Fig. 8. Location update average time

4.2 Tunneling Average Time

Figure 9 illustrates the tunneling average time for different data sizes.

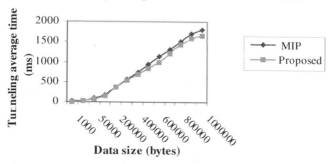

Fig. 9. Tunneling average time

The tunneling time is smaller in the proposed model compare to the Mobile IP model. The data processing is faster because of the multiprocessor architecture. The messages sent to the mobile users spend less time in the message queue on the HA. In the mobile IP model the message $i + 1$ will wait longer than the message i on the message queue, increasing the processing time. But in the proposed model the message $i + 1$ can be process in parallel on a different processor while processing the message i on an other one. As a result the total time spent in the system is reduced. Overall the tunneling time is reduces by 10 % to 30% in the proposed model which is above the targeted objective of 25%.

4.3 Task Scheduling and Assignment

Figure 10 shows the number of tasks missing their deadline for different number of processors with different speeds following Gaussian and exponential distributions.

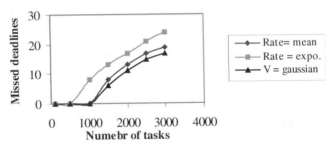

Fig. 10. Processors with different speeds distribution

The number of tasks missing their deadline in the Gaussian distribution is lower compare to the exponential distribution. The reason is that in the first case the speeds of the processors are close to the mean speed. It is the contrary in the exponential case where the distribution is larger with more low speeds. This leads to a higher ratio Execution Time/Processor Speed and number of missed deadlines. So, for configurations with different speeds, the speeds of the processors must follow a Gaussian distribution in order to have an optimal scheduling and assignment for the tasks.

5 Conclusion

In this paper we presented a Mobile IP architecture and mobility management algorithms in a real-time context. The implementation of the proposed architecture and algorithms gave better results for the location update and tunneling average times as well as the CMR compare to the existing architecture and algorithms. The location update time is reduced by 20% to 80% while the tunneling time is reduced by 10% to 30%. These results meet time constraint in real-time systems. The multiprocessor architecture is the core of the proposed model. It gives a faster parallel processing for the mobile users.

The scheduling and assignment algorithm is optimal for different number of processors with different speeds. This achievement is something new compare to actual real-time multiprocessor scheduling and assignment algorithms. In the current algorithms, the processors must have the same speed to guarantee an optimal scheduling and assignment.

Many investigations are on going in real-time mobility management for Mobile IP networks. The areas cover the implementation of real-time algorithms in real networks as well as proposition of new algorithms and architectures. Also, since the current protocols are designed for micro-mobility, the WAN and global roaming areas are some new domains of interest.

References

1. Akyildiz I. F., McNair J., Ho J. S. M., Uzunalioglu H., Wang W.: Mobility Management in Next-Generation Wireless Systems, *Proceedings of the IEEE*, vol. 87, no. 8, pp. 1347-1384, Aug. 1999.
2. Chrishna C.M., Kang G. S.: Real-Time Systems, *McGraw Hill*, 1997.
3. D. B. Johnson, C. Perkins: Mobility support in IPv6, *Inter-net Engineering Task Force*, Internet draft, draft-ietf-mobileip-ipv6-22.txt, May 2003.
4. James D. S.: Mobile IP, The Internet Unplugged, *Prentice Hall PTR*, 1998.
5. P. Calhoun, C. Perkins: Tunnel establishment protocol, *Internet Engineering Task Force*, Internet draft, draft-ietf-mobileip-calhoun-tep-01.txt, March 1998.
6. Perkins C.: IP Mobility Support, *Internet Engineering Task Force*, RFC 2002, Oct. 1996.
7. Perkins C. and Johnson D. B.: Route Optimization in Mobile IP, *Internet Engineering Task Force*, Internet drafts, draft-ietf-mobileip-optim-11.txt, Sept. 2001.
8. Perkins C.: Minimal Encapsulation within IP, *Internet Engineering Task Force*, RFC 2004, Oct. 1996.
9. R. Caceres and V. Padmanabhan: Fast and scalable handoffs for wireless networks, in *Proc. ACM/IEEE MOBICOM'96*, pp.56–66.

Proactive QoS Routing in Ad Hoc Networks

Ying Ge[1], Thomas Kunz[2], and Louise Lamont[1]

[1] Communications Research Center, 3701 Carling Ave, Ottawa, ON K2H 8S2
{ying.ge,louise.lamont@crc.ca}
[2] Dept. of Systems and Computer Engineering, Carleton University, Ottawa, ON K1S 5B6
tkunz@sce.carleton.ca

Abstract. In this paper, we analyze the advantages and disadvantages of the proactive QoS routing in ad-hoc networks. We discuss how to support bandwidth QoS routing in OLSR (Optimized Link State Protocol), a best-effort proactive MANET routing protocol. Using OPNET, we simulate the algorithm, exploring both traditional routing protocol performance metrics and QoS-specific metrics. Our analysis of the simulation results shows that the additional message overhead generated by the proactive QoS routing have a negative impact on the performance of the routing protocol. Given the negative results, we identified research areas that would be worthwhile investigating in order to obtain better performance results.

1 Introduction

QoS routing in Ad-Hoc network is difficult. To support QoS routing, the link state metrics such as delay, bandwidth, jitter, loss rate and error rate in the network should be available and manageable. However, getting and managing such link state information in a MANET is not trivial because the quality of a wireless link changes quite frequently due to mobility and variations in the surroundings. In addition, it is also complex to evaluate the QoS routing performance. Compared to the traditional best-effort routing, QoS routing has two additional overheads – "computational cost" and "protocol overhead" [2]. "Computational cost" comes from the more frequent path selection computations, since besides maintaining the source-destination connection, additional computations are needed to determine paths that satisfy the QoS demands. The additional "protocol overhead" comes from the need to distribute the frequently updated link state information. There is a trade-off between the QoS performance the QoS routing protocol achieves and the additional cost it introduces.

In on-demand QoS protocols such as [3] and [11], a route is found based on specific QoS requirements. However, the unpredictable nature of Ad-Hoc networks and the requirement of quick reaction to QoS demands make the idea of a proactive protocol more suitable. When a request arrives, the control layer can easily check if the pre-computed optimal route can satisfy such a request. Thus, waste of network resources when attempting to discover infeasible routes is avoided. These advantages of the proactive QoS routing motivate us to look into this area. However, similar to a proactive best-effort routing protocol, a proactive QoS routing may introduce "protocol" overhead. Do these additional overhead have a negative effect on the Ad-Hoc network? If yes, then how much additional overhead does a proactive QoS routing protocol introduce into the network? How does the additional overhead affect the performance of the routing protocol? Can we minimize the costs to achieve better

S. Pierre, M. Barbeau, and E. Kranakis (Eds.): ADHOC-NOW 2003, LNCS 2865, pp. 60–71, 2003.

performance? Or should we just give up on proactive QoS routing? The goal of this paper is to investigate the answers to these questions through the performance evaluation of a proactive bandwidth QoS routing algorithm that we have proposed.

In [5], we studied the approach of proactive QoS routing and proposed 3 heuristics that allow OLSR (Optimized Link State Protocol [8]) to pre-compute the best bandwidth route among all the possible routes. That work presents the performance of the heuristics in a static network. In this paper, we implement one QoS OLSR heuristic, which guarantees to find the best bandwidth path in the static network and has comparably low overhead, in OPNET and evaluate the routing algorithm's performance with node movements and data flows, and consequently, analyze the feasibility of proactive routing in MANET.

The rest of the paper is organized as follows: a brief description of OLSR and QoS versions of OLSR is given in Section 2. The detailed implementation of QoS OLSR in OPNET is discussed in Section 3. Section 4 lists the OPNET simulation parameters and discusses the simulation results in OPNET. Section 5 analyses whether proactive QoS routing is practical in an Ad-Hoc network and discusses future work.

2   OLSR and QoS OLSR

The IETF's MANET Working Group has introduced the Optimized Link State Routing (OLSR) protocol for mobile Ad-Hoc networks [8]. The protocol is an optimization of the pure link state algorithm. The key concept used in the protocol is that of multipoint relays (MPRs). The MPR set is selected such that it covers all nodes that are two hops away. A node's knowledge about its neighbors and two-hop neighbors is obtained from HELLO messages – the message each node periodically generates to declare the nodes that it hears. The node N, which is selected as a multipoint relay by its neighbors, periodically generates TC (Topology Control) messages, announcing the information about who has selected it as an MPR. Apart from generating TCs periodically, an MPR node can also originate a TC message as soon as it detects a topology change in the network. A TC message is received and processed by all the neighbors of N, but only the neighbors who are in N's MPR set retransmit it. Using this mechanism, all nodes are informed of a subset of all links – links between the MPR and MPR selectors in the network. So, contrary to the classic link state algorithm, instead of all links, only small subsets of links are declared. For route calculation, each node calculates its routing table using a "shortest hop path algorithm" based on the partial network topology it learned. MPR selection is the key point in OLSR. The smaller the MPR set is, the less overhead the protocol introduces. The proposed heuristic in [8] for MPR selection is to iteratively select a 1-hop neighbor that reaches the maximum number of uncovered 2-hop neighbors as an MPR. If there is a tie, the one with higher degree (more neighbors) is chosen.

Table 1. Node B's MPR(s), based on Fig. 1.

Node	1 Hop Neighbors	2 Hop Neighbors	MPR(s)
B	A, C, F, G	D, E	C

From the perspective of node B, both C and F cover all of node B's 2-hop neighbors. However, C is selected as B's MPR as it has 5 neighbors while F only has 4 (C's degree is higher than F).

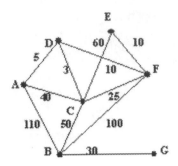

Fig. 1. Simple network. An edge between two nodes indicates that the two nodes connected by this edge are within reach of each other. The edge weight represents the QoS link attribute we are interested in, available bandwidth.

OLSR is a routing protocol for best-effort traffic, with emphasis on how to reduce the overhead. So in its MPR selection, the node selects the neighbor that covers the most unreachable 2-hop neighbors as MPR. However, in QoS routing, by such MPR selection mechanism, the "good quality" links may be "hidden" to other nodes in the network. As an example, we will consider the network topology in Fig. 1. In the OLSR MPR selection algorithm, node B will select C as its MPR. So for all the other nodes, they only know that they can reach B via C. Obviously, when D is building its routing table, for destination B, it will select the route D-C-B, whose bottleneck bandwidth is 3, the worst among all the possible routes. Also, when "bandwidth" is the QoS constraint, nodes can no longer use the "shortest hops path" algorithm as proposed in OLSR. Because of these limitations of OLSR in QoS routing, the QoS OLSR version revises it in two aspects: MPR selection and routing table computation.

The decision on how each node selects its MPRs is essential to determining the optimal bandwidth route in the network. In selecting the MPRs, a "good bandwidth" link should not be omitted. Based on this idea, we previously explored three revised MPR selection algorithms [5]. In this paper, we implement the best variant (OLSR_R2) in OPNET to compare its performance with the original OLSR protocol. The idea behind OLSR_R2 is to select the best bandwidth neighbors as MPRs until all the 2-hop neighbors are covered.

Table 2. Node B's MPR(s), using OLSR_R2.

Node	1 Hop Neighbors	2 Hop Neighbors	MPR(s)
B	A, C, F, G	D, E	A, F

Among node B's neighbors, A, C, and F have a connection to its 2-hop neighbors. Among them, link BA has the highest bandwidth. So A is first selected as B's MPR, and the 2-hop neighbor D is covered. Similarly, F is selected as MPR next and E is covered, so all 2-hop neighbors are covered and the algorithm terminates. This revised OLSR MPR selection algorithm improves the chance that a better bandwidth route is found. However, by using such algorithm, the overhead also increases because the number of MPRs in the network is increased.

Besides the MPR selection method, a node also needs to change the "shortest hops path" algorithm in its routing table computation so as to find the best bandwidth route. We use the "Extended BF" algorithm [6], which computes the best bandwidth paths from a source to any reachable destinations with minimum hop count (shortest-widest path).

With bandwidth constraint as QoS metric, it is reasonable to view the "bandwidth" as available bandwidth. Most probably, the devices in the Ad-Hoc network will be configured with the same wireless card, which means that all nodes in the network have the same maximum bandwidth. So we are only interested in how much of the remaining bandwidth is available for new traffic. However, in real networks, bandwidth computation is a complex issue. Many papers such as [9] discuss how to compute bandwidth in Ad-Hoc networks. Here, we use a rather simple and straightforward approach: measuring how much time a node monitors an idle channel and thus is available to transmit new messages over a link (node's idle time), which is similar to the approach suggested in [1].

3 QoS OLSR Implementation in OPNET

The Naval Research Laboratory (NRL) of the United States Department of Defense developed the original OLSR model in OPNET. To implement the QoS versions of the OLSR protocol, besides changing the MPR selection mechanism and the routing table calculation, the following revisions are made to develop the QoS OLSR model.

QoS OLSR uses the media idle time to reflect the available bandwidth over a link. Modifying the standard OPNET Wireless LAN model achieves this task. Each OPNET OLSR node connects to the wireless media. The OPNET Wireless LAN simulation model includes a transmitter, and a receiver. If a node is sending packets, its transmitter becomes busy. If there are other nodes beginning transmission within the interference range of the current node, its receiver senses the busy media and sends a media busy signal. As the OPNET Wireless LAN model already defines functionalities to capture changes of the media, the media idle time calculation, using a sliding window over the past 5 seconds, is straightforward.

Also, the QoS OLSR versions need to know the available bandwidth on the neighbor link to select MPRs, and the available bandwidth of the far-away links to compute the routing table. As idle time should be used to calculate the available bandwidth on the links, we revise the format of OLSR Hello and TC messages to include the idle time.

a. Hello message: in addition to the original information such as neighbor address and neighbor link type, a node also includes its own idle time in the Hello messages. Upon receiving a Hello message from its neighbor, a node reads the neighbor idle time, and selects MPRs using the QoS MPR selection algorithm.

b. TC message: the TC message originator not only puts its own idle time in TC messages, but also piggybacks its MPR selectors' idle times, which are obtained from the Hello messages. When a node receives TC messages, it knows the idle time information of both the TC message originator and the MPR selectors, thus gets information about the links and the link bandwidth between the TC message originator and its MPR selectors. In this way, it learns the partial network topology and the bandwidth condition of that partial network, and is ready to calculate the routing table.

Furthermore, QoS OLSR needs to decide when to originate a TC message. In the original OLSR, if a node detects changes in its MPR selector, it generates a new TC message to propagate the changes in the network topology. In QoS OLSR, however, changes in link bandwidth condition must also be propagated for the correct computation of the best bandwidth routes. If an MPR generates a TC message as soon as it detects a bandwidth change over the link between its MPR selector and itself, there will be many messages flooding into the network, causing extremely high overhead. So in our QoS OLSR version, a "threshold" of bandwidth change is defined. If an MPR finds there is "significant bandwidth change", it will generate a new TC message informing the whole network about the change, enabling other nodes to update their routing table reflecting such changes. There is a tradeoff in how to define the "threshold". On one hand, if the "threshold" is low, TC messages will be generated as soon as there is a small percentage change of the bandwidth. That will cause frequent generation of TC messages, introducing high overhead, although more accurate bandwidth information is obtained. On the other hand, if the "threshold" is high, TC messages will not be generated until there is a very large percentage change of the bandwidth. Thus, the overhead is reduced, but the nodes only obtain relatively inaccurate bandwidth information. In the rest of the paper, we will utilize different "threshold" values to compare the network performance, and analyze the performance tradeoffs.

4   OPNET Simulation

The following environment parameters are defined for OPNET simulations:

Movement Space: 1000m x 1000m flat space

Number of Nodes: 50 nodes

Simulation Time: 900 seconds.

Movement Model: each node randomly selects a destination in the 1000m x 1000m area, moves to that destination at a speed distributed uniformly between 0 and "maximum speed". After it reaches the destination, the node selects another destination and another speed between 0 and "maximum speed", and moves again. In the set of experiments reported here, we use 5 different "maximum speed" values: 20m/s, 10m/s, 5m/s, 1m/s, and 0m/s.

Communication Model: In each simulation, there are 20 communication pairs. Each source sends 64-byte UDP packets at a rate of 4 packets/second. So in total, 80 packets are sent each second.

OPNET Model Parameter: see Table 3.

Routing Protocol: 4 routing protocols – Original OLSR, QoS OLSR with 20% bandwidth updating threshold (20% OLSR), QoS OLSR with 40% bandwidth updating threshold (40% OLSR), and QoS OLSR with 80% bandwidth updating threshold (80% OLSR). All the QoS OLSR algorithms use the OLSR_R2 [5] mechanism to select MPRs, and the "Extended BF" algorithm to calculate the routing table.

Table 3. OPNET Model Parameter.

OLSR Parameters	Hello Interval	_0.5s_
	TC Interval	2s
Wireless LAN Parameters	Data Rate	2 Mbps
	Buffer Size	256000 bits
	Retry Limit	7
	Wireless LAN Propagation Range	250 M

The OPNET simulation results are grouped into two sets: Basic Performance and QoS Performance (The data shown in this section are the average value from multiple runs.)

Basic Performance – the basic performance is measured by a set of metrics used to evaluate most routing protocols: "Packet Delivery Ratio" and "End to End Delay".

**Packet Delivery Ratio:** percentage of packets that successfully reach the receiving nodes each second.

**End to End Delay**: the average time between a packet being sent and being received

QoS performance – metrics that relate to the bandwidth QoS routing studied in this paper: "Error Rate" and "Bandwidth Difference".

**Error Rate:** the percentage of times the routing algorithms do not find the optimal bandwidth path.

**Bandwidth Difference:** the average difference between the optimal bandwidth and the detected non-optimal bandwidth in percentage.

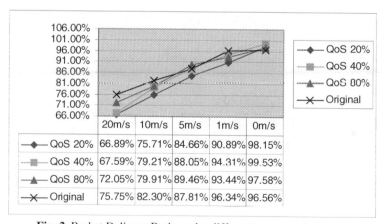

Fig. 2. Packet Delivery Ratio under different movement patterns.

Fig. 2 compares the packet delivery ratio of the 4 algorithms. From high movement to low movement, packet delivery ratio for all algorithms rises continuously. With lower movement, the established links between the nodes have a lower probability of breaking, thus, there are less stale routes in the node routing tables, which results in a higher ratio for correct packet delivery. In low movement scenarios (speed 5m/s, 1m/s and 0m/s), the 4 algorithms achieve similar packet delivery ratio. However, in high movement scenario, the original OLSR protocol has higher packet delivery ratio than the 3 QoS versions, especially with a speed of 20m/s where the performance difference between the QoS versions of OLSR and the original OLSR protocol are statistically significant. There are two main reasons:

a. High Overhead: The original OLSR protocol concentrates on how to reduce the overhead, and minimizes the MPR sets to reduce the TC messages flooding into the network. However, the QoS versions of OLSR select the best bandwidth path, so in their MPR selection mechanism, they select neighbors with high idle time as MPR, resulting in a larger MPR set than the original OLSR protocol. So more TC messages are generated and relayed into the network by QoS OLSR versions. (See Fig. 3)

For all algorithms, there are fewer TC messages sent at lower movement than at higher movement. This is because at lower movement, less TC messages are generated to reflect topology changes. Also, 20% OLSR has the highest number of TC messages generated and relayed, while the original OLSR protocol has the least number of TC messages. Under the same speed, the difference of TC messages sent between the original OLSR protocol and the 3 QoS OLSR versions comes from two aspects:

1. The original OLSR protocol only generates TC messages to reflect topology change, while QoS OLSR versions also need to generate TC messages to reflect bandwidth change; with a lower bandwidth update threshold, more TC messages are generated to reflect bandwidth change, causing the highest overhead in 20% OLSR

2. QoS OLSR versions have larger MPR sets than the original OLSR protocol, so more TC messages are generated and relayed by the larger MPR sets. Among the QoS OLSR algorithms, 20% OLSR may select more MPRs than 40% and 80% OLSR. With the possibly larger MPR set, more TC messages are generated and relayed by 20% OLSR than 40% OLSR and 80% OLSR.

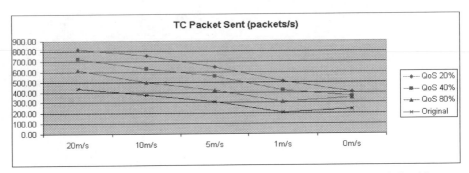

Fig. 3. Average TC message overhead in the network (in packets/s) for the 4 algorithms.

With higher overhead introduced into the network, especially for the 20% OLSR at higher movement, the wireless media is more heavily loaded.

b. Incorrect Routing Table: if there are overlapped two hop neighbors covered by multiple MPRs, there is a high probability that TC packets collide at these neighbors, resulting in inaccurate routing tables. This problem happens in all 4 OLSR algorithms. But because of the different MPR selection mechanism, the QoS OLSR algorithms have more overlapped two hop neighbors than the original OLSR protocol, causing more TC message collisions.

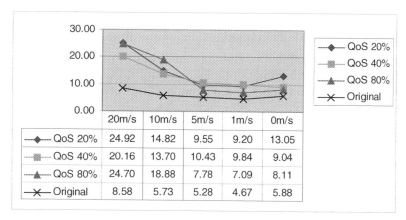

	20m/s	10m/s	5m/s	1m/s	0m/s
◆ QoS 20%	24.92	14.82	9.55	9.20	13.05
▓ QoS 40%	20.16	13.70	10.43	9.84	9.04
▲ QoS 80%	24.70	18.88	7.78	7.09	8.11
✕ Original	8.58	5.73	5.28	4.67	5.88

Fig. 4. End-To-End Delay (ms) of data packets for 4 OLSR algorithms.

Fig. 4 shows the End-to-End Delay for each algorithm under each movement pattern. Basically, for all movement patterns, the original OLSR has the lowest delay. Furthermore, in the high movement scenarios, the delay between the QoS versions of OLSR and the original OLSR protocol is statistically different. As the original OLSR has the lowest overhead, its network is the least congested, resulting in the least delay. Also, the original OLSR algorithm always computes the shortest hop path, while the QoS OLSR versions may compute longer paths because they target the best bottleneck bandwidth path, which also affects the end-to-end delay of the data packets.

For the three QoS OLSR algorithms, we can see that at higher movement speed (20m/s and 10m/s), the 80% threshold QoS OLSR has a higher delay, while at lower movement speed (5m/s, 1m/s and 0m/s), its delay is close to the original OLSR. To analyze this phenomenon, recall that the 80% threshold QoS OLSR has the most inaccurate bandwidth information of the network, which means that the routing algorithm may select a route that is still relatively congested. At higher movement, all the QoS OLSR algorithms have higher overhead because of the frequent updates due to topology change (see Fig. 3), causing the network to be congested. Working on the already congested networks, 20% QoS OLSR and 40% QoS OLSR do a better job in directing the traffic to the less congested routes, resulting in the lower packet delay. However, at lower movement speed, there are much less topology updates, so the more frequently sent bandwidth update messages in 20% and 40% OLSR tend to make the network busy, resulting in a larger delay than the 80% OLSR.

Fig. 5 and Fig. 6 show the "Average Difference" and "Error Rate" among the 4 algorithms under different movement patterns. All QoS OLSR outperform the original OLSR in both the "Error Rate" and "Bandwidth Difference". Among the QoS OLSR algorithms, 20% OLSR updates the bandwidth condition most frequently, introducing the highest overhead, but gets the most accurate bandwidth information. So the routes it calculates are closest to the optimal routes. The 40% and 80% OLSR, however, update bandwidth information less frequently, introducing less overhead, but their QoS performances are not as good as that of 20% OLSR.

The results for "Bandwidth Difference" and "Error Rate" of each algorithm are calculated based on its own network conditions – the bandwidth difference between the routes the routing algorithm calculated and the optimal paths in the network in

which the routing algorithm works. However, because the QoS OLSR versions introduce more overhead than the original OLSR protocol, the networks in which the QoS
OLSR versions work may have worse overall available bandwidth than a network that
runs the original OLSR algorithm. So one may question if the QoS OLSR versions
really improve the route bandwidth condition. To explore this question, for each scenario and OLSR version, the average available bandwidth over both the optimal
routes and the paths found by the routing algorithms are computed. In the following,
as available bandwidth is directly related to idle time in percentage, we report available bandwidth as percentage of idle time.

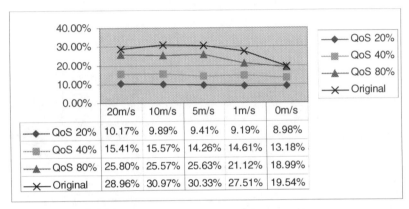

Fig. 5. Comparison of Average Bandwidth Difference.

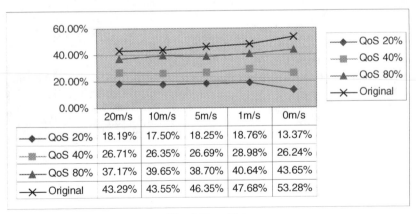

Fig. 6. Error Rate

To calculate the average available bandwidth on the routes the routing algorithms
find, first we obtain the average optimal route bandwidth (see Table 4.).

The results shown are consistent with our former analysis: The lower the movement speed, the less the overhead all the OLSR algorithms introduce into the network.
So from speed 20m/s to 0m/s, the optimal bandwidth conditions for all the OLSR
algorithms rise continuously. The original OLSR algorithm has the least overhead, so
the network that runs the original OLSR algorithm always has the best bandwidth

condition. Compared with 80% OLSR, 40% OLSR evenly directs traffic throughout the network, so under high movement (speed 20m/s, and 10m/s) where the wireless media are rather busy, 40% OLSR has better optimal bandwidth routes than that of the 80% OLSR, although it has more overhead than 80% OLSR. Under low movement (speed 5m/s, 1m/s, and 0m/s), the added overhead of 40% OLSR has a negative effect on the network bandwidth condition, thus the 40% OLSR has less optimal bandwidth than 80% OLSR. As the 20% OLSR has the highest overhead, its optimal bandwidth routes have the lowest available bandwidth.

From the results, we can also see that because the original OLSR has the lowest overhead, it provides the network with the best bandwidth condition – its best bandwidth paths have the highest bottleneck bandwidth among all the OLSR versions. However, as the original OLSR does not make efforts to find these optimal bandwidth paths, the actual path it finds may have a lower bandwidth than the paths the QoS OLSR versions find. In the following, we compare and analyze the actual bandwidth on the path the 4 versions of OLSR calculate.

Table 4. Avaliable bandwidth on the optimal paths (measured as idle time).

Algorithm	20m/s	10m/s	5m/s	1m/s	0m/s
QoS 20%	77.68%	80.93%	82.29%	84.69%	89.73%
QoS 40%	82.23%	84.92%	86.29%	87.46%	90.17%
QoS 80%	78.17%	84.27%	87.17%	90.08%	92.34%
Original	87.07%	87.28%	90.63%	91.14%	93.08%

The actual average available bandwidth the routing algorithms calculate
= the available bandwidth on the optimal paths x ((1- "Bandwidth Difference")
    x "Error Rate") + (1- "Error Rate"))
= the available bandwidth on the optimal paths x (1- "Bandwidth Difference"
    x "Error Rate")

Using the "Bandwidth Difference" and "Error Rate" values, the result for actual average available bandwidth the routing algorithms calculated is shown (see Fig. 7). We can see that although the QoS OLSR versions introduce more overhead, the routes they compute still have higher available bandwidth than the routes in a network running the original OLSR. In movement patterns with maximum speed 20m/s, 10m/s, 5m/s, and 1m/s, among all the OLSR algorithms, the 40% OLSR always computes the route with the best available bandwidth, as it has less overhead than 20% OLSR and more accurate bandwidth information than 80% OLSR. In the fixed network case, because of few topology updates, all the algorithms have low overhead. Thus, 20% OLSR finds the routes with highest bandwidth, for it has the most accurate bandwidth information. Based on these results, we conclude that the QoS OLSR versions do achieve bandwidth improvement over the original OLSR algorithm.

5   Analysis of QoS Routing and Future Work

As mentioned in Section 1, there is a trade-off between the QoS performance that the QoS routing protocol achieves and the additional cost it introduces. The QoS OLSR versions we study in this paper confirm this – QoS OLSR algorithms do enhance the network QoS performance. However, in order to achieve this improvement, additional "protocol overhead" is also introduced, which degrades the performance of these QoS routing protocols, especially with respect to "Packet Delivery Ratio" and "End-to-End Delay" in high mobility cases. Does this then imply that we should abandon proactive QoS routing and switch to on-demand QoS routing because of the cost? Not necessarily:

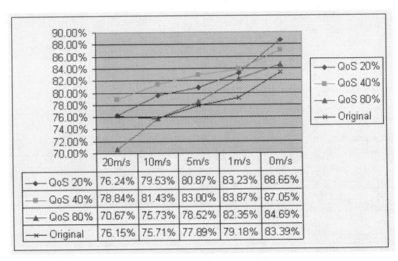

	20m/s	10m/s	5m/s	1m/s	0m/s
QoS 20%	76.24%	79.53%	80.87%	83.23%	88.65%
QoS 40%	78.84%	81.43%	83.00%	83.87%	87.05%
QoS 80%	70.67%	75.73%	78.52%	82.35%	84.69%
Original	76.15%	75.71%	77.89%	79.18%	83.39%

Fig. 7. Average available bandwidth (in idle time) on the routes of the 4 OLSR algorithms

– We do not know if on-demand routing algorithms have the same overhead problems. [3] discusses the performance of the "ticket-based probing" algorithm in a delay-constrained environment, calculating what percentage of routes that the algorithm finds meet the delay request. But it fails to analyze other aspects of the routing algorithm, such as control overhead, packet delivery ratio etc. [11] tests the CEDAR algorithm using bandwidth as the QoS parameter, giving a detailed performance evaluation. However, [11] does not experiment with node movement. Nor does it run the simulation in a real shared-channel environment, and the impact of channel interference and packet collision are not considered.

– Many proposed proactive QoS routing algorithm such as [10] and [7] just present a basic idea, without performance evaluation. So it is not clear whether the negative effect on the routing performance caused by the additional routing overhead is a common problem to proactive QoS routing.

Based on the above analysis, proactive QoS routing is still worth studying. As the added overhead is the main cost that affects the QoS routing algorithm's performance, the future work on QoS routing in Ad-Hoc networks may be focused on how to reduce the overhead. Our future work plans include the following:

– TC packet collisions at the 2-hop neighbors cause the problem of stale routing tables. To avoid this problem, we can add some jitter mechanism into the OLSR protocol – when an MPR receives a TC message, it waits for a random delay time before it relays that TC message, instead of relaying it immediately.

– Compared to the data packet load, the additional overhead the QoS OLSR versions introduce use a large amount of link bandwidth. This overhead is relatively independent of the nominal link bandwidth. We plan to explore whether the use of 802.11b, with up to 11 Mbps data rate, reduces the added network load and the resulting negative effect on the delivery ratio and delay.

Acknowledgements

This work is funded by the Defense Research and Development Canada (DRDC) and benefited from various discussions with members of the CRC RNS mobile networking group.

References

1. G. S. Ahn, A. T. Campbell, A. Veres and L. H. Sun, "SWAN: Service Differentiation in Stateless Wireless Ad-Hoc Networks", IEEE Computer and Communications Societies, 2002, pages 457-466, June 2002
2. G. Apostolopoulos, R. Guerin, S. Kamat, and S. K. Tripathi, "Quality of Service Based Routing: A Performance Perspective", Association for Computing Machinery's Special Interest Group on Data Communication '98, pages 17-28, September 1998
3. S. Chen and K. Nahrsted, "Distributed Quality-of-Service Routing in Ad-Hoc Networks", IEEE Journal on Selected Areas in Communications, Vol. 17, No. 8, August 1999, pages 1488-1505
4. S. Chen and K. Nahrstedt, "An Overview of Quality-of-Service Routing for the Next Generation High-Speed Networks: Problems and Solutions", IEEE Network Magazine, Vol.12, No.6, pages 64-79, November 1998
5. Y. Ge, T. Kunz and L. Lamont, "Quality of Service Routing in Ad-Hoc Networks Using OLSR", Proceeding of the 36th Hawaii International Conference on System Sciences (HICSS-36), Hawaii, USA, January 2003, ISBN 0-7695-1874-5, IEEE 2003.
6. R. Guerin and D. Willimas, "Qos Routing Mechanisms and OSPF Extensions", draft-qos-routing-ospf-00.txt", Internet-Draft, Internet Engineering Task Force, November 1996
7. Iwata, C. C. Chiang, G. Pei, M. Gerla and T. Chen, "Scalable Routing Strategies for Ad-Hoc Wireless Networks", IEEE Journal on Selected Areas in Communications, Vol.17, No.8, pages 1369-1379, August 1999
8. P. Jacquet, P. Muhlethaler, A. Qayyum, A. Laouiti, L. Viennot, T. Clauseen, "Optimized Link State Routing Protocol draft-ietf-manet-olsr-05.txt", INTERNET-DRAFT, IETF MANET Working Group
9. C. R. Lin and J. S. Liu, "QoS Routing in Ad-Hoc Wireless Networks", IEEE Journal On Selected Areas In Communications, Vol.17, No.8, pages 1426-1438, August 1999
10. R. Ramanathan and M. Steenstrup, "Hierarchically-Organized, Multihop Mobile Wireless Networks for Quality-of-Service Support", Mobile Networks and Applications, Vol.3, pages 101-119, 1998
11. P. Sinha, R. Sivakumar and V. Bharghanan, "CEDAR: a Core-Extraction Distributed Ad-Hoc Routing Algorithm", IEEE Journal on Selected Areas in Communications, Vol. 17, No. 8, August 1999, pages 1454-1465

Delivering Messages
in Disconnected Mobile Ad Hoc Networks

Ritesh Shah and Norman C. Hutchinson

Department of Computer Science, University of British Columbia
{rshah,norm}@cs.ubc.ca

Abstract. Many routing protocols for mobile ad hoc networks have been developed. These protocols find a route to a destination if such a route exists. We present a novel protocol that delivers messages between disconnected hosts, that is, when no route exists between them. Our protocol uses the nodes moving between the neighbourhoods of the source and destination nodes to act as carriers of the messages. We describe the protocol in detail, provide an initial simulation-based evaluation of its performance compared to both a naive scheme and the optimal scheme, and discuss some extensions to the protocol that we are exploring.

1 Introduction

An ad-hoc network is a self-starting network formed on-the-fly by a group of mobile nodes without the aid of any centralized administration or established infrastructure. Ad-Hoc networks find their use in situations where no fixed wired infrastructure is available or has been damaged by natural or man-made disaster.

Rapid advancement in wireless technology and increasingly affordable prices of wireless devices have made ad-hoc networks a reality. This has fuelled a lot of research activity in the field. Several protocols [1], [5], [6] have been developed to find and maintain routes between the nodes of an ad-hoc network. These routing protocols can be divided into three broad categories. First are the pro-active protocols, that use periodic advertisements to broadcast routing information, such as DSDV [2]. Second are on-demand protocols, that search for routes on-demand, such as AODV [4] and DSR [3]. Third are those routing protocols that use a hybrid approach, which is a combination of the first two approaches. While each approach has its advantages and disadvantages in finding a route between two mobile hosts when one exists, none of them handle the case of message delivery between two disconnected hosts.

Links in a MANET (Mobile Ad-Hoc Network) are susceptible to frequent breakage due to movement of nodes. This may cause some nodes to get disconnected from others. A message destined to such a disconnected node results in a delivery failure irrespective of the routing protocol used. Different protocols handle this situation differently but at most they invalidate the route, if there was one already in use and inform the source about the situation.

Why is the question of message delivery among disconnected hosts important? Consider a disaster relief scenario. Relief workers are working on several

S. Pierre, M. Barbeau, and E. Kranakis (Eds.): ADHOC-NOW 2003, LNCS 2865, pp. 72–83, 2003.

sites scattered in an area. The workers have mobile nodes to communicate among themselves. The sites may be separated by a distance that is several times the radio range of the devices. In such a case some of the sites might be disconnected from each other, forming multiple partitioned mobile ad-hoc networks in the area. While sporadic node movements between the sites may offer connectivity, it may be for brief periods of time. In such a situation it would be helpful to have some mechanism of delivering messages between disconnected hosts.

Consider a similar scenario on a long beach having several scenic-spots separated from each other by a distance several times the radio range. While there are nodes moving between the them, the scenic-spots may be disconnected for the majority of the time. It is easy to see that the nodes moving between these disconnected networks could be used as carriers of messages for other nodes.

Now the question is how to select the right carrier node. One option is to select all the nodes in the connected graph containing the source as carrier nodes. This could create unnecessary replication of messages and wastage of network bandwidth. So the goal is to find the right carrier node — the one that will come in contact with the destination within a certain period of time in the future. It is impossible for a source to choose the right carrier node without the knowledge of the present and future trajectories of all nodes. So a more refined goal could be to select those nodes as carrier nodes that have a higher probability of connecting to the destination in the future. Even this is difficult to ascertain without the knowledge of position and direction of movement of the disconnected destination and potential carrier nodes. A further refined goal could be to select carrier nodes in every direction (as the position and direction of movement of the disconnected destination is not known) and to minimize redundancy in doing so. In this paper we propose a completely decentralized protocol, Voilà, that replicates a message destined for a host disconnected from the source on selective nodes.

2   Routing Protocol

Our protocol will work with any proactive or on-demand routing protocol including DSDV, AODV and DSR. Besides routing messages, the only requirement that our algorithm places on the routing protocol is that it is capable of maintaining a neighbour list at each node. We assume that the routing protocol reports with an upcall to our protocol whenever a route to a destination cannot be found or is broken.

3   Algorithm

The algorithm is described from the points of view of each of the nodes that participate in it. The intuition behind this algorithm is that mobile nodes tend to exhibit correlated movement patterns. This correlated movement of nodes or "group mobility" has been studied in the past and several group based mobility models [9], [11] have been proposed. Based on this we propose that nodes that are close to each other need not store the same message, only one of them should be chosen to hold a message.

3.1   The Source Node

When a node X is unable to find a route to node Y or has lost the route to node Y, it buffers the message M meant for Y and requests the neighbour list of all of its neighbours. After the neighbour list is received from all its neighbours, node X selects the neighbouring node Z that has most neighbours that are not neighbours of X and adds it to its *picked* set. It then eliminates all of those neighbouring nodes that are also the neighbours of Z by putting them in the *eliminated* set. Node X repeats the same selection and elimination process with the neighbouring nodes set after removing node Z and those nodes that were just added to the eliminated set. The process ends when all the neighbouring nodes of X are either in the *picked* or in the *eliminated* set. The algorithm can be described in set notation as follows:

```
neigh(X) = neighbour list of node N
picked(X) = empty
eliminated(X) = empty

while(neigh(X) is not empty)
   Select a node Z : Z ∈ neigh(X)
                     and ∀ M ( M ∈ neigh(X) and M ≠ Z
                     |neighbour list of node Z - neighbour list of X| ≥
                     |neighbour list of node M - neighbour list of X| )

   picked(X) = picked(X) ⋃ {Z}
   eliminated(X) = eliminated(X) ⋃
              {M : M ∈ neigh(X) and M ∈ neighbour list of node Z}
   neigh(X) = neigh(X) - picked(X) - eliminated(X)
endwhile
```

After this process completes, the nodes in the *picked* set are sent the message M and the *eliminated* set in a HOLD control message. Each application message, M is uniquely identified by the tuple:

```
<message-id, source, destination>
```

A HOLD message sent by node X consists of the following fields:

```
<message-type, message-id, source, destination, M, eliminated(X)>
```

The *message-type* field indicates that its a HOLD message. The nodes in the *eliminated* set are sent a NACK for message M which they buffer in their NAKMSG queue. Node X and the nodes in the *picked* set buffer message M in their HOLDMSG queue.

3.2   Other Selected Nodes

When a node R receives a HOLD control message from another node S, it starts a similar selection process to select other nodes to hold the application message M. Like the original source node, node R requests the neighbour list from all of its neighbouring nodes. However, since some nodes are known to have already participated in the selection process for this message, node R can eliminate from consideration the node it received the message from (S), the original source node (X), and those nodes that are in the *eliminated* set sent by S.

A neighbour list request message, NBREQ, consists of the following fields:

`<message-type, message-id, source, destination>`

A node T receiving a NBREQ message from node R replies with a neighbour list reply message, NBREP. A NBREP control message sent by a node T consists of the following fields:

```
<message-type, message-id, source, destination,
                    neighbour list of node T, status>
```

The status field in the NBREP message reports on the status of this message at the node that sent the reply, and can be one of three values: NEW meaning that this node knows nothing of the message, NACKED, meaning that this node has previously been sent a NACK for the message, or HELD, meaning that this node is holding the message.

Nodes responding with NACKED or HELD in their status are eliminated from further consideration by R. After node R receives the neighbour list from the nodes, it uses the same algorithm as the source node X to select other nodes to hold message M. The only difference is in the initialization of the set *neigh(R)*. *neigh(R)* contains only those neighbouring nodes of R that have not been eliminated from consideration by R in any of the above steps. Again the nodes in the *picked(R)* set are sent a HOLD message containing the message M and the set of R's neighbours that were eliminated from further consideration. This set contains the final value of *eliminated(R)* together with the sending node (S) and those neighbours that were eliminated from consideration by R before. The nodes in the *eliminated(R)* set are sent a NACK for the message.

In order to bound the size of the HOLD control message, the *eliminated* set sent by any node contains only those nodes that are its neighbours. Therefore, the upper bound on the size of the *eliminated* set is the maximum number of neighbours a node can have. The alternative of accumulating eliminated nodes as the HOLD message propagates would eliminate some redundancy, at the expense of having HOLD messages that could be as large as the number of nodes in any partition.

Figure 1a shows an example network partition of 10 nodes. Suppose node 0 wants to send a message to a node outside the network partition. It initiates the process of selecting carrier nodes by first requesting the neighbour list from its neighbouring nodes 1, 2, 3, 4 and 5. After execution of the elimination process its *picked* set consists of nodes {2, 4} while its *eliminated* set consists of {1, 3, 5} which it sends to nodes in the *picked* set along with the message in a HOLD control message. When node 2 receives the HOLD message it starts the selection process by eliminating nodes 0, 1 and 3 from consideration and requesting the neighbour list from nodes 6 and 7. After execution of the algorithm its *picked* set consists of node {6} and its *eliminated* set consists of nodes {0, 1, 3, 7}. When node 6 receives a HOLD message it starts the selection process but ends up eliminating all nodes from its neighbour list and thus the process terminates at this node. Nodes 4 and 9 are the other nodes that hold the message.

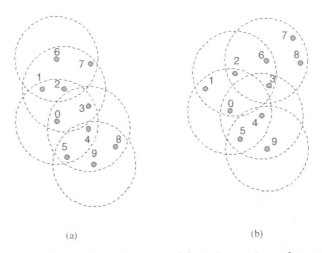

(a)   (b)

Fig. 1. (a) A network partition of 10 nodes (b) Node positions after some mobility

After a while if node 0 has another application message for an unreachable host, the nodes that are selected this time may be different from the ones previously selected as some of the nodes have moved to different positions as shown in Figure 1b. When node 0 executes the algorithm it again ends up with the *picked* set {2, 4} but this time the eliminated set does not contain node 3. This time the eliminated set is {1, 5}. When node 2 starts its selection process it requests the neighbour list from nodes 3 and 6. Node 3 is also being considered by node 4 at the same time. If node 3 receives a HOLD or a NACK message for the same application message from node 4 before the NBREQ message from node 2 arrives at node 3 then node 3 reports its status appropriately to node 2 either stating that it already holds the message (status=HELD) or it has been eliminated from consideration (status=NACKED) by some node which has selected one of its neighbours to hold the message (in the present scenario node 3 would receive a HOLD message from node 4). Ideally we would like a node to receive either a NACK or a HOLD control message for a particular application message and not both. This cannot be achieved because of distributed nature of the algorithm and changing neighbour lists of the nodes during the execution of the algorithm. Thus it is possible for a node to receive both a NACK and a HOLD message for the same application message and in any order. Again referring to Figure 1b, if node 3 receives the NBREQ message from node 2 before it receives the HOLD control message from node 4, then it will receive a HOLD message from node 4 and a NACK from node 2 for the same application message. In such a situation, the HOLD message always overrides the negative acknowledgement.

3.3 Message Delivery

Once every FIND_DST_INTERVAL, each node tries to find a route to the destinations of the messages it has stored in its HOLDMSG queue. If it is able to find a route to a destination then it delivers all the messages stored for the destination in its HOLDMSG queue. When a destination node receives a message, either from the original source or from an intermediate node, it sends an acknowledgement to the sending node. For each message stored in the HOLDMSG queue of a node, ids of the node that sent the HOLD message to the node and the nodes that were sent an HOLD message by the current node are also kept. When an acknowledgement is received for a message in the HOLDMSG queue the node forwards the acknowledgement to all such nodes. These forwarded acknowledgements help in removing messages that have been successfully delivered to the destination from the HOLDMSG queues of other nodes. This potentially prevents a lot of route search messages being broadcasted by all the nodes holding the message in the HOLDMSG queue. Some of the nodes to which the acknowledgements are forwarded may be disconnected from the forwarder node. In that case the acknowledgement is either dropped or a selection process similar to the one used for the data messages, is started depending on how important the acknowledgements are in a particular application. A message is removed from a HOLDMSG queue if an acknowledgement for the message is received or the message has been in the HOLDMSG queue for longer than MAX_HOLD_TIME.

3.4 Node Selection Metric

While selecting nodes for replicating a message, a node picks the one that has the largest number of neighbours that are not its neighbours. The intuition here is to select a node that is connected to the most nodes outside the range of the node currently involved in selection process and hence can potentially disseminate the message farthest in the connected graph containing the source.

4 Experimental Set Up

We implemented our protocol for the ns2 simulator [14] and have evaluated its performance for scenarios involving upto 120 nodes. The poor performance of ns2 in simulating large number of nodes has prevented us from experimenting with scenarios involving a larger number of nodes. We use the modified Random Way Point Model described in the next subsection of this paper for our experiments. The range of each wireless node is set to 100m. The traffic generator consists of an application running at each node that tries to send a message of 64 bytes once every five seconds to a random node (other than itself). The underlying routing protocol used is AODV. Table 1 shows the values of various parameters of AODV used in the simulations.

Table 1. AODV parameter values used in the simulations

Parameter	Value
EXPANDING_RING_SEARCH	ON
TTL_START	7
TTL_THRESHOLD	7
NETWORK_DIAMETER	30
RREQ_RETRIES	1
AODV_LOCAL_REPAIR	OFF
AODV_LINK_LAYER_DETECTION	OFF
HELLO_INTERVAL	1
ALLOWED_HELLO_LOSS	3

4.1   Mobility Model

The Random Way Point model is the most widely used mobility model in testing protocols in mobile ad-hoc networks. We have made two modifications to the Random Way Point model to make it more practical.

Random Way Point Model. The Random Way Point Model was first described in [3], since then it has been widely used to test MANET protocols in simulated environments. In this model nodes move around in a "room". Each node starts from an initial position selected randomly from within the simulated area. As the simulation progresses, each node pauses at its current position for a constant period of time and then randomly chooses a destination location from within the simulated region and moves there by selecting a random speed uniformly from the interval (0,V], where V is the maximum speed with which a node can move.

Modifications in Random Way Point Model. We believe that the applications developed for Mobile Ad Hoc environments will be mostly interactive and therefore people using them would be walking or strolling. Studies [10] have shown that the normal walking speed of an adult human being is around 1.5 m/s. We also consider the possibility of vehicles (particularly in the disaster relief scenarios outlined in section 1) moving at a slow speed. [13] shows that selecting a minimum speed greater than zero for mobile nodes is necessary to attain a meaningful steady state. Therefore we choose node speed from a normal distribution with mean speed set to 1.5 m/s. The minimum speed is set to 0.4 m/s while the maximum speed is set to 5.0 m/s.

The Random Way Point Model assumes uniform distribution of nodes, which is quite "unrealistic". The density of nodes would vary a lot in any of the scenarios described in section 1. There would be some hot-spots where nodes are clustered. In a disaster relief scenario these hot-spots could be sites where relief and rescue work is being carried out; on a beach these could be particularly scenic points, snack bars, or volleyball courts; in a military zone these could be various army camps. In our model we have five such hot-spots randomly distributed in a space of 3000m x 600m. Each hot-spot is a circular disc of radius 250m. Each node is initially placed at a random position in a randomly selected hot-spot. Nodes then, instead of selecting a destination uniformly from the whole area, select a random position within a randomly selected hot-spot. There is also a small probability (10%) of choosing a destination which is not in a hot spot.

4.2   Results

We compare the results from our protocol, Voilà, with those obtained from two other schemes, the Source-only scheme and the Oracle scheme. In the Source-only scheme, the source node buffers the messages that could not be delivered to a disconnected host. It tries to find a route to the destinations of buffered messages once every FIND_DST_INTERVAL. In the Oracle scheme every node is omniscient, i.e., it has the knowledge of the present and future trajectories of all the nodes in the network and hence can choose the optimal intermediate node to send the message to, if any such node exists. When a node wants to send an message to another node and cannot find a route to the destination, the routing algorithm makes an upcall to one of the three protocols. As the number of nodes increases from 30 to 120, the fraction of messages whose destination is disconnected from the sender decreases approximately linearly from 91% to 62%. Figure 2 shows the fraction of upcalled messages successfully delivered to the destinations in the three schemes as the number of nodes is varied (normalized to those delivered by the Oracle scheme). In this experiment, FIND_DST_INTERVAL is 30 seconds and MAX_HOLD_TIME is 65 seconds. Voilà performs better than the Source-only protocol all the time except when the number is nodes is 120. We believe that this is because the network traffic in Voilà is much higher than in the Source-only protocol. The traffic increases as the number of nodes increases, causing congestion in the network. We are working on various approaches to reduce the congestion when the number of nodes is large.

Figure 3 shows the fraction of upcalled messages delivered to the destination (normalized to those delivered by the Oracle scheme) as MAX_HOLD_TIME is varied from 35 to 155 seconds in a scenario of 70 nodes. The value of FIND_DST_INTERVAL is set at 30 seconds for this experiment. As MAX_HOLD_TIME increases, the probability of the source getting connected to the destination within MAX_HOLD_TIME also increases and hence we can see the rising bars for the Source-only scheme. We would expect Voilà to show similar behaviour but the increased network traffic arising from long HOLDMSG queues at high MAX_HOLD_TIME causes important packets to get dropped and hence we see a small decrease in perfomance of Voilà as MAX HOLD_TIME in creases. Figure 4 shows the fraction of upcalled messages delivered to the destination as FIND_DST_INTERVAL is varied from 30 to 150 seconds in a scenario of 70 nodes. The MAX_HOLD_TIME is calculated according to the formula ((2 * FIND_DST_INTERVAL) + 5) for this experiment. We see an expected drop in the number of upcalled messages delivered to the destination as FIND_DST_INT-

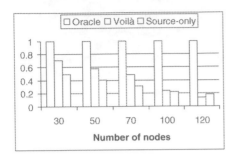

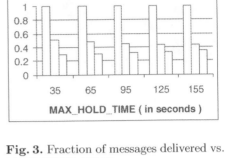

Fig. 2. Fraction of messages delivered vs. number of nodes

Fig. 3. Fraction of messages delivered vs. MAX_HOLD_TIME

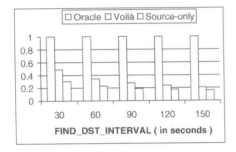

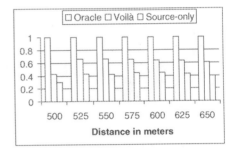

Fig. 4. Fraction of messages delivered vs. FIND_DST_INTERVAL

Fig. 5. Fraction of messages delivered vs. hot-spot separation

ERVAL increases. As in Figure 3, Voilà performs better than the Source-only protocol for all parameter values.

Any scheme for delivering messages across disconnected networks is sensitive to the sizes of the hot-spots and their relative positions. The size of a hot-spot determines the diameter of the network partition in the hot-spot and hence the number of nodes selected to hold a message. The relative positions of hot-spots determine how connected they are under the movement of nodes and hence how much a scheme like ours is warranted. It is easy to see that any such scheme is not useful for hot-spots that are completely connected at all times. To determine the sensitivity of the protocols to the positions of hot-spots we experimented with a scenario of 30 nodes and three circular hot-spots. The hot-spots are arranged such that their centers lie in a straight line. The centers of the adjacent hot-spots are separated by a distance which is varied from 500m to 650m. Since the radius of each hot-spot is 250m, the circumferences of adjacent hot-spots touch each other when distance is 500m. The number of messages whose destinations are not connected to the source increases as the distance between the hot-spots increases, from 27% at 500m to 38% at 650m. Figure 5 compares the performace of the three schemes.

5   Related Work

The problem of message delivery among disconnected Mobile Ad-Hoc Networks is not new to the research community. In particular, the idea of using intermediate nodes to relay messages among disconnected hosts has also been proposed earlier by Li and Rus [7]. Different research groups have approached the problem of delivery in a disconnected network from the perspective of different MANET applications and hence have come up with different solutions. Karumanchi et. al. [8] describe the problem of network partitioning in a MANET formed by a group of firefighters involved in a firefighting mission. Each firefighter is required to update its location information to servers in the network and must be able to obtain location information of other firefighters by querying the servers. Their solution employs quorum-based strategies to update and query information in a partitioned network. Wang and Li [9] describe the problem of network partitioning among group of mobile nodes that are requesting and downloading information on demand from a centralized service. Their goal is to provide service coverage even when the network is partitioned by replicating the service to appropriate nodes before a partition takes place. They employ a partition prediction scheme to predict partitioning in the network before it occurs [12]. Li and Rus [7] consider the problem of network partitioning in a domain where it is possible to instruct mobile nodes to change their trajectories, such as in a robotic network where a team of robots is deployed to perform sensing tasks in a remote or hazardous environment.

6   Future Work

We have not reseached the parameter space completely, we plan to run more simulations and determine the sensitivity of our protocol to more parameters. We are also considering several extensions and improvements to our basic scheme. These are described in the next subsections.

6.1   Reducing Overhead

The network overhead seems to override the benefit of replication in our protocol when the number of nodes is large. We are working on several approaches to solve this problem. One of them is to use a cache to buffer the neighbour lists of neighbouring nodes as it seems wasteful to ask for neighbour lists for each message. A cache entry containing the neighbour list of a mobile node would be valid for a short period of time and all neighbour list requests sent to the mobile node within that period will be serviced from the cache. This can help reduce neighbour list request and reply traffic in the network. But this would raise other issues; specifically, the status information provided by a neighbour in a NBREP message would be lost. It is to be evaluated how much of a trade-off this is between reducing overhead and accuracy. Another approach to reduce neighbour list request traffic is to broadcast the NBREQ message. One approach to reduce the route search messages is to limit the number of route search messages initiated by a node every FIND_DST_INTERVAL. Again it is to be evaluated how much of a trade-off this is between throughput and reducing overhead.

6.2   Nodes with GPS Receivers

In our algorithm when a node tries to select neighbourhood nodes to hold its message, it selects those that are not close to each other (separated by an angular distance of more than 60 degrees). But it is possible that two nodes positioned far apart in the neighbourhood could be heading towards the same destination. It is not possible to eliminate such cases unless the node running the algorithm knows the direction of the movement of the neighbouring nodes. If all the nodes are equipped with GPS receivers then it is possible for them to know their location information. The nodes can periodically relay this information to their neighbouring nodes by piggybacking it on the local broadcast messages and other unicast messages. The periodically arriving location information from the neighbours can help a node to compute a neighbour's speed and more importantly, its direction of movement. A node can then use this information to select appropriate nodes, i.e., the elimination criterion can change from nodes placed closely to each other to nodes moving in almost the same direction.

7   Conclusion

We have demonstrated a simple algorithm to deliver messages in disconnected Mobile Ad-Hoc networks. Our scheme does not entail any extra requirements on the present routing algorithms. The only requirement is a local broadcast mechanism used to discover the neighbours of nodes. Such a mechanism is used in some of the current routing protocols. We have presented results obtained from simulating our scheme.

References

1. E. M. Royer and C-K. Toh, "A Review of Current Routing Protocols for Ad-Hoc Mobile Wireless Networks", IEEE Personal Communications Magazine, April 1999, 46-55
2. C. E. Perkins and P. Bhagwat, "Highly Dynamic Destination-Sequenced Distance-Vector Routing (DSDV) for Mobile Computers," in SIGCOMM'94, 1994
3. D. B. Johnson and D. A. Maltz, "Dynamic Source Routing in Ad Hoc Wireless Networks," Mobile Computing, 1996. Kluwer Academic Publishers
4. C. E. Perkins and E. M. Royer, "Ad-hoc On-Demand Distance Vector Routing," in 2nd IEEE Workshop. Mobile Comp. Sys. and Apps., , Feb 1999, 90-100
5. J. Broch, D. A. Maltz, D. B. Johnson, Y.-C. Hu, and J. Jetcheva, "A performance comparison of multi-hop wireless ad hoc network routing protocols". In Mobile Computing and Networking, 1998, 85-97

6. C. E. Perkins, E. M. Royer, S. R. Das and M. K. Marina, "Performance Comparison of Two On-demand Routing Protocols for Ad Hoc Networks", IEEE Personal Communications Magazine, special issue on Mobile Ad Hoc Networks, Vol. 8, No. 1, Feb 2001, 16-29

7. Q. Li and D. Rus, "Sending Messages to Mobile Users in Disconnected Ad-hoc Wireless Networks," in Proceedings of the Sixth ACM/IEEE Internatinal conference on Mobile Computing and Networking (Mobicom 2000), Aug 2000, 44-55

8. G. Karumanchi, S. Muralidharan and R. Prakash, "Information Dissemination in Partitionable Mobile Ad Hoc Networks", in Proceedings of IEEE Symposium on Reliable Distributed Systems, Lausanne, Switzerland, Oct 2000

9. K. Wang and B. Li. "Efficient and Guaranteed Service Coverage in Partitionable Mobile Ad-Hoc Networks." in Proceedings of IEEE INFOCOM 2002, Vol. 2, Jun 2002, 1089-1098

10. P. S. Rodman and H. M. McHenry, "Bioenergetics and the origin of hominid bipedalism", Americal Journal of Physical Anthropology, Vol. 52, 1980, 103-106

11. X. Hong, M. Gerla, G. Pei, and C. Chiang, "A Group Mobility Model for Ad Hoc Wireless Networks," in Proceedings of the 2nd ACM InternationalWorkshop on Modeling and Simulation of Wireless and Mobile Systems, 1999

12. K. Wang and B. Li. "Group Mobility and Partition Prediction in Wireless Ad-hoc Networks", in Proceedings of IEEE International Conference on Communications (ICC 2002), Vol. 2, April-May 2002, 1017-1021

13. J. Yoon, M. Liu, and B. D. Noble, "Random waypoint considered harmful", in Proceedings of INFOCOM '03, April 2003

14. http://www.isi.edu/nsnam/ns/

Extending Seamless IP Multicast Edge-Coverage through Mobile Ad Hoc Access Networks

Pedro M. Ruiz[1], Antonio F. Gomez-Skarmeta[1],
Pedro Martinez[1], and David Larrabeiti[2]

[1] University of Murcia, Facultad de Informatica, Dept. of Information and
Communication Engineering, Campus de Espinardo,
E-30100 Espinardo (Murcia), Spain
{pedrom,skarmeta,pma}@dif.um.es
[2] University Carlos III of Madrid, Dept. of Telematic Engineering,
Campus de Leganes, Avda. Universidad, 30,
E-28911 Leganes (Madrid), Spain
dlarra@it.uc3m.es

Abstract. The provision of multicast communications in wireless and wired networks has followed different paths which have led to different solutions. Little has been accomplished to-date in bringing together the traditional IP multicast model used in fixed networks and multicast routing protocols for wireless ad hoc networks. We analyse the provision of an integrated IP multicast service in which mobile hosts can seamlessly participate in IP multicast sessions regardless of the currently underlying network type. We propose a multicast architecture in combination with a new ad hoc multicast routing protocol called MMARP. MMARP nodes are challenged with special IGMP-handling capabilities allowing our solution to combine the efficiency of multicast ad hoc routing protocols with the support of standard-IP nodes without an impairment in the performance of the protocol. Our empirical results demonstrate that such kind of multicast ad hoc access networks offer a good performance when compared with the traditional single-hop wireless multicast access.

1 Introduction

IP Multicast is suited for efficient multipoint communications among a group of nodes. It has emerged as one of the most researched areas in networking. The problem of efficient packet distribution to a specific group of destinations has been researched since the late 80's and most of the routers nowadays support IP multicast routing protocols. The main benefit of IP Multicast is that the bandwidth consumption for group communications is dramatically reduced compared to unicast-based group communications. This is of particular interest for 'all-IP' and 'beyond 3G' mobile networks consisting of a high number of user terminals using applications which are typically interactive, multiparty and bandwidth-avid.

Many projects like the IST project MIND (Mobile IP-based Network Developments) [1] have researched the extension of IP-based radio access networks to

S. Pierre, M. Barbeau, and E. Kranakis (Eds.): ADHOC-NOW 2003, LNCS 2865, pp. 84–95, 2003.

include ad-hoc wireless elements within the access infrastructure as a natural evolution towards 'beyond 3G' systems. In this ad hoc fringe, a user terminal employs those of other users as relay points to provide multi-hop paths between mobile nodes and the fixed access network architecture.

The provision of an integrated IP multicast service in such an heterogeneous scenario consisting on traditional IP core networks interconnecting a variety of wireless and wired access networks and technologies is extremely complex. There are specific solutions for wireless ad hoc networks, but the real challenge is their effective and efficient integration with (fixed) IP multicast protocols to achieve a seamless IP multicast service in which group members from any of these network types can take part in the same IP multicast session. Furthermore, mobile nodes should be allowed to move among these types of networks without any service disruption.

To our knowledge, for the specific problem of IP multicast interworking between IP access networks and wireless and mobile ad hoc networks, there are not satisfactory solutions so far. The typical intra-domain IP multicast protocols for fixed networks (i.e. IGMPv2[2] for multicast group membership and PIM-SM[3] for IP multicast routing) are not able to deal with the quick and unpredictable link changes which characterise ad hoc networks. They would consume too much overhead to keep updated distribution paths in such variable topologies. In addition, multicast ad hoc routing protocols like CAMP[4], ODMRP[5], and ADMR[6] among others, incorporate specific functionality which enables them to cope with the particular characteristics of ad hoc networks but they are only suitable for isolated ad hoc networks. These protocols do not provide any means to interoperate with the protocols used in the fixed IP networks and they do not support the attachment of standard IP multicast nodes to the ad hoc extension. In fact, the only few proposals to connect ad hoc networks to the Internet, like the one by Lei and Perkins[7] have only considered the case of unicast traffic.

In this paper we propose an integrated IP Multicast solution for ad hoc network extensions consisting of a novel IP multicast architecture and the Multicast MAnet Routing Protocol (MMARP). MMARP is a new multicast ad hoc routing protocol based on the same basic mechanisms as other ad hoc multicast routing protocols. However, it incorporates additional functionalities to deal with the complexity of supporting traditional IP nodes whilst interoperating smoothly with fixed IP networks. MMARP nodes are able to intercept and process standard IP multicast messages. They further permit standard IP nodes to seamlessly participate in IP multicast communications as they do when attached to a fixed IP network. The novelty of our approach is not only the provision of such an integrated IP multicast solution, but also the way in which the functions are divided among the fixed and ad hoc nodes so that the interworking is achieved without a noticeable impairment in the overall performance.

The remainder of the paper is organised as follows: section 2 comments on the problems, requirements, and proposed architecture for ad hoc access network extensions. A detailed description of the MMARP protocol is given in section 3.

Section 4 presents some empirical results. Finally, section 5 gives some conclusions.

2   Proposed Multicast Architecture

One of the most important design issues in the multicast architecture for seamless IP multicast provision in ad hoc network extensions is the separation of the functions between the different network boundaries. We followed a top-down approach which allowed us to derive the best design options from the particular requirements and related issues of a seamless and integrated multicast solution.

2.1   Requirements

The first step towards an integrated IP multicast solution is the identification of the requirements. As an objective for ad hoc network extensions we seek a trade-off in which at least the following requirements are met:

- Interoperability with IP Multicast mechanisms in fixed networks
- Efficiency, scalability and low signalling overhead
- Resilience and robustness (e.g. several points of attachment to the fixed network)
- Compatibility with inter-domain multicast routing
- Support of seamless moving of terminals among network types

2.2   Problems to Solve

Trying to map the traditional IP multicast model into the concrete scenario of ad hoc network extensions, allows us to identify specific problems which need to be solved. According to the IP multicast model for IP multicast hosts, the process of taking part in multicast communications is quite straightforward. When they wish to send multicast traffic they simply use a class-D address as a destination and send the datagrams. When they are interested in receiving multicast traffic, they use the Internet Group Management Protocol (IGMP[6]) to inform their First Hop Multicast Router (FHMR) about the group they wish to join. This simple operation is not automatically supported in ad hoc networks due to some of the problems presented below.

TTL Issues. IGMP uses IP datagrams with a time-to-live (TTL) of one hop for the communication between hosts and routers. Thus, by default, only directly connected hosts are able to join multicast groups since IGMP messages are unable to transit a multi-hop ad hoc network fringe.

Multihop Nature of MANETs. Packets sent by sources which are more than one hop away will not automatically be received by the FHMR. However, intermediate ad hoc nodes must ensure that these packets reach the FHMR as it is required by most IP Multicast routing protocols (e.g. PIM-SM). The support of standard IP nodes is an issue that requires that ad hoc nodes incorporate capabilities for the interception and processing of IGMP messages since these are the means by which hosts join IP multicast groups in fixed networks. To date, none of the proposed multicast ad hoc routing protocols is able to handle such types of messages.

Flat Addressing. An additional issue relates to the differences between the hierarchical addressing architecture which is used in fixed networks and the flat addressing architecture used in ad hoc networks. The problem is that multicast routers usually perform a process called an 'RPF-check' on every incoming packet. This process drops any packet which arrives at an interface which that router would not use to reach the source of the packet.

2.3   Proposed Architecture

There are several alternatives to achieve efficient network layer multicasting support between nodes within the ad hoc network extension and those in the access network. As we showed in [8], the most relevant are basically what we called a tunnel-based approach, and multicast ad hoc fringe. The former is based on the creation of a tunnel between receivers and the access routers. We have selected the multicast ad hoc fringe approach because, as we demonstrated in [8], it is much better in terms of scalability, simplicity and performance.

The key point in our proposed architecture is the idea of confining any new functionality to within the ad hoc fringe, challenging ad hoc nodes with the ability to process standard protocols (i.e. IGMP) to interact with non-ad hoc nodes. This mechanism exploits the anonymous nature of IP multicast because the FHMR does not need to know which node is interested in joining a particular multicast group, but only if there is any. So, when a standard-IP host generates an IGMP Report, internally ad hoc nodes will not need to transport that message. They use the MMARP protocol to create efficiently multicast paths within the ad hoc extension and any of the ad hoc nodes at a single hop from the FHMR will regenerate such an IGMP Report message. This shields the solution from the particular IP multicast routing protocol being used in the fixed network: we can interoperate in the same way with all of them just by sending IGMP Reports. So, our approach does not require any changes in standard IP nodes and routers. Mobile nodes will behave according to the standard IP Multicast model in which there is no requirement for senders and the only requirement for receivers is the use of the IGMP protocol to join multicast groups.

In addition, as the use of Standard-IP mechanisms (e.g. the ARP or IGMP protocols) within ad hoc networks is costly and usually offers limited performance, we propose a specific multicast ad hoc routing protocol called MMARP which incorporates particular path creation mechanisms to support standard-IP messages without an impairment in the protocol's performance. These specific MMARP extensions are described in the next section, whereas the proposed architecture is depicted in Fig. 1.

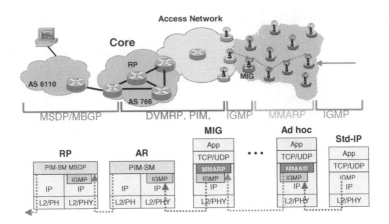

Fig. 1. Proposed multicast architecture

The AR and RP nodes in the figure represent standard multicast-enabled routers running PIM-SM. 'Ad hoc' represent pure ad hoc nodes and 'Std IP' represnts a standard IP multicast-enabled mobile host. The protocol providing efficient paths between the nodes within the ad hoc network fringe is the MMARP protocol presented below. From the point of view of the core network and the AR, the ad hoc fringe is seen just as another BMA subnet (i.e. group membership are being dynamically updated by IGMP Report messages received by the ARs).

3   The MMARP Protocol

The MMARP protocol is especially designed for mobile ad hoc networks (MANETs). It is fully compatible with the standard IP Multicast model and it allows standard IP nodes to take part in multicast communications without requiring any change because MMARP supports the IGMP protocol as a means to interoperate both with access routers and standard IP nodes. The interoperation with access routers is performed by the Multicast Internet Gateways (MIGs) which are the ad hoc nodes situated just one hop away from the fixed network. Every MMARP node may become a MIG at any time. The only difference between a MIG and a normal MMARP node is that the MIG is responsible for notifying the access routers about the groups memberships within the ad hoc fringe. The mechanism allows MMARP to work with any IP multicast routing protocol in the access network and, therefore, it shields the MMARP operation from the protocols performing the intra-domain or inter-domain multicast routing.

For the remaining text we use the terms standard IP source or standard IP receiver to refer to a traditional IP Multicast source or receiver and we use the term ad hoc sender or ad hoc receiver to refer to pure ad hoc nodes.

3.1   Overall Operation

MMARP uses an hybrid approach to build a distribution mesh similar to the one used by ODMRP[5]. Routes among ad hoc nodes are established on-demand, whereas routes towards nodes in the fixed networks are maintained proactively. This offers a good trade-off between efficiency, smooth interworking with the fixed network while still having a good protection against link breakages (see Fig. 2). However, the way in which the mesh is created is different from ODMRP due to the special requirements which MMARP nodes have to face. For example, MMARP nodes can participate in the mesh creation process on behalf of standard IP nodes or even on behalf of the access router (AR). In addition, they have behave so that the standard IP multicast model can be preserved (i.e. making all the traffic generated within the ad hoc fringe to be delivered to the AR). These specific differences are explained in the next subsections.

The reactive part consists of a request phase and a reply phase. When an ad hoc node has new data to send, it periodically broadcasts a MMARP_SOURCE message which is flooded within the entire ad hoc network to update the state of intermediate nodes as well as the multicast routes. These messages have an identifier which allows intermediate nodes to detect duplicates and avoid unnecessary retransmissions. When such a message is received by an ad hoc node for the first time, it stores the IP address of the previous hop and rebroadcasts the packet. When one of these messages arrives at a receiver, or at a neighbour of a standard IP receiver, it broadcasts a MMARP_JOIN message including the IP address of the selected previous hop towards the source. When an ad hoc node detects its IP address in an MMARP_JOIN message, it recognises that it is in the path between a source and a destination. It then activates its MF_FLAG (Multicast Forwarder Flag) and rebroadcasts a MMARP_JOIN message containing its previously stored next hop towards the source. In this way, a shortest multicast path is created between the source and the destination. When there are different sources and receivers for the same group, the process results in the creation of a multicast distribution mesh like the one presented in Fig. 2).

The proactive part of the protocol is simply based on the periodic advertisement of the MIGs as default multicast gateways to the fixed network. As the TTL of IGMP messages is fixed at one, the reception of an IGMP Query can be used by ad hoc nodes to detect that they are MIGs and activate its MIG_FLAG. MIGs periodically broadcast a MMARP_DFL_ROUTE message which is flooded to the whole ad hoc network. The reception of this message informs intermediate nodes about the path towards multicast sources in the access network. When the MMARP_DFL_ROUTE message reaches a receiver or a neighbour of a receiver, this node initiates a joining process similar to the one that we have just described for the reactive approach. When the MIG receives the MMARP_JOIN message, it then sends an IGMP Report towards the FHMR, ensuring the IP multicast data from sources in the fixed network reach the destinations within the ad hoc network extension.

The protocol incorporates local repairing mechanisms to overcome link breakages during the creation of the distribution mesh. Whenever a node is unable to

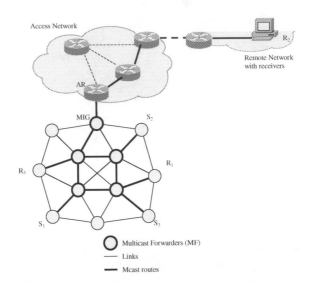

Fig. 2. Multicast mesh after request/reply phase

deliver a MMARP_JOIN message to its next hop after four retries, it broadcasts a MMARP_NACK message to its one-hop neighbours. Upon the reception of this message, the neighbours use their own route to reach that next hop. Should any of them not know an alternate path, they repeat the process until a path is found. Although this recovery process does not offer optimal routes, it offers a quick recovery before the next topology refresh.

Once the mesh is established, the data forwarding is very simple: data packets addressed to a certain multicast group are only propagated by ad hoc nodes which have their MF_FLAG active for that group. When such a data packet arrives at a node whose MF_FLAG for that group has not expired, it checks that it is not a duplicate and in that case retransmits the packet. In any other case the packet is dropped.

3.2   Support of Standard IP Multicast Protocols

The protocols used by standard IP nodes to perform their basic operation (such as ARP, or IGMP) were designed to operate in BMA (Broadcast Medium Access) networks. However, in multihop ad hoc networks, the link layer has a different semantics. The neighbours of a node are able to receive the frames it sends but it is not guaranteed that they are able to directly communicate among all of them. In traditional ad hoc routing protocols without explicit support for standard IP nodes this is not a problem because each ad hoc node sends its own source announcement or join message. In order to be compatible with the standard IP multicast model, MMARP nodes in the neighbourhood of a standard IP node have to send MMARP_SOURCE or MMARP_JOIN messages on behalf of the standard IP node. This means that messages generated by standard IP nodes, may be received by all neighbours and processed independently, creating unnecessary paths.

The MMARP protocol has been designed to avoid unnecessary generation of these messages. It includes a field in its header which facilitates the identification of the node which actually triggered the sending of the control message; this allows ad hoc nodes to identify all the MMARP packets which are triggered by a specific standard IP node, independently of the ad hoc neighbour which actually generated it. Thus, ad hoc neighbours of standard IP nodes and intermediate ad hoc nodes are able to detect these types of MMARP_SOURCE and MMARP_JOIN messages as duplicate and avoid the creation of unnecessary paths.

4   Empirical Results

We have set up an indoor 802.11b multicast wired-to-wireless ad hoc network testbed to evaluate the feasibility of our MMARP-based seamless IP multicast approach for wireless ad hoc access networks. Our target is to evaluate the benefits of MMARP-driven infrastructureless ad hoc access networks when compared to traditional single-hop wireless IP multicast in a realistic scenario.

4.1   Testbed Description

As it is shown in Fig. 3, the testbed consists of six x86-compatible PCs and a laptop. Different processor and memory configurations are used, since there are not any specific hardware requirements. In fact, all of these PCs are able to support the workload of the experiments. Three out of the six PCs are acting as MMARP-enabled nodes running Red Hat Linux 7.2 with the 2.4.17 kernel. They have a Lucent 802.11b pcmcia card as unique NIC. The nodes labelled as WR (Wireless Router) and WWR(Wired-to-Wireless Router) are PCs running FreeBSD 4.6 OS. The WWR node is equipped with two NICs, one of them being a Lucent WaveLan pcmcia card to provide coverage for the wireless area, while the other one is a 100 Mbps Ethernet NIC. The Wired Router (WR) contains two 100 Mb/s Ethernet NICs, one connected to the WWR and the other one to the rest of fixed networks. The Sender and Receiver nodes are both running Red Hat 7.2 with kernel 2.4.17. The sender is a x86-compatible desktop with a 100 Mb/s Ethernet card whereas the receiver is a laptop PC equipped with a Lucent 802.11b-compatible pcmcia wireless NIC. Wireless 2.422 GHz channel operating at the maximum capacity of 2 Mb/s has been used for the experiment. We have previously checked that this channel was not occupied by any other equipment. All the WaveLan NICs are operated in ad hoc mode.

4.2   Description of the Experiments

To assess the effectiveness of our proposal, we have performed two different tests: single-hop IP multicast and multihop ad hoc IP multicast. The former consists of

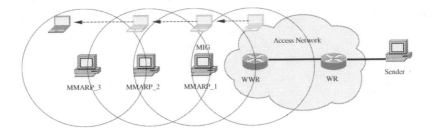

Fig. 3. Topology of the testbed

the network depicted in Fig. 3, in which every MMARP node is switched off, so that there is a dedicated IEEE 802.11b wireless link between the receiver and the WWR. The wired part of the network is running the PIM-SM routing protocol to create the multicast path between the source and the WWR, which acts as an IGMP designated router forwarding multicast datagrams to the receiver when it joins the source.

The multihop tests are exactly the same regarding the wired part of the network. However, we deploy a self-organising ad hoc network extension with nodes running MMARP rather than a single-hop link between the receiver and the WWR. The receiver joins the multicast source in the fixed network through this multihop access network.

We use CBR traffic generator to measure the end-to-end bandwidth and packet delivery ratio. This application generates UDP packets with a payload of 900 bytes (i.e. 942 bytes including the IPv4 and UDP headers) which are then accounted at the destination. For each of the tests we have performed several measurements at increasing distances (7m, 15m, 24m, 30m, 42m) between the receiver and the WWR. At each distance, we have repeated the measurements using three different data rates of 100 packets/s (753.6 Kb/s), 50 packets/s (376.8 Kb/s) and 25 packets/s (188.4 Kb/s) respectively. The results of the different trials are described in the next section.

As expected, in our indoor scenario the performance depends not only on the distance but on the node's position as well. This is mainly due to random noise caused by traversing walls, obstacles, etc. The results are calculated as the mean values over quite a huge number of measurements per experiment. In addition, the measurements are performed in the same positions for each distance. So, random noise is expected to be nearly the same in all the trials, not affecting the validity of our experiments.

4.3 Experimental Results

To be sure about the cause of the packet losses in our analysis, we empirically checked that there were not packet losses within the wired part of the network. So, all the packet losses perceived at the receiver will occur in the wireless part of the network.

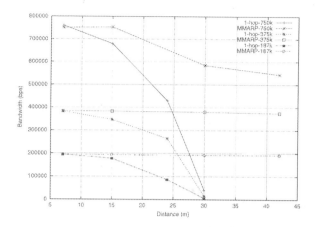

Fig. 4. Effective data rate achieved at increasing distance

As it is shown in Fig. 4, both approaches are able to deliver the transmitted bandwidth at short distances. For those cases the link-layer contention is low and the signal quality is good enough.

In the single-hop trials, due to the degradation of the signal strength with increasing distance, the achieved bandwidth is lower as the distance increases. This is clearly assessed both for the achieved bandwidth and the packet delivery ratio in the '1-hop' cases of Fig. 4, and Fig. 5. These results are basically the expected behaviour as long as it is commonly known that (particularly in indoor scenarios) the signal strength usually decreases at a rate inversely proportional to d^2, d^3 and even in some cases d^4 in really bad indoor conditions. In our case, it is clearly shown that the bandwidth and packet delivery ratios rapidly drop to zero for distances around 30 m and beyond. As expected for the 1-hop dedicated link, given a fixed distance, the difference between the achieved bandwidth and the one being used at the source is bigger at higher data rates.

In the case of the multihop MMARP-based multicast ad hoc access network, it can be noticed that the performance at increasing distances degrades much slower than $1/d^2$. This is because the average distance in each of the intermediate hops is lower than in the single-hop trial. Thus, the mean signal strength is higher and the achieved bandwidth and packet delivery ratio are higher as well.

However, as Fig. 4 shows, MMARP only manages to achieve a 100% delivery ratio at distances at which only one or two of the MMARP nodes are needed. At a distances higher to 30m some packet losses come up. These losses are mainly due to the well-known hidden terminal problem which happens among the nodes MMARP_1, MMARP_2 and MMARP_3. As long as IEEE 802.11b does not implement layer 2 acknowledgements of multicast frames (as it does for unicast traffic), each time a collision happens the packet is lost.

It is also particularly noticeable that the trial with the higher bandwidth in the multihop case performs much worse than the others. This is because, due to contention, the effective bandwidth, even in the ideal case, is lower than the

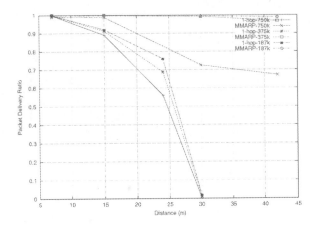

Fig. 5. Packet delivery ratio at increasing distance

753,6 Kb/s generated by the source. When MMARP_1 receives a packet from WWR and forwards it to MMARP_2, the effective bandwidth is reduced to half of the original. One half of the channel is used for receiving the packet and the other half for sending it. When MMARP_2 sends the packet to MMARP_3, the effective bandwidth is further reduced to a third of the original (in optimal channel conditions). Leaving thus an effective bandwidth of 667 Kb/s (2/3 Mb/s) which is lower than the 753 Kb/s that the source is using.

However, in the trials without that bandwidth limitation the MMARP protocol has demonstrated to be able to deliver mostly 100% of the packets (even in non-optimal channel conditions) without an impairment in the overhead or the scalability of the protocol. The differences in the packet delivery ratio between these two multihop cases are mainly due to the hidden terminal problem. At higher data rates, the probability of two packets actually colliding is higher. However, as the figure shows, the performance has not been severely degraded for that reason. So, it is clear that the real limitation towards multicast ad hoc access networks is mostly the IEEE 802.11b MAC layer, which is known not to be very adequate for ad hoc networks. This demonstrates that, regarding the protocol's behaviour, having a higher number of nodes in the same radio link is not an issue. Only nodes with the MF_FLAG active will forward packets, and only the best of all those nodes would be selected as a forwarder.

5   Conclusions and Future Work

Currently there is not a real solution to seamlessly support efficient IP multicast communications in future heterogeneous wireless scenarios. We present our solution for ad hoc networks extending fixed IP access networks. It consists of a novel architecture and a new multicast ad hoc routing protocol called MMARP. This approach is the first to our knowledge being able to support seamless roaming from multicast nodes (including traditional IP multicast hosts) between traditional IP multicast networks and ad hoc network extensions. In the authors' opinion, in addition to the proposed solution, it is also an important contribution the demonstration through empirical experimentation that this kind of extensions driven by MMARP are able to easily extend IP multicast edge-coverage in a cost-effective way, without an impairment in the overall throughput. The results show that even at distances which the traditional single-hop approach is not able to cover, the multihop option offers more than a 98% packet delivery ratio.

For future work, we are working towards the analysis of the approach in hybrid ad hoc networks (e.g. mixed WLAN, Bluetooth scenarios), and with different layer 2 protocols to improve the performance of the IEEE 802.11b MAC layer.

Acknowledgements

This work has been partially funded by Spanish MCYT by means of the projects ISAIAS(TIC2000-0198-P4-04) and SAM(TIC2002-04531-C04-04).

References

1. IST-MIND Official Web site. [On-line] http://www.ist-mind.org/
2. Fenner, W.: Internet Group Management Protocol, Version 2. IETF Request For Comments, RFC 2236, November, 1997.
3. Estrin, D., Farinacci, D., Helmy, A., Thaler, D., Deering, S., Handley, M., Jacobson, V., Liu, C., Sharma P, Wei, L.: Protocol Independent Multicast Sparse Mode (PIM-SM): Protocol Specification. RFC 2362, June 1998.
4. Garcia-Luna-Aceves, J.J., Madruga, E.L.: The Core Assisted Mesh Protocol. IEEE JSAC, Vol 17, No. 8, August 1999, pp.1380–1394.
5. Lee, S.-J., Su, W., Gerla, M.: On-Demand Multicast Routing Protocol in Multihop Wireless Mobile Networks. ACM/Kluwer Mobile Networks and Applications, 2000.
6. Jetcheva, J.G., Johnson, D.B.: Adaptive Demand-Driven Multicast Routing in Multi-Hop Wireless Ad Hoc Networks. Proceedings of the 2001 ACM International Symposium on Mobile Ad Hoc Networking and Computing, ACM, Long Beach, CA, (October 2001): 33–44
7. H. Lei and C.E. Perkins. Ad Hoc Networking with Mobile IP. In Proceedings of the Second European Personal Mobile Communications Conference, October 1997, pp. 197–202.
8. Ruiz, P.-M., Brown, G., Groves, I.: Scalable Communications for Ad hoc Extensions connected to Mobile IP Networks. Proceedings of the (PIMRC'2002). Lisbon, September, 2002. Vol. 3, pp. 1053–1057.

A Uniform Continuum Model for Scaling
of Ad Hoc Networks

Ernst W. Grundke and A. Nur Zincir-Heywood

Faculty of Computer Science, Dalhousie University
6050 University Avenue, Halifax, Nova Scotia, Canada B3H 1W5
{grundke,zincir}@cs.dal.ca

Abstract. This paper models an ad-hoc network as a continuum of nodes, ignoring edge effects, to find how the traffic scales with N, the number of nodes. We obtain expressions for the traffic due to application data, packet forwarding, mobility and routing, and we find the effects of the transmission range, R, and the bandwidth. The results indicate that the design of scalable adhoc networks should target small numbers of nodes (not over 1000) and short transmission ranges. The analysis produces three dimensionless parameters that characterize the nodes and the network: α, the *walk/talk ratio*, or the ratio of the link event rate to the application packet rate; β, the *forwarding overhead*, or the average number of hops required for a packet to travel from source to destination; and γ, the *routing overhead*. We find that the quantity $\alpha\gamma/\beta$ characterizes the relative importance of routing traffic and user data traffic. These quantities may be useful to compare the results of various simulation studies.

Keywords: Ad-hoc networks, mobile, scaling, continuum, model.

1 Introduction

Several features distinguish ad-hoc networks [5] from their traditional wired counterparts: (1) Ad-hoc networks consist of mobile nodes that communicate by relatively low-powered radio signals. (2) Nodes act both as hosts for application software and as routers to forward incoming packets to other nodes. (3) Ad-hoc networks need to be highly dynamic: the nodes should be able to move, including entering and leaving the network, without manual configuration. (4) Finally, since nodes rely on batteries, power is a scarce resource.

Some recent papers have explored algorithms and protocols for this combination of constraints. Santivanez et al. [11] model the scaling of ad-hoc routing protocols. Gupta and Kumar [6] analyze the capacity of wireless networks under a sophisticated model, although mobility and routing are not considered. Hong, Xu and Gerla [7] analyze the scalability and operational features of routing protocols for mobile ad-hoc networks. They divide routing protocols into three categories: flat routing, hierarchical routing and geographical (GPS-augmented) routing.

This paper investigates the scaling of a very simple model with minimum *a priori* assumptions; the effects of mobility and (flat) routing are included. To this end, we construct an ad-hoc network model by specifying just the average density of nodes (the number of nodes per unit area): we are not concerned with discrete nodes, their

S. Pierre, M. Barbeau, and E. Kranakis (Eds.): ADHOC-NOW 2003, LNCS 2865, pp. 96–103, 2003.

exact positions or movements, or their exact links with other nodes. Thus, our model focuses on a *continuum* approximation rather than a graph-theoretic view of the network.

In order to simplify the analysis, we assume that all nodes see similar traffic conditions. In other words, we ignore the *edge effects* resulting from nodes near the extremity of a network having fewer neighbors than centrally located nodes. In this sense our model is *uniform*.

Our goal is to model the traffic in an ad-hoc network using simple and optimistic assumptions. Our simple assumptions lead to a mathematically tractable model, which in turn reveals several dimensionless parameters that characterize the operation of an ad-hoc network; these may be helpful in bridging the gaps between the parameters of various simulations. Our optimistic assumptions lead to upper bounds on the performance of real networks. We avoid an exclusively asymptotic analysis for a large number of nodes [11] in order to deal with practical finite cases; therefore our results are derived in *cf* rather than $\Theta(f)$ format, although the values of constants are approximate at best.

In Sections 2 and 3 we define the parameters to describe two-dimensional network geometry and node behavior, respectively. An expression for the traffic due to user applications is derived in Section 4. In Section 5 we find how much traffic results from mobility to support routing. The user data traffic and routing traffic are combined in Section 6 to find the total traffic and the power requirement. The two-dimensional results are extended to m dimensions in Section 7, and conclusions are drawn in Section 8.

2 Network Geometry in Two Dimensions

We consider an ad-hoc network whose nodes lie in a plane. With each node we can associate a Voronoi cell, which is the set of points closer to that node than any other. We approximate cells by circles with radii r_1 on average, giving an average distance $d_1 = 2r_1$ between neighboring nodes, and we suppose that the cell area is approximately πr_1^2. (See [6], who investigate feasible Voronoi tessellations in detail.) Thus the *node density* (the number of nodes per unit area) is approximately $4/\pi d_1^2$.

We assume that the network consists of N nodes occupying a circular region of diameter $D+d_1$, and that the maximum distance between any pair of nodes is D. The node density must be approximately $4N/(\pi(D+d_1)^2)$, so that $Nd_1^2 = (D+d_1)^2$, or $D = d_1(\sqrt{N}-1)$.

3 Node Model

The essential features of nodes are that (a) they generate user data, (b) they forward packets, (c) they move, (d) they have a finite radio transmission range, and (e) they have a finite transmission bandwidth.

(a) We assume that each node is a random source of user data, being characterized by a rate p_T, the number of new data packets created and transmitted by a node per unit time to a randomly chosen destination node. (The subscript T is meant to suggest an application transmitting, or *talking*.).

(b) Packet forwarding is discussed in Section 4.

(c) When a node moves, it may enter or leave the radio range of one or more other nodes. We assume that nodes are able to detect the making and breaking of radio links, and that such *link events* occur at each node at a rate p_W per unit time. The symbol W is meant to suggest a user *walking* while carrying a mobile device.

We define $\alpha = p_W/p_T$ to be the *walk/talk ratio*, the ratio of a node's rate of link events to its rate of producing user data packets. Notice that α depends only on the nodal behavior and not on the network configuration. The value α is a useful dimensionless measure of the impact of node mobility.

(d) It is important to model the radio transmission *range* of a node because of a fundamental tradeoff in ad-hoc networks: a large range can reduce the number of hops required to transport a packet to its destination, but it also reduces the number of nodes that can transmit simultaneously. We assume that the transmission range of a node is R: at distances exceeding R, a node's signal cannot be received and does not interfere with other reception, either because the signal is too weak or because some aspect of the physical layer restricts the nodes' participation. (For example, a frequency hopping scheme may form a logical small-scale network of this size.) We assume that $d_1 \leq R \leq D$, since (i) for $R < d_1$ the network becomes largely disconnected, and (ii) for $R > D$ all N nodes are already within range. The number of nodes within range of any transmitting node is a group of approximately $g = (R+r_1)^2/r_1^2$ nodes, including the transmitting node.

(e) The quantities p_T and p_W cannot be arbitrarily large because the packet transmission rate for each node is finite. Let b be the maximum possible value for p_T, realized when a node transmits new user data packets continuously and handles no other traffic. (Assuming one packet per link event, p_W must also satisfy $p_W \leq b$.) Then b is node's *bandwidth* expressed in packets/second. However, because of the nature of a typical physical layer, a node cannot attain a packet transmission rate of b. If we assume that the g nodes within radio range share a channel at any moment, the average maximum packet transmission rate per node is only b/g.

4   User Data Traffic

Our network is assumed to have only two types of traffic: application (*user*) data packets and routing packets. We assume that there is no gateway to other networks. We begin by using the above network geometry and node model to find the traffic due to user data alone.

First we define β (>1), the *forwarding overhead*, to be the average number of hops required for a packet to travel from source node to destination node. (We ignore data traffic between local applications.) With the assumption of uniformity, this implies that for every user data packet injected into the network by a node, the node must perform on average β packet transmissions. This is a cost of participating in an ad-hoc network: in order to originate data packets at a rate p_T, a node must transmit data packets at a rate of βp_T.

The average distance from the source to the destination [4] is about $D/2$. We assume that routing is optimal, i.e. a packet travels roughly in a straight line in hops of length R. Then the distance $D/2$ can be covered in $\beta = D/(2R)$ hops. Since $D = d_1(\sqrt{N}-1)$, we have

$$\beta = \frac{r_1}{R}(\sqrt{N}-1) .$$ (1)

We note that β is dimensionless, and, because of the limits on R, satisfies $1 \le \beta \le (\sqrt{N}-1)/2$. (Following the algebra strictly, $R \le D$ would give $0.5 \le \beta$, but it is precisely as R approaches D that the edge effect begins to matter, and $\beta<1$ makes no sense.) In our model, the forwarding overhead β depends only on the network configuration and not on the nodal behavior. In a more realistic model, it would depend on the routing effectiveness and on traffic patterns.

The overhead β is $\Theta(\sqrt{N}/R)$, which appears to favor large R (but see below). Although dependence on N is not strong, it may still impose restrictions on the number of nodes. For instance, if the overhead is to be no greater than 5 (e.g. for reasons of power consumption), and if the range is as short as possible, N must not exceed 121.

The user data packet rate βp_T cannot exceed the limit b/g established earlier: $\beta p_T \le b/g$, or

$$p_T \le \frac{br_1R}{(R+r_1)^2} \frac{1}{\sqrt{N}-1} .$$ (2)

This result shows that small values of R (that is, $R \approx 2r_1$) are preferable. In that case we obtain

$$p_T \le \frac{2b}{9(\sqrt{N}-1)} .$$ (3)

The limiting case $p_T = \beta b/N$ is obtained for a low walk/talk ratio, where routing traffic can be ignored because the nodes are nearly stationary. For example, if $N=100$ nodes have a bandwidth of 1Mbps each and a range $R = 2r_1$, then each node can transmit only 2Mbps/(9×9) $\approx$ 25 Kbps on average (including bandwidth overheads such as packet headers), and then only if all nodes are stationary. Throughput drops further if the range increases beyond the minimum of $2r_1$.

We emphasize that these results are independent of any mobility and any routing algorithm. In fact, they would apply to wired networks if hosts were required to perform routing. We note that the finite *minimum traffic load* of Santivanez et al. [11] must already include the effect that $p_T \to 0$ as $N \to \infty$.

5 Routing Traffic

Link events are generated by each node at a rate p_W. We assume non-hierarchical (flat) proactive routing, that is, a link event is promptly sent as one routing packet to all nodes in order to keep routing tables updated.

We define γ (>1), the *routing overhead*, to be the number of packets generated by one link event. As with user data traffic, uniformity implies that every node must transmit on average γ routing packets per link event, or γp_W packets per unit time. To estimate γ, we assume that by transmitting one packet a node can broadcast to $g-1$ other nodes. However, the information will be new to only about half of those nodes. In total we must reach N nodes, so that

$$N = \gamma \frac{g-1}{2},$$

(4)

or

$$\gamma = \frac{2N}{\frac{R^2}{r_1^2} + \frac{2R}{r_1}}.$$

(5)

If the range is minimized (that is, $R \approx 2r_1$), the routing overhead is $\gamma = (N/4) - 1$.

Clearly this estimate needs to be refined to take specific routing protocols into account. The assumption of reaching $(g-1)/2$ nodes with a single packet may be quite optimistic, since it requires a receiving node to "know" whether its position justifies rebroadcasting a given packet (e.g. whether it is at the edge of the previous transmitter's range in the "forward" direction). On the other hand, some protocols [1,2,9,10] effectively combine multiple events into a single packet to improve efficiency.

The routing overhead is $\Theta(N/R^2)$, and is potentially much more serious than the forwarding overhead, which is only $\Theta(\sqrt{N}/R)$. It is interesting that γ is $\Theta(\beta^2)$.

The routing packet rate γp_W cannot exceed the limit b/g established earlier: $\gamma p_W \leq b/g$, or

$$p_W \leq \frac{b}{\gamma g} = \frac{b}{2N} \frac{g-1}{g} = \frac{b}{2N} [1 - \frac{r_1^2}{(R+r_1)^2}] \approx \frac{b}{2N}.$$

(6)

Therefore the maximum rate at which a node can generate link events is $\Theta(1/N)$. It is almost independent of R because, in our model, γ is $\Theta(N/R^2)$ while the bandwidth reduction factor, $1/g$, is $\Theta(R^2)$.

The limiting case $p_W = b/(\gamma g)$ is obtained for a high walk/talk ratio, where the user data traffic is starved to zero by frequent link events. This sets a fundamental limit on the product Np_W, the network link event rate. Fortunately, the consequences are not numerically serious for networks of modest size. For example, if $N=100$ nodes have a bandwidth of 1Mbps each, then each node would have to generate almost 5 Kbps of link event packets in order to saturate the network. Similarly, if p_W = 1 event/second and routing packets contain 500 bits, then saturation occurs at about $N=1000$.

6   Total Traffic and Power

The combined effect of user data traffic and routing traffic is that each node transmits $\beta p_T + \gamma p_W$ packets per unit time, and the bandwidth constraint for the combined traffic is $\beta p_T + \gamma p_W \leq b/g$. Relative to transmitting p_T packets of user data per unit time, a node incurs an overhead of $\beta + \alpha\gamma$.

Depending on whether α, the walk/talk ratio, is greater or less than β/γ, routing traffic or forwarding traffic, respectively, dominates in the network. This suggests using values of $\alpha\gamma/\beta$ in order to compare the results of various simulation studies; see, for example, [1-3,8,10]. In the previous sections, a low (high) walk/talk ratio should be taken to mean $\alpha\gamma/\beta \ll 1\ (\gg 1)$.

Assuming that the range R is limited by a threshold of received signal strength, the average antenna power is proportional to R^2 and to the rate of packet transmission. The average power requirement per node is

$$P = k\,R^2\,(\beta p_T + \gamma\,p_W) \tag{7}$$

where k is a constant. (The antenna power for continuous transmission is kR^2b.) The power is $\Theta(N)$, with the leading term arising from mobility. P is the radiated antenna power only, and does not include the power consumption of the node's circuitry itself.

7   Other Dimensions

This model has been built for the most practical case of two dimensions, although we could equally well have chosen m dimensions. For $m=1$ we have a model of N nodes spread in a line (e.g. nodes in vehicle on a road), and for $m=3$ we have a model of N nodes spread in a volume (e.g. nodes carried by users in a multi-storied building).

In m dimensions, the node density is measured in units of nodes per (unit length)^m. The total number of nodes, N, and the network diameter (a distance), D, are related by $Nd_1{}^m = (D+d_1)^m$. The number of nodes in a radio group is $g = (R+r_1)^m/r_1{}^m$. The data transmission overhead β becomes

$$\beta = \frac{r_1}{R}(\sqrt[m]{N}-1), \tag{8}$$

the routing overhead γ becomes

$$\gamma = \frac{2N}{(\dfrac{R}{r_1}+1)^m - 1}. \tag{9}$$

The bandwidth constraints in m dimensions become

$$p_T \leq \frac{br_1{}^{m-1}R}{(R+r_1)^m}\frac{1}{\sqrt[m]{N-1}} \tag{10}$$

and

$$p_W \leq \frac{b}{2N}[1-\frac{r_1{}^m}{(R+r_1)^m}]. \tag{11}$$

For $m=3$, from (10) the maximum p_T is $\Theta(1/R^m N^{1/\ m})$, which makes it especially important to keep the range small, while N can grow larger than was feasible in two dimensions.

The R^2 factor in the power requirement is unchanged because it arises from the three-dimensional spreading of radio signals, regardless of m.

8 Conclusion

The simple continuum model without edge effects has yielded a number of analytic results. In two dimensions the user data traffic is $\Theta(\sqrt{N}/R)$, and routing traffic is $\Theta(N/R^2)$, where N is the number of nodes and R is the transmission range. The maximum (bandwidth-limited) user data traffic per node is $\Theta(1/R\sqrt{N})$, and the maximum link event rate is $\Theta(1/N)$. It will be interesting to see how closely simulations and real networks follow these scaling trends. Our results confirm that the design of flat ad-hoc networks should target small numbers of nodes (100's, not 1000's), and should strive for short transmission ranges.

This analysis has produced three dimensionless parameters that characterize an ad-hoc network. The node behavior is characterized by α, the *walk/talk ratio,* which is the ratio of the link event rate to the application packet rate. The network is characterized by β, the *forwarding overhead*, and by γ, the *routing overhead*. We find that the quantity $\alpha\gamma/\beta$ characterizes the relative importance of routing traffic and user data traffic; the two are equal when $\alpha\gamma/\beta = 1$. These dimensionless parameters may prove useful to compare the results of various simulation studies and to scale ad-hoc networks.

References

1. Amin K., Mayes J., Mikler A..: Agent-based Distance Vector Routing. IEEE/ACM MATA 2001: 3rd International Workshop on Mobile Agents for Telecommunications Application, Canada, August (2001). Retrieved from http://students.csci.unt.edu/~amin/

2. Braginsky D., Estrin D.: Rumor Routing Algorithm for Sensor Networks, WSNA '02 September 28 (2002) Atlanta, Georgia, USA

3. Celebi E.: Master's Thesis: Performance Evaluation of Wireless Mobile AdHoc Network Routing Protocols (2001). Retrieved May 16, 2003 from http://cis.poly.edu/~ecelebi/

4. Contla, P. A.., Stojmenivoc, M.: Estimating Hop Counts in Position Based Routing Schemes for Ad Hoc Networks. Telecommunication Systems, Vol. 22 (2003) 109-118

5. Corson, S., Macker, J.: Mobile Ad hoc Networking (MANET): Routing Protocol Performance Issues and Evaluation Considerations. IETF RFC 2501 (1999) Retrieved May 21, 2003, from http://www.ietf.org/rfc/rfc2501.txt

6. Gupta P., Kumar P. R.: The Capacity of Wireless Networks. IEEE Transactions on Information Theory, Vol. 46 Issue 2 (2000) 388-404

7. Hong X., Xu K., Gerla M.: Scalable Routing Protocols for Mobile Ad Hoc Networks, IEEE Network, July/August (2002) 11-21

8. Johansson P., Larsson T., Hedman N., Mielczarek B., Degermark M.: Scenario-based Performance Analysis of Routing Protocols for Mobile Adhoc Networks. ACM Mobicom '99, Seattle Washington USA (1999)

9. Liang, S., Zincir-Heywood, N., Heywood, M.: The Effect of Routing under Local Information Using a Social Insect Metaphor. IEEE 2002 World Congress on Computational Intelligence, Congress on Computation (2002) 1438-1443

10. Minar N., Kramer K. H., Maes P.: Cooperative Mobile Agents for Dynamic Network Routing. In: Software Agents for furutre Communications Systems, Springer-Verlag, New York (1999)

11. Santiváñez, C., McDonald, B., Stavrakakis, I., Ramanathan, R.: On the Scalability of Ad Hoc Routing Protocols. Proc. IEEE INFOCOM 3 (2002) 1688-1697

Probabilistic Protocols for Node Discovery in Ad Hoc Multi-channel Broadcast Networks

G. Alonso[1], E. Kranakis[2], C. Sawchuk[2], R. Wattenhofer[1], and P. Widmayer[1]

[1] Department of Computer Science, Swiss Federal Institute of Technology, ETH Zurich, Switzerland
[2] School of Computer Science, Carleton University, Ottawa, ON, K1S 5B6, Canada

Abstract. Ad hoc networks consist of wireless, self-organizing nodes that can communicate with each other in order to establish decentralized and dynamically changing network topologies. Node discovery is a fundamental procedure in the establishment of an ad hoc network, as a given node needs to discover what other nodes are in its communication range. Existing multi-channel node discovery protocols are typically constrained by the network configuration that will be imposed on the nodes once they are discovered. We present a communication model that is independent of the network configuration that will be established after node discovery. We present a pair of node discovery protocols for $k \geq 2$ nodes in a multi-channel system and analyze them using the given communication model.

1 Introduction

Ad hoc networks consist of wireless, self-organizing nodes that can communicate with each other in order to establish decentralized and dynamically changing network topologies. Since these networks are an integral part of the new wireless solutions sought for home or personal area networks, sensor networks, and various other commercial and educational networks, eliminating the shortcomings of ad hoc networks is an important goal in network research [12].

Before a node can communicate with the other nodes in its communication range, it must be aware of those nodes and thus node discovery is an essential part of the rendezvous layer for any node that engages in ad hoc network formation [16]. Efficient network formation requires that the rendezvous layer be able to find all nodes in communication range in the shortest time and with the smallest energy expenditure possible. Obviously, the complexity of node discovery is a function of both the number of nodes present and the number of communication channels available to these nodes. Until recently, nearly all ad hoc networks were formed by nodes that used single channel technology such as 802.11 or IR LANs and thus most of the research about node discovery in ad hoc networks assumes there is a single broadcast channel [15]. The introduction of Bluetooth [8], however, has boosted interest in node discovery in multi-channel systems with frequency-hopping. Such research is especially important since the

S. Pierre, M. Barbeau, and E. Kranakis (Eds.): ADHOC-NOW 2003, LNCS 2865, pp. 104–115, 2003.
© Springer-Verlag Berlin Heidelberg 2003

node discovery protocol in the Bluetooth standard [8] does not scale well and is both time and energy intensive [16].

The node discovery protocol in the Bluetooth standard [8] is asymmetric in that it assigns different roles and different frequency-hopping speeds to various nodes. Salonidis et al [14] point out that when two or more Bluetooth users want to form an ad hoc network, they cannot explicitly assign roles. They need a symmetric protocol for node discovery, i.e., one that does not depend on pre-assigned roles for the nodes. Salonidis et al [14][15], Law et al [10], and Siegemund and Rohs [16] have subsequently developed symmetric node discovery protocols for Bluetooth.

Naturally, these protocols are constrained by the configuration requirements of Bluetooth, e.g., scatternets are comprised of connected piconets where the latter contains one master and seven slave nodes. There exist few multi-channel node discovery protocols that are independent of any network configuration. Since the performance of existing multi-channel node discovery protocols is inextricably linked to the resulting network configuration, it is difficult to compare the performance of protocols that execute in different network configurations. In this paper, however, we present a communication model that is independent of any network configuration that may be imposed on nodes once they are discovered. The model is an extension of the work by Alonso et al [1] to the multi-channel case for $k \geq 2$ nodes. We present a pair of node discovery protocols for $k \geq 2$ nodes and analyze them using the multi-channel communication model.

1.1   Multi-channel Communication Model

Consider a collection of $k \geq 2$ nodes and $f \geq 2$ broadcast channels or frequencies. At each point in time, a given node must either talk (T) or listen (L) on one of the f channels. A node cannot talk and listen at the same time. The state of a node is denoted by (S, i) where $S = T$ or $S = L$ and i represents the chosen frequency, $i = 1, \ldots, f$.

A node a hears the broadcast of another node b if, at the given time, nodes a and b choose the same frequency i, node a listens (L) and node b talks (T), and no other node talks on frequency i. If a node other than node b also talks on frequency i, then collision occurs on frequency i and no node listening on that frequency hears a broadcast. (In this model, spatial frequency reuse, like that used in cellular phones, is not possible.) The nodes are unable to distinguish between collision and noise when listening to a given frequency. *Node discovery* occurs when node a hears the broadcast of node b and, in the next step, node b hears the broadcast of node a.

An event E describes the states of the k nodes at a given point in time:

$$E = \begin{pmatrix} S_1 & i_1 \\ S_2 & i_2 \\ \vdots & \vdots \\ S_k & i_k \end{pmatrix} \tag{1}$$

where (S_m, i_m) is the state of the m-th node.

A node discovery protocol dictates how a node should choose its state at each point in time. A run of a given protocol is the sequence of events generated by the node's choices. Let $E \rightarrow E'$ denote that event E is immediately followed by event E' in a given run of the protocol. A run terminates when the k nodes have discovered each other. In the two node case, the last two events of the run, $E \rightarrow E'$, are 1) in event E, the first node hears the second node talk, and 2) in the last event, E', the second node hears the first node talk. Thus node discovery with two nodes occurs under the events

$$
\begin{pmatrix} T\ i \\ L\ i \end{pmatrix} \rightarrow \begin{pmatrix} L\ j \\ T\ j \end{pmatrix} \text{ or } \begin{pmatrix} L\ i \\ T\ i \end{pmatrix} \rightarrow \begin{pmatrix} T\ j \\ L\ j \end{pmatrix} \tag{2}
$$

The relationship between frequencies i and j depends on the protocol's frequency allocation method and is discussed below.

Let a node be represented by the random variable X that assumes the values of (S, i), the possible states of the node. When a node must randomly choose whether to talk or listen, let p denote the probability that the node will talk (T) and let $q = 1 - p$ denote the probability that the node will listen (L). When a node must randomly choose a frequency, let F_i denote the probability that the node will choose frequency i. Thus p_i, the probability that a given node will talk on frequency i, equals $\Pr[X = (T, i)] = pF_i$ and q_i, the probability that a given node will listen on frequency i, equals $\Pr[X = (L, i)] = qF_i$. Since $p + q = 1$ and $\sum_{i=1}^{f} F_i = 1$, then $\sum_{i=1}^{f} p_i + \sum_{i=1}^{f} q_i = 1$.

After certain events, e.g., one node hears the broadcast of another node, a node may have to decide whether to stay with the same frequency i in the next step, i.e., **static** frequency allocation, or to again randomly choose a frequency, i.e., **dynamic** frequency allocation. If the initial contact occurred on a given frequency i, one might argue that frequency i is a natural choice for further communication and thus static frequency allocation should occur. One can also argue, however, that chances for continued contact may be just as good if the next frequency is again randomly chosen and thus dynamic frequency allocation can be used.

We assume that nodes are synchronized so that they start an algorithm at the same time, choose their respective states at the same time, and maintain those states for the same amount of time. While it is unlikely that all the nodes that want to participate in a given session of node discovery will start the node discovery protocol at the same time, Salonidis et al [15] demonstrate that node synchronization can be accomplished in a reasonable amount of time. They show that if the times at which the respective nodes start the protocol are modelled as a carefully chosen Poisson process then, after a first node has started the node discovery protocol, the remaining nodes have start times that are identically and independently distributed according to a truncated exponential distribution. Given this distribution, a timeout value can be estimated and incorporated into the beginning of the node discovery protocol so that node synchronization occurs before the nodes engage in discovery. The size of the timeout is usually small relative to the time required for node discovery.

We also assume that the nodes know the value of k, the number of nodes in the system. If the number of nodes k was unknown, then a node might need to estimate k in the course of a node discovery protocol, but we leave the study of such cases to a later date.

1.2   Our Contribution

As mentioned earlier, the multi-channel communication model just described for $k \geq 2$ nodes is **independent** of any network configuration that might be imposed on the nodes once they are discovered. We present two node discovery protocols for $k \geq 2$ nodes and analyze them using the multi-channel communication model.

In the random protocol **RP**, each node randomly chooses whether to talk or listen and also randomly chooses a channel or frequency. The nodes' respective choices of actions (talk or listen) and frequencies over time can be represented by a string of symbols. If the nodes' respective choices of actions and frequencies in a given time t are such that one node can hear the other node's broadcast, then the subsequence of symbols representing that event is called a success pattern. By analyzing the occurence of these success patterns, we determine that the expected run time of the random protocol **RP** for two nodes is

$$\frac{1 + \sum_{j=1}^{f} p_j q_j}{2 \left(\sum_{j=1}^{f} p_j q_j \right)^2}$$

We also analyze another node discovery protocol for the two node case. In the conditional protocol **CP**, a node randomly chooses to talk or listen until 1) the node talks or 2) the node listens *and* hears the other node's broadcast.

If a given node talked at time t, it will listen at time $t + 1$ in an attempt to determine if the other node heard its broadcast, while if the given node listened at time t and heard another node's broadcast, it will talk at time $t + 1$ in an attempt to answer the other node. The nodes will choose their respective frequencies according to either static or dynamic frequency allocation.

The **CP** protocol has two phases. The first phase ends for a given node when that node either talks or hears the other node talk. The second phase consists of one step and the node's behaviour in that step is determined by whether it talked or listened at the end of phase 1. A single execution of the two phases is called a subrun. If, at the end of a subrun, node discovery has not occurred, then another subrun is executed. The length of a subrun of the **CP** protocol is an identically distributed random variable with a finite mean and the number of subruns in the **CP** protocol is a random variable with non-negative integer values and a finite mean. Since the length of a subrun is independent of the number of subruns for the **CP** protocol, Wald's identity implies that the expected run time of the **CP** protocol for two nodes is the product of the expected length of a subrun and the expected number of subruns.

A node's choice of frequency is random for each step in phase 1, but the frequency choice in the phase 2 (one step) depends on whether static or dynamic frequency allocation is used. With static frequency allocation, the frequency used in phase 2 is the same frequency used in the final step of phase 1, while with dynamic frequency allocation, the frequency for phase 2 is randomly chosen.

The expected run time for the **CP** protocol with two nodes is

$$\frac{2p(1-p)+1}{(2p(1-p))^2(\sum_{i=1}^{f} F_i^2)}$$

with static frequency allocation and

$$\frac{2p(1-p)+1}{(2p(1-p))^2(\sum_{i=1}^{f} F_i^2)^2}$$

with dynamic frequency allocation. The expected run time for the **CP** protocol with two nodes is longer under dynamic frequency allocation, as opposed to static frequency allocation, by a factor of $\phi = 1/\sum_{i=1}^{f} F_i^2$. For example, if there are f equally likely frequencies such that $F_i = F_j$ for all i, j, then the expected run time of the **CP** protocol with two nodes is f times greater under dynamic frequency allocation than under static frequency allocation.

Having analyzed the **RP** and **CP** protocols for the two node case, we turn to the $k \geq 2$ node case. In the random protocol **RP** for $k \geq 2$ nodes, each node again decides at random whether to talk or listen and also randomly chooses a frequency. The expected run time for the **RP** protocol with $k \geq 2$ nodes is

$$\frac{1 + \sum_{j=1}^{f} p_j q_j (1-p_j)^{k-2}}{2\binom{k}{2} \left(\sum_{j=1}^{f} p_j q_j (1-p_j)^{k-2}\right)^2}$$

Unfortunately, calculating the expected run time of the **CP** protocol for $k > 2$ nodes is not as straightforward. At any time $t > 0$ in the **CP** protocol, a given node can be both in phase 1 relative to one subset of nodes and in phase 2 relative to another subset of nodes. Tracking the potential overlap of phases across the nodes becomes more complicated as the number of nodes increases. Our analysis of the expected run time for the **CP** protocol with $k \geq 2$ nodes, therefore, relies on simulation methods rather than a closed-form solution.

1.3   Outline of the Paper

In section 2, we present and analyze the random protocol **RP** and the conditional protocol **CP** for the two node case. In section 3, we present and analyze the $k \geq 2$ nodes case for the **RP** and **CP** protocols. The paper ends in section 4 with some summary remarks and a brief description of open problems. Due to space limitations, only outlines of the proofs are given.

2   Random Protocol for Two Node Multi-channel System

In the random protocol **RP**, each node decides at random whether to talk (T) or listen (L). The two nodes thus generate an event at each time t

$$E = \begin{pmatrix} S & i \\ S' & i' \end{pmatrix}$$

such that S and S' are either T or L, and i and i' are the frequencies chosen.

We use the technique described in [3,11] to analyze the **RP** protocol. The random protocol **RP** succeeds when, for some $i, j = 1, 2, \ldots, f$, either

$$\begin{pmatrix} T & i \\ L & i \end{pmatrix} \rightarrow \begin{pmatrix} L & j \\ T & j \end{pmatrix} \text{ or } \begin{pmatrix} L & i \\ T & i \end{pmatrix} \rightarrow \begin{pmatrix} T & j \\ L & j \end{pmatrix} \tag{3}$$

Define the events A_i and B_i as follows:

$$A_i = \begin{pmatrix} T & i \\ L & i \end{pmatrix}, B_i = \begin{pmatrix} L & i \\ T & i \end{pmatrix} \tag{4}$$

A success pattern is a pair of events such that the two nodes discover each other, e.g., $A_i B_j$, $i, j \in 1, \ldots, f$, and thus there are $2f^2$ success patterns:

$$A_1 B_1, \ldots, A_1 B_f, \ldots, A_f B_1, \ldots, A_f B_f, B_1 A_1, \ldots, B_1 A_f, \ldots, B_f A_1, \ldots, B_f A_f.$$

For each $i, j \in \{1, 2, \ldots, f\}$, the pattern $A_i B_j$ (respectively $B_i A_j$) may either overlap itself, or the last event of $A_i B_j$ (respectively $B_i A_j$) may overlap with the first event of $B_j A_k$ (respectively $A_j B_k$) for $k \in \{1, 2, \ldots, f\}$. The former case occurs with probability $\frac{1}{p_i q_i p_j q_j}$ while the latter case occurs with probability $\frac{1}{p_j q_j}$.

The resulting system of $2f^2$ linear equations is

$$\left[\begin{array}{c|c} D & U \\ \hline U & D \end{array} \right] \left[\begin{array}{c} \Pi \\ \Pi \end{array} \right] = \begin{bmatrix} E[N] \\ E[N] \\ \vdots \\ E[N] \\ E[N] \\ E[N] \\ \vdots \\ E[N] \end{bmatrix} \tag{5}$$

where

$$\Pi' = [\pi_{1,1}, \ldots, \pi_{1,f}, \pi_{2,1}, \ldots, \pi_{f,1}, \ldots, \pi_{f,f}]$$

and $\pi_{i,j}$ (respectively $\pi_{i \mid f,j}$) is the probability that $A_i B_j$ (respectively B_i, A_j) occurs before any other pattern. N is the run time for exactly one subrun of RP. However, because RP always executes exactly one subrun, N is also the run time for the entire protocol.

The $2f^2$ x $2f^2$ matrix $\left[\begin{array}{c|c} D & U \\ \hline U & D \end{array} \right]$ is defined as follows:

- D is a diagonal $f^2 \times f^2$ matrix with the $((i,j),(i,j))$-th entry equal to $\frac{1}{p_i^2 q_i^2}$.
- U is an $f^2 \times f^2$ matrix formed by a column of f matrices, i.e., $U = [V, V, \ldots, V]^T$ where V is defined as

$$
V = \begin{bmatrix}
\frac{1}{p_1 q_1} & \frac{1}{p_2 q_2} & \cdots & \frac{1}{p_f q_f} & 0 & 0 & \cdots & 0 & \cdots & 0 & 0 & \cdots & 0 \\
0 & 0 & \cdots & 0 & \frac{1}{p_1 q_1} & \frac{1}{p_2 q_2} & \cdots & \frac{1}{p_f q_f} & \cdots & 0 & 0 & \cdots & 0 \\
\vdots & \vdots & \vdots & \vdots & \vdots & \vdots & \vdots & \vdots & \vdots & \vdots & \vdots & \vdots & \vdots \\
0 & 0 & \cdots & 0 & 0 & 0 & \cdots & 0 & \cdots & \frac{1}{p_1 q_1} & \frac{1}{p_2 q_2} & \cdots & \frac{1}{p_f q_f}
\end{bmatrix}
$$

With the condition

$$
\sum_{i=1}^{i=f}\sum_{j=1}^{j=f} \pi_{ij} = 1 \tag{6}
$$

the resulting system of linear equations has $2f^2 + 1$ unknowns and $2f^2 + 1$ equations. Solving this system of equations gives us $E[N]$, the expected runtime of **RP**.

Theorem 1 (RP). *The expected run time for the **RP** protocol is:*

$$
\frac{1 + \sum_{j=1}^{f} p_j q_j}{2 \left(\sum_{j=1}^{f} p_j q_j \right)^2}. \tag{7}
$$

3   Conditional Protocol for Two Node Multi-channel Systems

The conditional protocol **CP** is implemented as a series of two-phase subruns where phase 1 consists of a finite number of random steps and phase 2 consists of a single step.

In phase 1 of a subrun of the **CP** protocol, a node follows the random protocol **RP** until 1) the node talks (T) or 2) the node listens (L) *and* hears the other node's broadcast.

Phase 2 of a subrun of the **CP** protocol consists of a single step. The behaviour of a node in phase 2 is conditional on the way in which phase 1 ended. If a node talked (T) at the end of phase 1, then it will listen (L) in the phase 2 in an attempt to determine if the other node heard its broadcast. If a node listened (L) and heard the other node's broadcast at the end of phase 1, then it will talk (T) in phase 2 in an attempt to answer the other node's broadcast.

If a subrun is successful, then node discovery occurs in phase 2 and the **CP** protocol terminates. If a subrun is unsuccessful, however, then another subrun is executed, i.e., phase 1 and phase 2 are repeated, until node discovery occurs.

Let the probability of success in a subrun of the **CP** protocol be denoted by $Pr[\text{success in subrun}]$. Since the subruns of the **CP** protocol are independent trials, the number of subruns of the **CP** protocol is a geometric random variable with parameter $Pr[\text{success in subrun}]$. The expected number of subruns of the **CP** protocol is therefore

$$E[\text{number of subruns}] = \sum_{k=1}^{\infty} Pr[\text{failure in subrun}]^{k-1} Pr[\text{success in subrun}]k$$

$$(8)$$

3.1 Wald's Identity

If, for the **CP** protocol, the expected number of subruns and the expected length of a subrun are known, then Wald's identity can be used to calculate the expected run time of the protocol.

Wald's identity can be stated as follows [13]. Let $W_i, i \geq 1$ be independent and identically distributed random variables with a finite mean, $E[W] < \infty$. Let N be a stopping time for $W_1, W_2, \dots$ such that $E[N] < \infty$, i.e., the event $N = n$ is independent of $W_{n+1}, W_{n+2}, \dots$, for all $n \geq 1$. Then

$$E\left[\sum_{i=1}^{N} W_i \right] = E[W]E[N].$$

$$(9)$$

To apply Wald's identity to the present problem, let W_i be the length of a subrun of the **CP** protocol and let N be the number of subruns for the protocol. Defined in this manner, the W_i are identically distributed random variables with a finite mean and N is a random variable with non-negative integer values and a finite mean. The length of a subrun is independent of the number of subruns for the **CP** protocol, so W_i is independent of N. Wald's identity thus implies that the expected run time for the **CP** protocol is the product of the expected length of a subrun and the expected number of subruns, i.e.,

$$E[\text{run time for CP protocol}] = E[\text{length of subrun}]E[\text{number of subruns}].$$

$$(10)$$

If we calculate the expected length of a subrun and the expected number of subruns for the **CP** protocol, then equation 10 allows us to calculate the expected run time for the protocol.

3.2 Expected Length of a Subrun of CP Protocol

Phase 1 of the **CP** protocol ends when a node talks or when a node listens and hears the broadcast of the other node. Phase 1 therefore ends when at least one of the nodes talks so the only event that does not bring an end to phase 1 is the event where both nodes listen. Therefore $Pr[\text{phase 1 ends}]$ equals

$$1 - Pr[\text{both nodes listen}] = 1 - (1 - p)^2 = 2p - p^2$$

This implies that the expected length of phase 1 in the **CP** protocol is

$$E[\text{length of phase 1}] = \sum_{i=1}^{\infty} (2p - p^2)(1 - (2p - p^2))^{i-1} i = \frac{1}{2p - p^2} \qquad (11)$$

and therefore, because phase 2 has only one step, the expected length of a subrun is:

$$E[\text{length of subrun}] = \frac{1}{2p - p^2} + 1 \qquad (12)$$

As mentioned earlier, the **CP** protocol allows for static or dynamic frequency allocation. With static frequency allocation, a node that uses a frequency i at the end of phase 1 will use the same frequency i in the single step that makes up phase 2. With dynamic frequency allocation, a node randomly chooses a frequency in all steps of either phase.

Substituting the appropriate expressions into equation 10, we obtain the following results.

Theorem 2. *The expected run time for the* **CP** *protocol with static frequency allocation is*

$$\frac{2p(1-p)+1}{(2p(1-p))^2 \sum_{i=1}^{f} F_i^2}. \qquad (13)$$

Theorem 3. *The expected run time for the* **CP** *protocol with dynamic frequency allocation is*

$$\frac{2p(1-p)+1}{(2p(1-p) \sum_{i=1}^{f} F_i^2)^2}. \qquad (14)$$

The expected run time for the **CP** protocol with two nodes is longer under dynamic frequency allocation, as opposed to static frequency allocation, by a factor of $\phi = 1/\sum_{i=1}^{f} F_i^2$. With a uniform probability distribution for the frequencies, $\phi = f$.

3.3   Comparison of Two Node Protocols

To make a simple comparison of the two node protocols, assume that the probability of talking equals the probability of listening, and that the f frequency choices are uniformly distributed. The **CP** protocol with static frequency yields the best expected run time, $E[\text{run time}] = 6f$, followed by the **CP** protocol with dynamic frequency with an expected run time of $E[\text{run time}] = 6f^2$. The **RP** protocol has the poorest performance, with an expected run time of $E[\text{run time}] = 8f^2 + 2f$.

4   Random Protocol for $k \geq 2$ Node Multi-channel System

In the node discovery problem with $k \geq 2$ nodes, node discovery occurs when **two** of the k nodes discover each other. Consider the event

$$A_i^{ab} := \begin{pmatrix} \vdots & \vdots \\ T\,i \\ \vdots & \vdots \\ L\,i \\ \vdots & \vdots \end{pmatrix}, \quad \text{respectively,} \quad B_i^{ab} := \begin{pmatrix} \vdots & \vdots \\ L\,i \\ \vdots & \vdots \\ T\,i \\ \vdots & \vdots \end{pmatrix},$$

such that $a < b$ and

1. the state of the ath node is (T, i), (respectively, (L, i)),
2. the state of the bth node is (L, i), (respectively, (T, i)), and
3. for all $c \neq a, b$, the node c is either talking at a frequency other than i, or listening at any frequency, i.e., no other node talks at frequency i.

Clearly, there are $2f\binom{k}{2}$ such events A_i^{ab}, B_i^{ab}, where $i = 1, 2, \ldots, f$, $a < b$, and $a, b = 1, 2, \ldots, k$. Let the random variable X be defined as the current state of the $k \geq 2$ node system. Suppose that the current state of the system is described by the event A_i^{ab}, i.e., $X = A_i^{ab}$. Given the definition of A_i^{ab}, the event $X = A_i^{ab}$ is the intersection of the three events listed above and thus

$$\Pr[X = A_i^{ab}] = p_i q_i \prod_{c \neq a, b} \left(\sum_{j=1}^{f} q_j + \sum_{j=1, j \neq i}^{f} p_j \right) = p_i q_i (1 - p_i)^{k-2}. \quad (15)$$

Similarly, $\Pr[X = B_i^{ab}] = p_i q_i (1 - p_i)^{k-2}$.

In the random protocol for $k \geq 2$ nodes, node discovery occurs if for some frequencies $i, j \in \{1, 2, \ldots, f\}$ and two nodes $a < b$, event A_i^{ab} (respectively, B_i^{ab}) is followed by event B_j^{ab} (respectively, A_j^{ab}). It follows that the success patterns are $A_i^{ab} B_1^{ab}, A_i^{ab} B_2^{ab}, \ldots, A_i^{ab} B_f^{ab}$, and $B_i^{ab} A_1^{ab}, B_i^{ab} A_2^{ab}, \ldots, B_i^{ab} A_f^{ab}$. As in the two-node case, we calculate the expected runtime of the **RP** protocol by solving a system of linear equations.

Theorem 4. *The expected run time of the* **RP** *protocol for* $k \geq 2$ *nodes is*

$$\frac{1 + \sum_{j=1}^{f} p_j q_j (1 - p_j)^{k-2}}{2\binom{k}{2} \left(\sum_{j=1}^{f} p_j q_j (1 - p_j)^{k-2} \right)^2}. \quad (16)$$

5 Conditional Protocol
for $k \geq 2$ Node Multi-channel System

Like the two node case, the conditional protocol **CP** for $k \geq 2$ nodes is implemented as a series of two-phase subruns. In phase 1 of the protocol, a node follows the random protocol **RP** until the node talks (T), or it listens (L) and hears the broadcast of another node.

Phase 2 of a subrun of the **CP** protocol is still a single step and the behaviour of a node in this phase is conditional on the way in which phase 1 ended, i.e., a node that talked (T) at the end of phase 1 will listen (L) in phase 1 while a node that listened at the end of phase 1 will talk (T) in phase 2.

If a subrun of the protocol is successful, node discovery occurs, i.e., **two** of the k nodes discover each other and the protocol terminates. Otherwise, another subrun of the protocol is executed.

While calculating the expected run time of the **CP** protocol for two nodes was straightforward, the corresponding exercise for the $k \geq 2$ node case is quite complicated. For example, a given node can be in phase 1 relative to one set of nodes yet, at the same time, it can be in phase 2 relative to another set of nodes. Our analysis of the expected run time for the **CP** protocol with $k \geq 2$ nodes, therefore, relies on simulation results.

5.1   Comparison of Multi-node Protocols

Once again we made a simple comparison of the $k \geq 2$ node protocols by assuming that the probability of talking equals the probability of listening, and that the f frequency choices are uniformly distributed. The expected run time for the random protocol **RP** quickly became astronomical compared to the expected run time for either version of the **CP** protocol. The latter run times were estimated through simulation results. For a given number of nodes k in the **CP** protocol with static frequency allocation, and $f < k$ frequencies, the expected run time fell as the number of frequencies increased and f approached k. The expected run time then increased as f continued to increase. The behaviour of the expected run times for the **CP** protocol with dynamic frequency allocation was less predictable although, when $f < k$, the dynamic version of the protocol often outperformed the static version.

6   Conclusion

We presented a new communication model for node discovery and used it to compare the behaviour of random **RP** and conditional **CP** node behaviour protocols. We found closed form solutions for the expected run times of the protocols with the exception of the conditional protocol when $k > 2$ and were able to compare the performances of the protocols. In the future, it would be useful to explore protocols that use an estimate of the number of nodes within communication range, rather than assume that the number of nodes is known. It is unlikely that the exact number of nodes is known and the effects of poor estimation are likely to be significant.

Acknowledgements

Research of E. Kranakis and C. Sawchuk was supported in part by NSERC (Natural Sciences and Engineering Research Council of Canada) and MITACS (Mathematics of Information Technology and Complex Systems) grants. C. Sawchuk was also supported by an Ontario Graduate Scholarship.

References

1. G. Alonso, E. Kranakis, R. Wattenhofer, and P. Widmayer, Probabilistic Protocols for Node Discovery in Ad-hoc Single Broadcast Channel Networks, Workshop on Mobile AdHoc Networks (WMAN), International Parallel and Distributed Processing Symposium (IPDPS 2003), April 22 - 26, 2003, Nice, France.

2. D. Bertsekas and R. Gallager, Data Networks, Prentice Hall, 1992.
3. G. Blom and D. Thoburn, How Many Random Digits Are Required Until Given Sequences Are Obtained, J. Applied Probability, 19, 518-531, 1982.
4. BlueHoc: An Open-Source Simulator,
   http://oss.software.ibm.com/developerworks/opensource/bluehoc.
5. Ericsson Microelectronics: ROK 101 007 Bluetooth Module Datasheet Rev. PA5, April 2000.
6. R. Garcés, J.J. Garcia-Luna-Aceves, Collision Avoidance and Resolution Multiple Access for Multichannel Wireless Networks, IEEE Infocom 2000, March 26 - 30, 2000, Tel-Aviv, Israel.
7. W. Feller, An Introduction to Probability Theory and its Applications, Vol. II, Wiley, 1966.
8. J. Haartsen, Bluetooth Baseband Specification v. 1.0, www.Bluetooth.com.
9. O. Kasten, M. Langheinrich, First Experiences with Bluetooth in the Smart-Its Distributed Sensor Network, Workshop on Ubiquitous Computing and Communications, PACT 2001, October 2001.
10. C. Law, A.K. Mehta, K.-Y. Siu, Performance of a Bluetooth Scatternet Formation Protocol, The Second ACM Annual Workshop on Mobile Ad Hoc Networking and Computing (mobiHoc 2001), October 4-5, 2001, Long Beach, California, USA.
11. S. R. Li, A Martingale Approach to the Study of Occurrence of Sequence Patterns in Repeated Experiments, Annals of Probability, 8, 1171- 1176, 1980.
12. C.E. Perkins, editor, Ad Hoc Networking, Addison Wesley, 2001.
13. S. Ross, Stochastic Processes, John Wiley and Sons, 2nd edition, 1996.
14. T. Salonidis, P. Bhagwat, L. Tassiulas, Proximity Awareness and Fast Connection Establishment in Bluetooth, The First ACM Annual Workshop on Mobile Ad Hoc Networking and Computing (MobiHoc 2000), August 11, 2000, Boston, Massachusetts, USA.
15. T. Salonidis, P. Bhagwat, L. Tassiulas, R. LaMaire, Distributed Topology Construction of Bluetooth Personal Area Networks, In Proceedings of the Twentieth Annual Joint Conference of the IEEE Computer and Communications Societies (INFOCOM 2001), April 22 - 26, 2001, Anchorage, Alaska, USA.
16. F. Sicgcmund and M. Rohs, Rendezvous Layer Protocols for Bluetooth-Enabled Smart Devices, Technical Report, 1st International Conference on the Architecture of Computer Systems, ARCS - Trends in Network and Pervasive Computing, 2002.

Towards Adaptive WLAN Frequency Management Using Intelligent Agents

Fiorenzo Gamba[1], Jean-Frédéric Wagen[1], and Daniel Rossier[2]

[1] University of Applied Sciences of Western Switzerland
Bd de Pérolles 80, 1705 Fribourg, Switzerland
{gamba,wagen}@eif.ch
[2] Swisscom Innovations AG, Ostermundigenstrasse 93, 3006 Bern, Switzerland
daniel.rossier@swisscom.com

Abstract. Private, corporate and public *Wireless Local Area Networks* (WLAN) Hot-Spots are emerging. In this rapidly evolving environment, the configuration of WLAN access points raise the classical problem of re-using limited radio resources. In this paper, the problem of dynamic frequency allocation of WLAN access point in a highly competitive multi-provider Hot-Spots environment is addressed. Our solution aims at working in locations where planned and ad-hoc deployments might be side by side. An on-line adaptive optimization process is proposed and relies on available information delivered only by the local access points in order to maintain the quality of service as high as possible. This optimization process is implemented on a scalable and highly flexible agent-based framework. The easy deployment of intelligent agents in a real WLAN network and their integration in a simulation context allows us to perform extensive tests for small and large-scale networks. The proposed approach has been tested on a limited but practical demonstrator that showed encouraging results.

Keywords: WLAN Network, Frequency Optimization, Software Agent.

1 Introduction

The evolution of telecommunications is characterivzed by an increase in bandwidth availability and an increase in user mobility.

Mobile networks are evolving to support flexible access services and offer increasingly higher bandwidths as well as attractive pricing. For example, several mobile operators are about to offer new complementary services named here "Hot-Spot" access [1–3]. Hot-Spot solutions provide broadband mobile public access to the Internet and to corporate intranets. The coverage of a WLAN Hot-Spot is typically poor compared to a 2G and 3G mobile cellular solution but this limitation can be seen as an advantage to provide end-users with high bandwidth capacity. Current Wireless Local Area Network (WLAN) technologies deliver services access around 100 meters. Several mobile operators currently present WLAN Hot-Spot access as a complement to their GPRS and future UMTS offerings. User applications will have to be able to roam from one local Hot-Spot to another. For example: a user can access to her company's Email from the hotel lobby and access it again at the airport gate.

Despite many shortcomings with respect to security and inter-network or inter-operator roaming, WLAN IEEE802.11b products are increasingly becoming very

S. Pierre, M. Barbeau, and E. Kranakis (Eds.): ADHOC-NOW 2003, LNCS 2865, pp. 116–127, 2003.

popular in many countries. Many reasons can be mentioned for such a success: low price, software support and a very successful interoperability under the WiFi logo.

WLAN products continue to raise new expectations since they can demonstrate efficient LAN access at rates above 2 Mbps (IEEE 802.11), at 11 Mbps for the IEEE 802.11b specifications, or even several tens of Mbps in the near future (IEEE 802.11a and HIPERLAN)[1].

In this paper, we propose a novel approach to address the dynamic frequency allocation of WLAN *Access Point* (AP) in a multi-provider Hot-Spots environment, in order to maintain high level of *Quality of Service* (QoS) taking into account users' traffic. In Section 2, a description of the WLAN channel allocation problem and management issues are outlined. Our intelligent agent based approach to tackle the channel allocation problem of the WLAN access points is described in Section 3. Section 4 briefly presents the fundaments of frequency optimization. Section 5 shows some preliminary results in our test environment, and Section 6 concludes this paper.

2   Channel Allocation in WLAN Networks

Corporate users and, as price decreased, campus communities [4], have been the first to exploit the benefits of WLAN based on the IEEE802.11x family of standards. Private users are also using the technology to avoid new wires in their homes and sometimes to share Internet access with other users. An increasing number of telecommunication operators and visionary companies have identified new business models for the deployment of public WLAN access at popular locations (Hot-Spots) or in their business premises [5]. As a single example close to the authors, the Swiss operator Swisscom Mobile has launched commercial services based on public WLAN Hot-Spot access based on GSM subscriptions or special value cards since the end of 2002[2].

In this rapidly evolving environment, the deployment of WLAN has to face typical issues regarding optimal utilization of radio resources that can be provided within the allocated radio spectrum. The complexity of the optimization problem is amplified by the fact that all WLAN stakeholders, from private to business entities, share the same spectrum allocation without any cooperation unless proprietary bi-lateral agreements could be arranged. For example, a public WLAN Hot-Spot operator might find difficulties to offer its service at the bus stop near a popular restaurant, which has installed its own WLAN coverage. Adding private WLAN users in the flats above the restaurant further illustrates the need for an autonomous management of the WLAN access points. A manual and static configuration of each access points, besides being tedious, could only provide an acceptable solution for the conditions found during the measurements survey. A better solution would be to centralize some information about all the access points potentially interfering. Based upon a centralized database, the best frequency allocation could be computed. However, this solution is not possible in practice because it does not scale to a very large number of access points. Furthermore, the conditions can be quite complex since the access points are deployed in planned or ad-hoc manners depending on operators and service providers.

---

[1] http://www.alcatel.com/atr

[2] http://www.swisscom-mobile.ch/sp/4EGAAAAA-fr.html, visited March, 2, 2003.

In current deployment of WLAN hot spots, the frequency channel is fixed manually [6] and will be changed only if consumers complain or if a new survey is undertaken by WLAN operators.

We developed an autonomous management system that will detect the new situation and adapt the radio resource allocation so that the quality of service is maximized.

2.1 Towards Agent-Based Management

This paper describes the architecture of an autonomous and adaptive management system dedicated to on-line optimization of access point channel assignment; the process mainly relies on environmental information issued from the surrounding access points and information exchange between enhanced access points.

The target architecture resorts to autonomous software agents which can run on physically distributed or centralized platforms. From a logical viewpoint, each software agent is delegated to a single access point. Technical information such as the currently used frequency channel, the number of users, the number of rejected packets, etc. are gathered by each software agent and constitute their internal knowledge, i.e. their internal representation of the *local* environment. The implementation of advanced mechanisms to exchange internal knowledge between agents will enable the enhanced access point to perform a *local* optimization.

One of the major characteristics of the proposed solution resides in its ability to deal with currently deployed WLAN networks in concordance with the established IEEE 802.11 standards [7][8] and related management systems. Thus, management techniques directly rely on standardized protocols and information models making a vendor-independent implementation possible. *Simple Network Management Protocol* (SNMP) has become the de facto standard for IP network management. SNMP is commonly use to manage IP based elements and also for wireless elements.

3 Intelligent Agent Approach

The management of future WLAN networks will have to cope with highly competitive environment where several operators will deploy their own infrastructure in the same geographical areas. Furthermore, the diversity of WLAN equipments will make such a network very heterogeneous. As a consequence, centralized network management systems will be not sufficient to control fine-grained resources allocation with respect to the available frequency spectrum. On the other hand, legacy management systems such as SNMP-based systems can not be ignored in the overall architecture. The elaboration of hybrid centralized and de-centralized network management systems therefore constitute a general trend, not only for WLAN networks but also for other mobile networking technology like ad-hoc networks [9].

In this context, *intelligent agents* can be considered as one of the most promising approaches addressing issues related to distributed applications in the rapidly expanding communication industry [10]. Intelligent agents can be seen as a software program that can perform specific tasks for a user and possesses a degree of intelligence that permits it to perform parts of its tasks autonomously and to interact with its envi-

ronment in a useful manner [11]. An intelligent agent exhibits the following properties: *autonomy* - the agent is capable of following its goal autonomously that is, without interactions or commands from the environment - *reactivity* - the agent is capable
of reacting appropriately to influences or information from its environment - *pro-
activity* - under specific circumstances, the agent can take the initiative in performing
appropriate actions - *social ability* - the agent is able to communicate with other
agents and to interact with its environment in order to fulfill its tasks.

Intelligent agents have been considered for network management in numerous research projects[3]. These various projects have led to multi-layer agent architectures in
which each layer implements different abstraction views; examples of such layers are
the co-operation layer, the planning layer and the reactive layer. In this context, the
Belief-Desire-Intention [12] probably constitutes one of the most popular agent architecture and has also been considered, under different forms, in agent-based network
management systems.

The development of agent standards in telecommunication is obviously a *sine qua
non* condition for the successful deployment of software agents in large-scale networks. The most popular agent standard at the moment is the *Foundation for Intelligent Physical Agents* (FIPA)[4].

The framework we propose in this paper consists of a simple architecture in which
we mainly exploit the message and communication facilities provided by the agent
platform on the one hand, and the capability of an agent to implement different parallel behaviors on the other hand. Details about agent behavior are given in Section 3.2.

Our agent-based framework is therefore composed of intelligent agents, called
AWM_agents, which exchange information concerning their local environment in
order to perform on-line optimization by (re-)configuring the access point frequency
(or channel). The enhanced access point consequently exhibits an autonomous and
adaptive behavior.

3.1 Jade and LEAP Agent Platforms

Jade[5] is a freely downloadable *Java* agent platform, which is fully compliant with the
last revision of FIPA specifications; the intra-agent activity model defined in *Jade* is
based upon a non-pre-emptive concurrency model. A *Jade* agent is implemented with
a *Java* thread, which enables asynchronous inter-platform communication as specified
by FIPA. The *Jade* agents can implement one or several *behaviors*: while intra-agent
activities are synchronous, inter-agent communication relies on an asynchronous
process. The behaviors are executed in a thread-per-agent concurrency model, in
which there is no stack to be saved: they are managed by an internal scheduler implementing a round-robin non-pre-emptive policy among all the behaviors available
in the ready queue of an agent [13]. The synchronous characteristic of intra-agent
activity and related cooperative processes makes *Jade* an attractive agent platform for
the study of the agent behavior in the context of telecommunication applications, so
that our *AWM_agents* have been implemented into the *Jade* environment. Still, *Jade*

---

[3] An excellent overview of project activities concerning intelligent agents for network management can be found in [13].
[4] http://www.fipa.org
[5] http://sharon.cselt.it/projects/jade

is made up of numerous classes and can thus be difficult to implement into embedded systems.

The *Lightweight and Extensible Agent Platform* (LEAP)[6] is a project aiming at the realization of a FIPA platform that can be deployed seamlessly on any *Java*-enabled device endowed with sufficient resources and with a wired or wireless connection, such as PDAs and smart phones [15]. *LEAP* significantly contributes to providing network devices with an embedded agent platform.

LEAP appears to be particularly interesting because it can easily be deployed in a simple Java processor based platform, which can be connected to any vendor-independent access point. We are now investigating the deployment of our *AWM_agent* into a *LEAP*-based Java processor based platform.

3.2 Agent Behavior

The *Jade* agent platform [16] provides a novel approach towards task design with generic behaviors. Based on message exchanges between agents, several behavior schemes corresponding to various task types are defined in order to enable multiple interactions with other agents.

The behaviors are divided into two main categories, respectively *simple* and *composite* behaviors. A simple behavior consists in a task that is activated only once and cannot be blocked - *oneShotBehaviour* - or in a cyclically activated task. A composite behavior is made up of several behaviors according to a parent-child relationship; it may consist of a sequential behavior - *SequentialBehaviour* - which executes the sub-behaviors sequentially and terminates when all sub-behaviors have been executed; on the contrary, parallel behavior - *ParallelBehaviour* - allows the developer to implement sub-behaviors which can be executed in a non-deterministic order. Finally, a behavior can be described with a *finite state machine* (FSM); the parent behavior controls the transitions between the FSM states and activates the behaviors corresponding to the current state.

From the communication point of view, the agents can interact via *intra-platform* communication: all the agents participating in the interaction are managed by the same platform; they *reside* in the same environment. The agents can also be distributed over several platforms, in which case they interact via an *inter-platform* communication mechanism. In both cases, agents communicate via ACL messages. For example, if an agent platform is dedicated to one and only one access point, the hosted *AWM_agent* endowed with an SNMP manager can be logically perceived as the access point itself and only inter-platform communication will take place. On the contrary, if an agent platform hosts several *AWM_agents*, the agent platform is responsible for managing several access points, and intra-platform communication will take place between the *AWM_agents* residing in the platform.

3.3 AWM Agent Architecture

In our framework, an *AWM_agent* is dedicated to an access point and performs three basic tasks: at first, the agent continuously monitors its local environment by querying the SNMP agent of the access point; the retrieval of particular values from MIB vari-

---

[6] http://leap.crm-paris.com

ables will give information about the number of frames with errors (MIB:FCS), the number of frames delivered correctly (MIB:InUPkt), the number of associated stations (MIB:NAS) and the frequency channel (MIB:Channel), so that the agent can have an internal representation of the environment. Secondly, the agent handles incoming messages issued from other agents. The message contents are examined and processed accordingly. The exchange of information between neighboring agents improves the channel assignment. The agent finally has to perform local computation to determine the most appropriate channel.

The overall agent architecture and processes are depicted in Figure 1.

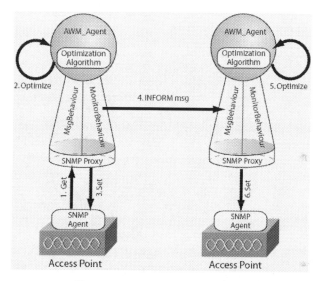

Fig. 1. AWM Agent Architecture

We now briefly introduce the general scenario involving our *AWM_agents* and related access points. The optimization algorithm we have implemented is currently being patented and is therefore not detailed in this paper. It is however important to mention that the overall architecture perfectly suits a wide range of distributed optimization algorithms.

The monitoring task and the message processing are implemented by means of two *Jade* cyclic behaviors. The monitoring process is implemented into the *MonitorBehaviour*, while message handling is implemented into the *MsgBehaviour*.

The *AWM_agent* queries the SNMP agent in order to retrieve the information from the access point, which is controlled by the agent (1). In case of interference, the agent activates the optimization algorithm (2) and computes the new channel to be assigned to the access point. The new channel is set via a SNMP request (3). The agent then sends the new channel to the neighboring agents (4). The receiving agent reads the message contents, makes sure that it fits the AWM ontology and in turn activates the optimization algorithm if necessary (5) to compute the new channel based upon the updated information. Finally, the new channel is assigned by means of a SNMP request (6).

4 Basic Principles for Channel Assignment Optimization

Classical frequency optimization in cellular networks is based on simple rules regarding frequency channel allocation. Usually, the same frequency and even an adjacent frequency cannot be repeated at the same location or neighbored locations.

The particular definition [8] of overlapping frequency channels in WLAN IEEE802.11b with DSSS (*Direct Sequence Spread Spectrum*) and the CSMA/CA (*Carrier Sense Multiple Access/Collision Avoidance*) technique lead that a more complex rules regarding frequency channel allocation. Indeed, measurements as depicted on Figure 2 shows that a better user throughput is obtained when there is either a total overlapping or, as expected, no overlapping of the interfering channels allocated to different access points.

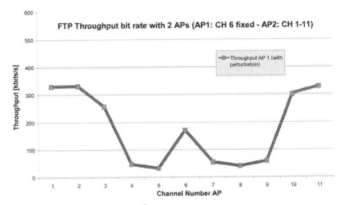

Fig. 2. Throughput measurements versus channel separation

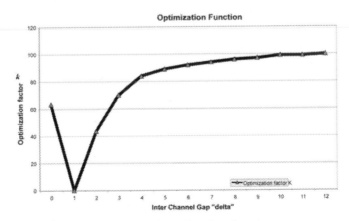

Fig. 3. Optimization function

Figure 2 reports FTP total throughputs measured in the following conditions: two users, each linked to its own access point: one with frequency 6 (AP1) and the other access point (AP2) with the frequency channel *x* (value on the *x*-axis). The result shows that partial overlapping is worst than a complete overlapping of the frequency channels. This can be explained from the effectiveness of the *collision avoidance* (CA) when the two channels are equal. Error rate measurements not reported here also shows that partial overlapping of frequency channels lead to a larger number of errors, while total overlapping or non-overlapping channels lead to negligible numbers of errors.

The results presented here have been taken into account in the design of our optimization function represented on Figure 3: a value between 0 and 100 is assigned to the difference between 2 channels.

The optimization function is not monotone. Partially overlapping channels lead to low numbers. However, re-using the same channel on competing access points is better that choosing partially overlapping frequency channels. As expected, choosing non-overlapping channel leads to the highest score.

The optimization function provided in Figure 3 can be adapted if necessary to take into account other functions if desired.

5 Experiments and Results

5.1 Test Bed Environment

Our test bed environment is based on four access points (AP) representing two *Wireless Internet Service Providers* (WISPs). An *Autonomous WLAN Management* (AWM) agent is connected to each AP and each AP is configured with a *Service Set Identifier* (SSID) that characterizes the WISP. Since The AWM agent must communicate between WISPs, then it is assumed that WISPs have to be inter-connected. At least, WISPs must allow their agents to exchange messages. It is recalled that the software agent platform chosen in this work simplify greatly this exchange of messages. Figure 4 shows the test environment, its architecture and the exchange of messages. Each of the 4 access points has its own PC acting as a proxy for the access point. The proxy runs the software agent platform and the software agents that have been designed to implement the AWM system.

Access Points are Cisco Aironet 350 products. The AWM Agent platform is based on Jade platform and runs on Pentium-III PCs. Wireless LAN clients are laptops with PCMCIA WLAN cards. To test traffic congestion, we have implemented a client emulator in the AWM agent. Thus associated terminals can be emulated by this feature on each AP.

This practical test environment has a limited size and can be used to demonstrate the feasibility of our approach and determine the user experience under different the frequency adaptation algorithm. Simulation environment has also deployed using the *Generic Network Management Tool* (GNMT) [17] described in the next sub-section. In this case, larger network with several tens of access points have been simulated. Comparisons with the practical test environment can also be performed.

5.2 Preliminary Results

In this section, we briefly present the first results we obtain with our experimental environment. Figure 5 presents the four access points with virtual interference links. It is recalled that a *Virtual Interference Link* (VIL) is defined as a communication channel between two access points which are subject to interfere each with other. Currently, VIL topology is determined and configured manually by editing a property file for each *AWM_agent*. Automatic VIL discovery mechanism is currently being investigated.

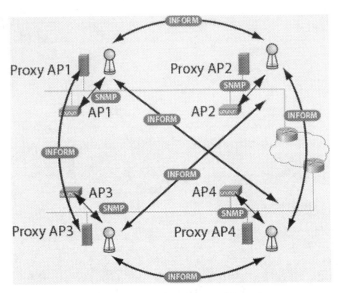

Fig. 4. Test environment architecture

The number appearing on each VIL corresponds to the difference of frequency channel between APs. For example, AP1 is configured on channel 13 and AP2 is also on channel 13, therefore the number 0 (=|freq(AP1)-freq(AP2)|=|13-13|) is circled on the VIL (AP1, AP2).

MIB Parameters have been introduced in Section 3.3. In the beginning of our experiment, we have configured each access point on the channel 1. We have then associated a certain number of stations (mobile users) to the access points according to the following scheme: two stations (users) are associated to AP1, three to AP3, five to AP2 and no station is associated to AP4. The numbers depicted on the figure shows the final (and stable) configuration we obtain after less than 10 minutes. It is important to see that the optimization algorithm takes into account the possibility to have two access points configured with the same channel (AP1 and AP2). As explained in Section 4, having two neighboring access points on the same channels may be considered as a better solution than having a small difference of frequency.

Figure 6 shows the adaptive process over time and, hence, the evolution of the assigned frequency channel for each access point. A different line profile (size and style) is given to each AP.

The optimization algorithm must obviously ensure that the process will converge and avoids cycles. The algorithm we implemented becomes stable after a few minutes. This algorithm is being patented, thus no more details are provided. During the optimization process, the APs may change several time their frequency channel. A user associated with a particular access point loose a few packets during the 1 to 3 seconds break occurring at each change of frequency channel. Practically, the end users do not feel these losses especially if the newer operating systems are used.

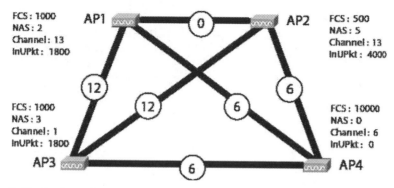

Fig. 5. Final values when the adaptive process becomes stable (circle number is on VIL)

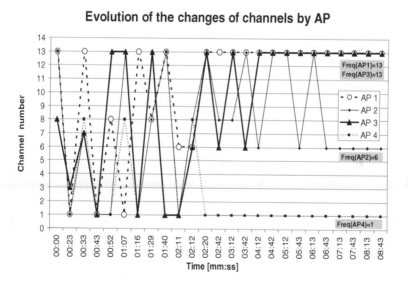

Fig. 6. Evolution of the adaptive process over time

6   Conclusions

An approach to address the problem of dynamic frequency allocation for WLAN access point in a multi-provider Hot-Spots environment has been presented. The operating frequency channel at each access point is modified in order to increase the quality of service measured at the access points. This channel (re-)configuration is performed via SNMP. Several parameters are combined in a metric defined to objectively measure the performance at each access point and compare the results to its neighbors. In our demonstrator, four parameters obtained from the MIB access points have been used: the number of frames with errors (FCS), the number of frames delivered correctly (InUPkt), the number of associated stations (NAS) and the frequency channel.

The exchange of information between access points regarding performance measures and newly assigned channel plays a central role in the solution presented here. Advanced communication mechanisms rely on the *Jade* agent-platform. Our agent-based framework is composed of intelligent agents, called *AWM_agents*, which closely interact in order to keep an up-to-date internal representation of their local environment and therefore to perform on-line optimization by (re-)configuring the frequency channel of the access point.

The architecture and functionalities of our solution has been explained. Each access point is controlled by an *AWM_agent*. The *AWM_agent* queries a *SNMP* agent in order to retrieve the information from the access point. In case of interference, the agent activates an optimization algorithm and computes the new channel to be assigned to the access point.

Preliminary results have been presented to illustrate the feasibility of our approach. The stability of the optimization process has been tested on the demonstrator and the simulator. Measurements demonstrated that frequency channels can be modified with little perturbation to the users associated to a given access point.

Future work will focus on extending the demonstrators to a large number of access points in a campus environment. Further tests and measurements will also be performed on the simulator. Furthermore, improvements of the optimization algorithm will be investigated.

References

1. Levillain, P., Wireless LAN for Entreprise. Alcatel Telecommunications Review, Q2002
2. Juha Ala-Laurila, Jouni Mikkonen, Jyri Rinnemaa, "Wireless LAN Access Network Architecture for Mobile Operators", IEEE Commun. Mag,. November 2001, pp 82-89.
3. Shidong Zhou, Ming Zhao, Xibin Xu, Jing Wang, Yan Yao, "Distributed Wireless Communication System: A New Architecture for Futur Public Wireless Access", ", IEEE Commun. Mag, .March 2003, pp 108-113.
4. David Kotz, Kobby Essein, "Analysis of a Campus-wide Wireless Network", MOBICOM'02, September 23-26, 2002 Atlanta, Georgia, USA
5. Heinz Luediger, Sven Zeisberg, "User and Business Perspectives on an Open Mobile Access Standard", IEEE Commun. Mag,. September 2000, pp 160-163.
6. Alex Hills, "Large-Scale Wireless LAN Design", IEEE Commun. Mag,. November 2001, pp 98-104.
7. IEEE Std 802.11, Part 11: Wireless LAN Medium Access Control (MAC) and Physical Layer (PHY) Specifications, including ieee802.11-mib, edition 1999
8. IEEE Std 802.11b, Part 11: Wireless LAN Medium Access Control (MAC) and Physical Layer (PHY) specifications: Higher-Speed Physical Layer Extension in the 2.4 GHz Band, ISBN 0-7381-1812-5.
9. Chien-Chung Shen, Chavalit Srisathapornphat, and Chaiporn Jaikaeo, "An Adaptive Management Architecture for Ad Hoc Networks", IEEE Commun. Mag, February 2003
10. Nicholas R. Jennings, "An Agent-based Approach for Building Complex Software Systems", Communications of the ACM, Vol.44, No.4 (April 2001).
11. Walter Brenner, Rüdiger Zarnekow, Hartmut Wittig, "Intelligent Software Agents", Springer-Verlag (Berlin Heidelberg, 1998).

12. Wooldridge and N.R. Jennings, "Agent Theories, Architectures, and Languages: a Survey", In M. Wooldridge and N.R. Jennings, editors, Intelligent Agents, number 890 in LNCS, pages 1-39 (Springer Verlag, 1995).
13. Alex L. G. Hayzelden, Rachel A. Bourne, "Agent Technology for Communication Infrastructures", Wiley & Sons Ltd, 2001.
14. Fabio Bellifemine, Agostino Poggi, Giovanni Rimassa, "Jade - a FIPA-compliant agent framework", in Proc. of the 4th International Conference on the Practical Application of Artificial Intelligence and Multi-agent Technology (PAAM'99), pp.97-108, (London, UK, 1999).
15. Federico Bergenti and Agostino Poggi, "LEAP: A FIPA Platform for Handheld and Mobile Devices", in Proc. of 8th Intl. Workshop on Agent Theories, Architecture and Languages (ATAL'2001) (Seattle, USA, August 2001).
16. Fabio Bellifemine, Giovanni Caire, Tiziana Trucco, Giovanni Rimassa, "Jade Programmer's Guide", available at http://sharon.cselt.it/projects/jade, 2002.
17. Daniel Rossier, "A Description of the Generic Network Management Tool", Technical Report, Department of Informatics, University of Fribourg (Switzerland, August 2002).

Analyzing Split Channel Medium Access Control Schemes with ALOHA Reservation[*]

Jing Deng[1], Yunghsiang S. Han[2], and Zygmunt J. Haas[3]

[1] The CASE Center and the Dept. of Electrical Engineering and Computer Science
at Syracuse University, Syracuse, NY, USA
jdeng01@ecs.syr.edu
[2] The Dept. of Computer Science and Information Eng.
National Chi Nan University, Taiwan, R.O.C.
yshan@csie.ncnu.edu.tw
[3] School of Electrical & Computer Engineering at Cornell University
Ithaca, NY, USA
haas@ece.cornell.edu

Abstract. In order to improve the throughput performance of Medium Access Control (MAC) schemes in wireless communication networks, some researchers proposed to split the single shared channel into two subchannels: a control subchannel and a data subchannel. The control subchannel is used for access reservation to the data subchannel over which the data packets are transmitted, and such reservation can be done through the use of the dialogues such as RTS/CTS (Ready-To-Send/Clear-To-Send) dialogue. In this paper, we evaluate the maximum achievable throughput of split-channel MAC schemes that are based on RTS/CTS dialogues with pure ALOHA contention resolution mechanism. We derive and calculate numerically the probability density function (pdf) of the contention resolution periods on the control subchannel. We then apply these results to calculate the throughput of the split-channel MAC schemes, which we then compare with the performance of the corresponding single-channel MAC schemes. Our results show that, when radio propagation delays are negligible, the maximum achievable throughput of the split-channel MAC schemes is lower than that of the corresponding single-channel MAC schemes in the scenarios that we have studied. Consequently, our results suggest that splitting the single shared channel of the MAC scheme in a wireless network should be avoided. Simulation results are presented to support our analytical results.

[*] This work was supported in part by the SUPRIA program of the CASE Center at Syracuse University, by the National Science Council of Taiwan, R.O.C., under grants NSC 90-2213-E-260-007 and NSC 91-2213-E-260-021, and by the DoD Multidisciplinary University Research Initiative (MURI) program administered by the Office of Naval Research under grant number N00014-00-1-0564. Part of Han's work was completed during his visit to the CASE Center and Dept. of Electrical Engineering and Computer Science at Syracuse University, USA.

1   Introduction

In wireless communication networks, Medium Access Control (MAC) schemes are used to control the access of active nodes to the shared channel [1]. As the throughput of the MAC schemes may significantly affect the overall performance of the wireless networks, some researchers proposed to split, either in time or in frequency, the single shared channel into two subchannels: a control subchannel and a data subchannel. The control subchannel is used for reservation of access to the data subchannel over which the data packets are transmitted, and such reservation can be done through the use of the RTS/CTS (Ready-To-Send/Clear-To-Send) dialogue. Examples of such split-channel MAC schemes can be found in [2], [3], [4], [5], [6], and [7].

In this paper, we analyze the performance of a generic split-channel MAC scheme, which is based on the RTS/CTS dialogue and with pure ALOHA [8] contention resolution on the control subchannel. A ready node sends an RTS packet on the control subchannel to reserve the use of the data subchannel. When the RTS packet is received, the intended receiver replies with a CTS packet to acknowledge the successful reservation of the data subchannel [9] [10].

Based on the previous work [11], we calculate the probability density function (pdf) of the contention resolution periods on the control subchannel. This pdf is then used to calculate the expected waiting time on the data subchannel and the throughput of the split-channel MAC schemes. We determine the maximum achievable throughput of the split-channel MAC scheme as a function of the ratio of the bandwidths of the control subchannel and the entire channel and compare the result to that of the corresponding single-channel MAC schemes. We show that, when pure ALOHA technique is used for contention resolution on the control subchannel and radio propagation delays are negligible, the throughput of the split-channel MAC schemes is inferior to that of the single-channel MAC schemes.

For notational convenience, we term single-channel MAC scheme as MAC-1 and split-channel MAC scheme as MAC-2. We further define MAC-2R as MAC-2 with parallel reservations; i.e., in the MAC-2R scheme, contention resolutions take place on the control subchannel in parallel with the transmission of data packets on the data subchannel.

The paper is organized as follows: Section 2 summarizes the related work. In Section 3, we present our main comparison results of comparing the MAC-1, the MAC-2, and the MAC-2R schemes. In Section 4, our numerical and simulation results are derived. We then conclude this work in Section 5.

2   Related Work

A dynamic reservation technique called split-channel reservation multiple access (SRMA) was introduced for packet switching radio channels in [2]. In SRMA, the available bandwidth was divided into three channels: two used to transmit control information and, one used for message transmission. Message delay of

SRMA was studied in that paper and it was shown that SRMA out-performs other MAC schemes under some network settings.

Split channel MAC scheme was compared with single channel MAC scheme in [12]. The authors categorized "scheduling epochs," the periods of time needed to schedule the next data transmission, into two groups: bandwidth-dependent component (e.g., contention resolution of reservation packets) and bandwidth-independent component (e.g., radio propagation delay). It was found that, if a system has no bandwidth-independent component in its scheduling epochs, the split-channel schemes may achieve the same performance as the single-channel schemes do. However, the analysis in that paper considered the average contention resolution period only, rather than the random distribution of these periods.

Similarly, [5] compared the performance of the single-channel MAC schemes and that of the split-channel MAC schemes by considering only the expected value of the contention resolution periods. In [5] and [6], the authors further proposed to use partial pipelining technique to solve the problem of unbalanced separation of the control channel and the data channel. This approach is similar to the generalized MAC-2R scheme, even though busy signals but not RTS/CTS dialogues are transmitted on the control subchannel.

In [11], the authors studied the contention resolution period of the pure ALOHA channel and the CSMA channel. They derived the Laplace transform of the pdf of the contention resolution periods of the two channels. The expected value and the variance of the resolution periods were calculated. Our work differs from [11], in that we study the throughput of the split-channel MAC schemes and compare it to that of the single-channel MAC schemes. We analyze the contention resolution periods numerically and use these results to determine the maximum achievable throughput of the split-channel MAC schemes.

In [3], RTS/CTS dialogue packets are transmitted on a separate signaling (control) channel. The protocol conserves battery power at nodes that are not actively transmitting or receiving packets by intelligently powering them off. A Power Controlled Dual Channel (PCDC) scheme for wireless ad hoc networks was proposed in [7]. By transmitting RTS/CTS dialogues on the control channel with maximum power and data packets on the main channel with adjustable (lower) power, interference-limited simultaneous transmission can take place in the neighborhood of a receiving node. However, these studies used separate channels mainly to achieve energy efficiency and low interference between neighboring transmissions in multi-hop networks.

3   Throughput Comparisons

3.1   Assumptions and Notations

In order to compare the throughput of the MAC-1, the MAC-2, and the MAC-2R schemes, we make the following assumptions. The wireless communication network we study is assumed to be fully-connected, i.e., all nodes are in the transmission range of each other. We also assume that the packet processing delays and the radio propagation delays are negligible and that the traffic generated by active nodes (including retransmissions) is Poisson with rate λ.

We establish the following notation:

- L_c, L_d: the length of a control packet and that of a data packet, respectively
- k: the ratio of data packet length to the control packet length; i.e., $k = \frac{L_d}{L_c}$
- R, R_c, and R_d: the data rate of the entire shared channel, the control subchannel, and the data subchannel, respectively; i.e., $R = R_c + R_d$
- r: the ratio of the data rate of the control subchannel to the data rate of the entire channel in the MAC-2 and the MAC-2R schemes; i.e., $r = \frac{R_c}{R} = \frac{R_c}{R_c + R_d}$
- γ_1, δ_1: the transmission time of a control packet and the transmission time of a data packet in the MAC-1 scheme, respectively; i.e., $\gamma_1 = \frac{L_c}{R}$ and $\delta_1 = \frac{L_d}{R} = k\gamma_1$
- γ_2, δ_2: the transmission time of a control packet and the transmission time of a data packet, respectively, in the MAC-2 or the MAC-2R schemes; i.e., $\gamma_2 = \frac{L_c}{R_c} = \frac{\gamma_1}{r}$ and $\delta_2 = \frac{L_d}{R_d} = \frac{k\gamma_1}{1-r}$
- δ: normalized data packet transmission time in the MAC-2 and the MAC-2R schemes; i.e., $\delta = \frac{\delta_2}{\gamma_2} = \frac{kr}{1-r}$.

3.2 Comparing the Throughput of the MAC-1 and the MAC-2 Schemes

Fig. 1 depicts an example of the operations of the MAC-1, the MAC-2, and the MAC-2R schemes. We treat the packet transmission on the channel as a renewal process. To send a data packet successfully, two control packets and a data packet need to be transmitted on the shared channel after the contention resolution period, which is the time between the end of the previous successful data transmission and the beginning of current successful RTS/CTS dialogue. According to [11], the expected value of the normalized contention resolution period in ALOHA channels ($\overline{w}$) is a constant, when normalized Poisson traffic arrival rate is fixed. In the MAC-1 scheme, the expected time of a data packet transmission cycle is:

$$\overline{t_1} = \overline{w}\gamma_1 + 2\gamma_1 + \delta_1 = (\overline{w} + 2 + k) \cdot \gamma_1 .$$

Thus, according to the property of renewal processes, the throughput of the MAC-1 scheme can be expressed as

$$S_1 = \frac{\delta_1}{\overline{t_1}} = \frac{k\gamma_1}{\overline{t_1}} = \frac{k}{\overline{w} + 2 + k} . \tag{1}$$

In the MAC-2 scheme, the available bandwidth is split into two subchannels. Channel requests can only be transmitted after the current data transmission ends. The throughput of the MAC-2 scheme is a function of r. The expected time of a renewal cycle is

$$\overline{t_2} = \overline{w} \cdot \gamma_2 + 2 \cdot \gamma_2 + \delta_2 = \left(\frac{\overline{w} + 2}{r} + \frac{k}{1-r} \right) \cdot \gamma_1 .$$

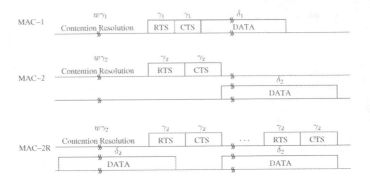

Fig. 1. Comparison of MAC-1, MAC-2, and MAC-2R

Therefore, the throughput of the MAC-2 scheme is

$$S_2(r) = \frac{\delta_2}{t_2} \cdot (1-r) = \frac{k}{\frac{\overline{w}+2}{r} + \frac{k}{1-r}}, \qquad (2)$$

where the term $1-r$ in the first equation represents the portion of the entire available bandwidth of the shared channel that is used as the data subchannel. Comparing (1) and (2), we conclude that $S_2(r) < S_1$ when $0 < r < 1$. Thus, the throughput of the MAC-2 scheme is always lower than that of the MAC-1 scheme.

3.3   Calculating the Throughput of the MAC-2R Scheme

For notational convenience, we normalize all variables with respect to γ_2 when calculating the throughput of the MAC-2R scheme, e.g., the data packet transmission time is now $\delta = \frac{\delta_2}{\gamma_2} = \frac{kr}{1-r}$.

As opposed to the operation of the MAC-2 scheme, in the MAC-2R scheme, contention resolutions take place on the control subchannel in parallel with the transmission of data packets on the data subchannel. A contention resolution period (W) begins on the control subchannel when the transmission of the data packet (for which the data subchannel was reserved in the previous reservation epoch) starts on the data subchannel. The contention period lasts until there is a successful RTS/CTS dialogue (see Fig. 2):

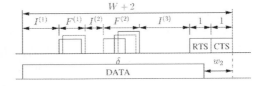

Fig. 2. An example of the contention resolution period in MAC-2R

$$W = \sum_{i=1}^{K-1} \left[I^{(i)} + F^{(i)} \right] + I^{(K)} , \qquad (3)$$

where K is the number of busy periods[1] during the $W + 2$ contention resolution periods, of which the last one is successful, $I^{(i)}$ is the i-th idle period, and $F^{(i)}$ is the i-th failed busy period, in which RTS packet collisions occur. In (3), the summation term includes $K - 1$ failed busy periods and the idle periods leading them. The second term, $I^{(K)}$, represents the idle period leading the successful busy period.

Let p_s be the probability that a successful RTS/CTS dialogue starts after an idle period on the control subchannel. The Laplace transform of the pdf of the contention resolution period W is [11]

$$W^*(s) = \frac{p_s}{\frac{1}{I^*(s)} - (1 - p_s) F^*(s)} , \qquad (4)$$

where $W^*(s)$ is the Laplace transform of $g(w)$, the pdf of W, $I^*(s)$ is the Laplace transform of $i(t)$, the pdf of the individual idle periods, and $F^*(s)$ is the Laplace transform of $f(t)$, the pdf of the individual failed periods[2].

Since the inter-arrival times of packet reservations for the control subchannel (newly generated and those scheduled for retransmission) are identical, independent, and exponentially distributed with mean $1/G$ in time units of γ_2, where $G = \lambda \gamma_2$, the Laplace transform of the channel idle time (I) is

$$I^*(s) = \frac{G}{G + s} .$$

The probability of a successful transmission of a packet after an idle period is given by

$$p_s = e^{-G} .$$

The duration of an unsuccessful transmission period F is given in [11] as

$$F^*(s) = \frac{G e^{-(s+G)} \left[1 - e^{-(s+G)} \right]}{(1 - e^{-G}) \left[s + G e^{-(s+G)} \right]} .$$

Thus,

$$W^*(s) = \frac{G e^{-G} \left[s + G e^{-(s+G)} \right]}{s^2 + sG \left[1 + e^{-(s+G)} \right] + G^2 e^{-2(s+G)}} \qquad (5)$$

and consequently,

$$\overline{w} = E[W] = -\left. \frac{\partial W^*(s)}{\partial s} \right|_{s=0} = \frac{1}{G} e^{2G} - 1 .$$

---

[1] We denote those contention periods with a packet transmission on the control subchannel as busy periods.

[2] We assume that all pdfs exist.

In the MAC-2R scheme, when the value of W (say, w) satisfies $w + 2 \leq \delta$, the RTS/CTS dialogue succeeds before the end of the current data packet transmission on the data subchannel. Thus, the next data packet transmission can start immediately after the current one ends. However, when $w + 2 > \delta$ (as shown in Fig. 2), the data subchannel will be left idle for a period of time, which we define as the waiting time on data subchannel (w_2). The expected value of this waiting time ($\overline{w_2}$) can be calculated as

$$\overline{w_2} = \int_{\delta-2}^{\infty} [w - (\delta - 2)] \cdot g(w) \, dw \ . \tag{6}$$

Note that the above equation holds even when $\delta - 2 < 0$.

Therefore, the throughput of the MAC-2R scheme can be expressed as

$$S_{2R}(r) = \frac{\delta}{\delta + \overline{w_2}} \cdot (1 - r) = \frac{1}{\frac{1}{1-r} + \frac{\overline{w_2}}{kr}} \ . \tag{7}$$

Note that the control subchannel access scheme is ALOHA for RTS packets. To maximize the throughput of the control subchannel, the RTS packet arrival rate in unit time on the control subchannel, $G = \lambda \gamma_2$, should be 0.5. In this case, the delay from when the control subchannel becomes available for reservation until a successful RTS/CTS dialogue takes place is minimized [11]. Thus, this value of G minimizes w.

Before we proceed to calculate $\overline{w_2}$, it is worthwhile to evaluate the throughput if we only consider the average delay of contention resolution on the control subchannel. In this case, the average time of each reservation cycle on the control subchannel is $E[W] + 2 = \overline{w} + 2$ and the time of each transmission cycle on the data subchannel is δ. The optimal throughput of the MAC-2R scheme occurs when $\delta = \overline{w} + 2$; i.e., the data packets are placed back-to-back and there is no waiting time needed on the data subchannel for conclusion of the contention resolution on the control subchannel. Thus,

$$\delta = \frac{kr^*}{1 - r^*} = \overline{w} + 2 \ ,$$

and the optimal r, which we label as r^*, based on the expected value of contention resolution delay is

$$r^* = \frac{\overline{w} + 2}{k + \overline{w} + 2} \ . \tag{8}$$

However, by substituting r^* into (7), we obtain that

$$S_{2R}(r^*) = \frac{k}{\overline{w} + 2 + k} \left(\frac{1}{1 + \frac{\overline{w_2}}{\overline{w}+2}} \right) \ ,$$

which is lower than S_1 for $\overline{w_2} > 0$.

In order to calculate $\overline{w_2}$, we need to derive $g(w)$ explicitly. Instead of deriving a closed-form for $g(w)$, we use a numerical inversion of Laplace transforms presented in [13]. The value of $g(w)$ for a specified value of w can be estimated as follows. First, $g(w)$ can be represented by a sequence of discrete values, $s_n(w)$,

$$g(w) = s_n(w) - e_d \text{ as } n \to \infty \ ,$$

where $e_d = \sum_{i=1}^{\infty} e^{-iA} g((2i+1)t)$ is the discretization error. Then, $g(w)$ can be approximated by the $s_n(w)$ sequence as:

$$g(w) \approx s_n(w) = \frac{e^{A/2}}{w} \left\{ \frac{1}{2} W^*(\frac{A}{2w}) + \sum_{i=1}^{n} (-1)^i Re(W^*) \left(\frac{A + 2i\pi j}{2w} \right) \right\} \ , \quad (9)$$

where A is a positive constant s.t. $W^*(s)$ has no singular points on or to the right of the vertical line $s = A/(2w)$ and $Re(W^*)(s)$ is the real part of $W^*(s)$ when s is substituted by a complex number $x + yj$. In (9), n represents the degree of discretization of $g(w)$, i.e., the larger the value of n is, the more accurate the estimation of $g(w)$ by $s_n(w)$ is. In the numerical results shown later, we found that $n = 30$ provides accurate enough results when compared with our simulation results.

If $|g(w)| \leq 1$, the error is bounded by [13]

$$|e_d| \leq \frac{e^{-A}}{1 - e^{-A}} \ .$$

When $A \geq 18.5$, the discretization error is 10^{-8}. The constant A can be further increased to improve the accuracy of the result.

4   Numerical and Simulation Results

We present our numerical and simulation results in this section. The available channel data rate is 1 Mbps and the control packet length is 48 bits[3]. Our simulation, written in C language, implements a network with 50 nodes, which are in the range of each other.

Fig. 3 depicts our numerical results of $g(w)$ for pure ALOHA-based MAC schemes and according to (5) and (9). We observe from this figure that, when normalized traffic load (G) is small, $g(w)$ decreases with the increase of w. As G increases, there is a knee in $g(w)$ around $w = 2$, where the decline of $g(w)$ suddenly slows down.

These numerical results can be verified at $w = 0$. The pdf of the contention resolution period w at $w = 0$ can be calculated as the pdf that exactly one RTS packet is sent out at $w = 0$ multiplied by the probability that no other RTS packets are transmitted on the control subchannel in the next unit time, i.e., $g(0) = Ge^{-Gw}|_{w=0} \cdot e^{-G \cdot 1} = Ge^{-G}$. For $G = 0.25, 0.50, 0.75, 1.00$, and 2.00, $g(0)$ is $0.1947, 0.3033, 0.3543, 0.3679$, and 0.2707, respectively. These results match exactly those shown in Fig. 3.

Fig. 4 depicts our numerical results of expected waiting time on the data subchannel, $\overline{w_2}$, of the pure ALOHA-based MAC-2R scheme. These results are

---

[3] Of course, these system parameters may be changed. However, our results suggest that the conclusions for different parameters' values remain unchanged.

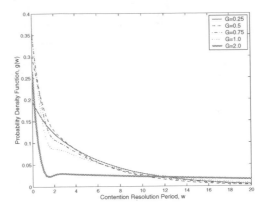

Fig. 3. Probability density function of W, $g(w)$, with different G for MAC schemes

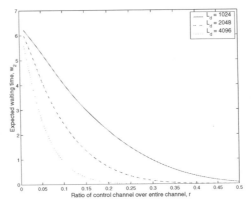

Fig. 4. Expected waiting time on the data subchannel, $\overline{w_2}$, for MAC-2R

calculated according to (6) and the pdf obtained through numerical calculations for different network settings. To minimize $\overline{w_2}$ and maximize the throughput of the MAC-2R scheme, we choose normalized traffic load $G = 0.5$ in the calculation of $g(w)$. In these results, the control packet length (L_c) is fixed at 48 bits, while the data packet length (L_d) takes on the values of: 1024, 2048, and 4096 bits to illustrate different operational overheads of the control packets.

As shown in Fig. 4, the expected waiting time on the data subchannel decreases exponentially as r increases. Furthermore, this decrease is much faster when $k = \frac{L_d}{L_c}$ is larger. Thus, for the same value of r, the expected waiting time on the data subchannel is significantly shorter in networks with larger k. This is due to a much longer data packet transmission time, δ. From this figure, we can also confirm the non-zero expected waiting time when r is chosen as the optimal value of $r^* = \frac{\overline{w}+2}{\overline{w}+2+k}$, as shown in (8). The non-zero expected waiting time on the data subchannel leads to an inferior performance of the MAC-2R scheme, compared to the performance of the MAC-1 scheme.

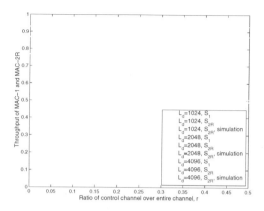

Fig. 5. Throughput comparisons between MAC-1 and MAC-2R

In Fig. 5, we compare the throughput performance of pure ALOHA-based MAC-1 and MAC-2R schemes for different data packet lengths. The straight lines represent the throughput of the MAC-1 scheme. The throughput of the MAC-2R scheme increases as r increases until the throughput reaches the maximum achievable value and then degrades. When r is too small, the control subchannel needs much longer time to come up with a successful RTS/CTS dialogue. However, when r is too large, the fraction of the entire available channel used to transmit data is too small, limiting the throughput of the MAC-2R scheme.

Comparing the throughput performance of the MAC-1 and the MAC-2R schemes, we observe that the MAC-1 scheme always out-performs the MAC-2R scheme, due to the non-zero waiting time on the data subchannel in the MAC-2R scheme. As expected, the throughput increases as L_d (or k) becomes larger, approaching 1 as L_d (or k) increases. In the same figure, Fig. 5, we also draw the simulation results of the MAC-2R scheme, demonstrating that our simulation results closely match those obtained by our analysis.

In Fig. 6, we show the ratio of the throughputs of the MAC-2R and the MAC-1 scheme, S_{2R}/S_1, as a function of r for different data packet lengths L_d. It can be observed that the maximum achievable throughput of the MAC-2R scheme is closer to the throughput of the corresponding MAC-1 scheme as L_d increases. Thus, the penalty for splitting the single channel is lower when data packet length is larger. As L_d increases, the optimum r that achieves the maximum throughput for the MAC-2R scheme becomes smaller.

In Fig. 6, we also draw symbols representing the performance of the MAC-2R scheme, when the single channel is split according to the expected value of the contention resolution periods. In these cases, r is set to $r^* = \frac{\overline{w}+2}{\overline{w}+2+k}$, as shown in (8). As shown in the figure, the throughput of the MAC-2R schemes is offset from the optimum operation point of the MAC-2R scheme. Interestingly, we find that such a non-optimum scheme would operate at the same relative performance S_{2R}/S_1 for the different values of L_d, as the three symbols are all

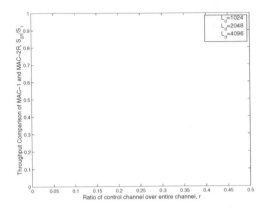

Fig. 6. Throughput comparisons between MAC-1 and MAC-2R

at 0.78 [4]. When the MAC-2R scheme is optimized according to the expected value of the contention resolution periods, i.e., setting r to r^*, we conclude that the throughput degradation of the MAC-2R scheme over the MAC-1 scheme can be as high as 22%.

5   Conclusions

In wireless communication networks, the Medium Access Control (MAC) scheme can significantly affect the performance of the network system. To improve the throughput performance of MAC schemes on random access channels, some researchers proposed to split the single shared channel into two subchannels: a control subchannel and a data subchannel. Control packets are sent on the control subchannel, while the data subchannel is used solely to transmit data packets. Therefore, separation of control packet transmission and data packet transmission is achieved.

Some previous publications in the literature claimed that the split-channel MAC scheme may achieve the same or better throughput as the corresponding single-channel MAC scheme does. However, as we show in this paper, these optimistic results were derived by considering only the expected value of the contention resolution periods, without taking into the account the random distribution of these periods. When the randomness of the contention resolution periods is considered, the split-channel schemes are inferior to the single-channel scheme in fully-connected networks and for the scenarios that we have studied here. According to our analysis, this result holds even if the split-channel schemes are optimized with respect to the ratio of the bandwidth of the control subchannel to the bandwidth of the entire channel.

---

[4] In fact, when $r = r^* = \frac{\overline{w}+2}{\overline{w}+2+k}$, $S_{2R}/S_1 = 1/\left[\left(\frac{1}{1-r} + \frac{\overline{w_2}}{kr}\right)\frac{k}{\overline{w}+2+k}\right]\Big|_{r=r^*} = 1/\left(1 + \frac{\overline{w_2}}{\overline{w}+2}\right)$. Since $\delta = \overline{w} + 2$ and it is not related to k, $\overline{w_2}$ is not related to k according to (6). Therefore, the ratio S_{2R}/S_1 is not related to k.

The inferior throughput performance of split-channel schemes is due to the fact that the control subchannel cannot generate a successful channel reservation dialogue during the period of time when data packets are transmitted on the data subchannel. The randomness of these contention resolution periods requires a larger portion of the available bandwidth to be allocated to the control subchannel, so that long waiting time on the data subchannel would be unnecessary. However, as the overall throughput of split-channel schemes is limited by the capacity of the data subchannel, such allocation of a larger bandwidth to the control subchannel results in significant loss of performance of the data subchannel.

Even though our results are derived for MAC protocols that are based on the RTS/CTS dialogue, these results can be applied to other split-channel MAC schemes as well. In particular, these results can be useful for system engineering in evaluating the advantage and disadvantage of splitting a single shared channel.

References

1. Gallager, R.G.: A perspective on multiaccess channels. IEEE Trans. on Information Theory **IT-31** (1985) 124–142
2. Tobagi, F.A., Kleinrock, L.: Packet switching in radio channels: Part III-polling and (dynamic) split-channel reservation multiple access. IEEE Trans. on Communications **COM-24** (1976) 832–845
3. Singh, S., Raghavendra, C.S.: PAMAS - power aware multi-access protocol with signaling for ad hoc networks. ACM Computer Communications Review **28** (1998)
4. Hung, W.C., Law, K.L.E., Leon-Garcia, A.: A dynamic multi-channel MAC for ad hoc LAN. In: Proc. 21st Biennial Symposium on Communications. (2002) 31–35 Kingston, Canada.
5. Yang, X., Vaidya, N.H.: Pipelined packet scheduling in wireless LANs. Research report, University of Illinois at Urbana-Champaign (2002)
6. Yang, X., Vaidya, N.H.: Explicit and implicit pipelining for wireless medium access control. In: Proc. of Vehicular Technology Conference (VTC). (2003) Orlando, Florida, USA.
7. Muqattash, A., Krunz, M.: Power controlled dual channel (PCDC) medium access protocol for wireless ad hoc networks. In: Proceedings of the 21st International Annual Joint Conference of the IEEE Computer and Communications Societies (INFOCOM 2003). (2003)
8. Abramson, N.: The ALOHA system - another alternative for computer communications. In: AFIPS Conference Proceedings of Fall Joint Computer Conference. Volume 37. (1970) 281–285
9. Karn, P.: MACA - a new channel access method for packet radio. In: ARRL/CRRL Amateur Radio 9th Computer Networking Conference. (1990) 134–140
10. IEEE 802.11: Wireless LAN MAC and physical layer specifications (1999)
11. Takagi, H., Kleinrock, L.: Output processes in contention packet broadcasting systems. IEEE Trans. on Communications **COM-33** (1985) 1191–1199
12. Todd, T.D., Mark, J.W.: Capacity allocation in multiple access networks. IEEE Trans. on Communications **COM-33** (1985) 1224–1226
13. Abate, J., Whitt, W.: Numerical inversion of Laplace transforms of probability distributions. ORSA J. Computing **7** (1995) 36–43

Preventing Replay Attacks for Secure Routing in Ad Hoc Networks

Jane Zhen and Sampalli Srinivas

Dalhousie University, Halifax, NS, Canada, B3H 1W5
{zhen,srini}@cs.dal.ca

Abstract. The design of secure routing techniques is a crucial and challenging requirement in mobile ad hoc networking. This is due to the fact
that the highly dynamic nature of the ad hoc nodes, their limited transmission range, and their reliance on an implicit trust model to route
packets make the routing protocols inherently susceptible to attacks. We
propose a solution to prevent two important types of replay attacks on
the Ad Hoc On-Demand Distance Vector (AODV) routing protocol. Our
technique is based on strengthening the neighbor authentication mechanism by a simple extension to the AODV protocol. Analysis of the
technique indicates that it achieves security with little overhead.

1 Introduction

Ad hoc networking is currently becoming a popular wireless technology for many
applications such as personal area networking, disaster relief and rescue operations, and a variety of military, business and scientific applications. The attractive features of such mobile ad hoc networks (MANET's) include automatic
self-configuration and self-maintenance, quick and inexpensive deployment, and
the lack of the need for fixed network infrastructures or centralized administration [1]. However, along side the advantages, a number of design challenges in
MANET's have emerged. One such crucial requirement is the design of secure
routing protocols. In such networks, the highly dynamic nature of the nodes
can cause the network's topology to change rapidly and unpredictably. Furthermore, wireless transmissions from each node are limited in their range. Due to
these reasons, the nodes must cooperate amongst themselves to exchange routing information and most routing algorithms for ad hoc networks rely on an
implicit trust model to exchange information between neighbors. As a consequence, MANET routing protocols are vulnerable to a variety of attacks such
as eavesdropping, denial of service, packet injection, traffic analysis and replay
attacks [2]-[18].

In this paper, we propose a solution to prevent two types of replay attacks
on the Ad Hoc On-Demand Distance Vector (AODV), which is currently on the
verge of becoming a standard routing protocol for ad hoc networks. The first
type of replay attack is the wormhole attack, in which attackers tunnel Route
Request (RREQ) packets from one node to another through a fast link such that

S: Pierre, M. Barbeau, and E. Kranakis (Eds.): ADHOC-NOW 2003, LNCS 2865, pp. 140–150, 2003.

the route passing through this tunnel will appear to be the shortest and thus gets selected. Consequently, attacker nodes at either ends of the tunnel can drop, delay or modify packets. We also identify a new type of replay attack that can occur on the AODV protocol - the *RREQ Flooding* attack. In this type, attackers can generate extra route discoveries by taking advantage of the "expanding ring" propagation of RREQ's in that not all nodes have the knowledge that the RREQ has been processed. If performed massively, these packets will result in a number of unnecessary resource-consuming route discoveries.

Our technique is based on strengthening the neighbor authentication mechanism by a simple extension to the AODV protocol in order to determine if the source nodes of RREQ packets are really in the neighborhood. By measuring the Round Trip Time (RTT) between two nodes and comparing the RTT value with an adaptive threshold, we can choose to discard or process the received RREQ. The threshold is calculated by requiring special Hello packets being replied immediately instead of periodically. Analysis of the technique indicates that it achieves security with little overhead.

The rest of the paper is organized as follows. Section 2 provides an overview of attacks on the AODV protocol and describes the two types of replay attacks. Section 3 surveys related work in this area. Section 4 describes the proposed approach. Section 5 gives a probabilistic analysis of the technique to provide an estimation of the overhead. Section 6 provides a discussion of the proposal and concluding remarks.

2   Attacks on the AODV Protocol

2.1   Overview of AODV

Briefly, the AODV routing protocol works as follows [2]. A node broadcasts a Route Request (RREQ) if it wants to communicate with another node and no valid route is found in its routing table. The RREQ has the latest sequence number of the originator, an RREQ_ID (or broadcast_id) to mark that it has not been processed, and the latest sequence number of the destination that the originator has in its routing table. Each intermediate node increments the hop count field in RREQ by one and broadcasts this RREQ until the RREQ reaches the destination or a node that has a higher destination sequence number than the one in the packet. Multiple replies (Route Replies - RREP's) may be generated and transmitted along the reverse path. Each intermediate node increments the hop count in RREP and updates its routing table if the RREP has a higher sequence number of the destination or a shorter hop count. This continues until the RREP arrives at the originator.

2.2   Known Attacks on AODV

A variety of known attacks on the AODV protocol have been identified [4,7,11,12,15,16].

Traffic Analysis: Resource limitations make it difficult to incorporate strong encryption mechanisms into wireless data transmissions. Furthermore, mutable fields in the routing packets such as hop count are not authenticated. This may result in exposure of information by traffic analysis.

Routing Loop: By impersonating other hosts' Medium Access Control (MAC) addresses and falsifying favored packets (such as those with higher sequence numbers or shorter hop counts), attackers can make routes to form a loop. This type of attack has become less likely as a result of node authentication mechanisms proposed recently [4].

Black/Gray Hole: By falsely claiming they have optimal routes to multiple destinations, attackers can manage to make relative amount of routes pass by them so as to manipulate packets later on. This attack has been solved by Deng et. al. [18].

Detour: Malicious nodes operate on packets illegally, such as by changing hop counts and sequence numbers arbitrarily, to poison routing tables and make it impossible for optimal routes to be chosen. Some solutions have been proposed but they only solve part of the problem [4].

Fake RERR(Route Error): A malicious node claims that an actually well-connected node is now unreachable by forging RERR packets. This RERR may have a high sequence number (fresher than any other) such that nodes will not accept any opposite information (such as RREQ to/from the isolated node).

Injection of Extra Control Packets: Injected control packets will result in unnecessary network operations. For example, injected RREQ will cause the network to be flooded without the need for data transmission, thus resulting in a denial of service attack. This can be solved by authenticating the source of the RREQ packets [15].

General Replay Attacks: Intruder nodes can launch attacks on the ad hoc network by replaying routing packets. While general authentication mechanisms cannot prevent replay attacks, the sequence number and the RREQ_ID fields are designed to reduce their possibility. However, there are two types of replay attacks that are particularly challenging to defend against. We describe these in the next section.

2.3   Two Special Replay Attacks

RREQ Flooding Attack. We identify a potential replay attack on the AODV protocol. The RREQ packets are broadcast in an incrementing ring to reduce the overhead caused by flooding the whole network. The packets are flooded in a small area (a ring) first defined by a starting TTL (time-to-live) in the IP headers. After RING_TRAVERSAL_TIME, if no RREP has been received, the flooded area is enlarged by increasing the TTL by a fixed value. The procedure is repeated until an RREP is received by the originator of the RREQ, i.e., the route has been found [2].

Nodes in the network avoid their packets from being replayed by incrementing a sequence number and RREQ_ID for each new packet. From the description in the last paragraph, we can observe that the RREQ's only exist within an area. The nodes outside the area will never know the freshest sequence numbers and RREQ_ID's. It is trivial to simply record the RREQ of one node and then broadcast it in another area of the network. If the area where the packets are broadcast to has up-to-date information, the packets will be simply discarded. If the information is out-of-date, the packets will provoke extra unnecessary rounds of route discoveries. By performing this attack massively, a denial of service attack can be launched.

This attack is illustrated in Fig.1. At time t0, attacker M overhears the RREQ from A. At time t1, M replays the RREQ to B. Because B has not heard the freshest RREQ, it will start processing the RREQ so an unnecessary round of route discovery is launched in the area.

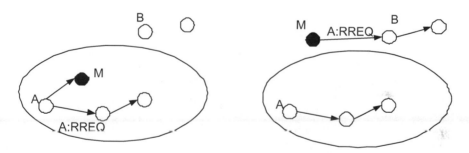

Fig. 1. RREQ Flooding Attack

Wormhole Attack. Wormhole attack has been described in [16]. Here we outline it just for a review. In this kind of replay attack, a tunnel is formed between two nodes through which attackers can transmit packets in a speed faster than the normal hop-by-hop propagation through legitimate wireless links by using long range directional wireless links or even wired links.

Figure 2 illustrates this attack. Node A wants to find out a route to node B. It broadcasts an RREQ which first reaches X and M1 (an attacker could be transparent to the network). While X relays RREQ to its neighbors W and Y, M1 tunnels this request to M2 (another attacker) using a fast link. M2 broadcasts this RREQ to Z, which in turn relays it to B. Since this is faster than the valid route, the valid RREQ is suppressed, and this route is shorter, the malicious route A-M1-M2-Z-B is selected. Furthermore, if nodes near A are about to

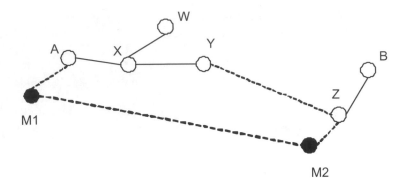

Fig. 2. Wormhole Attack

communicate with nodes near B, they will also choose this route passing M1-M2. Then M1 and M2 can drop, delay or modify packets in transit.

One severe problem is that performing both kind of attacks does not even require attackers to have legitimate keys in the network since only packet recording, replay and MAC spoofing are needed and these can easily be achieved without identification.

3   Related Work

There have been many proposals for securing routing protocols in MANET's. Marti *et. al.* [8] propose a "watch-dog" mechanism by implementing an overhearing module at each node to check if the forwarding of the packet at the next node has been changed illegitimately. Zapata and Asokan [3][4] present a secure extension to the AODV protocol to protect packets from malicious modification. Papadimitratos and Haas [5][9] propose the Secure Routing Protocol (SRP) to set up a secure association between two nodes, and a secure link state routing algorithm. Hu *et. al.* [6]propose SEAD (Secure Efficient Ad Hoc Distance Vector Routing) to secure the algorithm using a one-way hash chain. In [10], Yi *et. al.* define security levels for each node to avoid routing through un-trusted nodes. In [12] by Dahill *et. al.*, packets are signed hop-by-hop using PKI (Authenticated Routing for Ad Hoc Networks-ARAN). Other approaches are proposed to enhance co-operations among nodes in [13] by Buttyan and Hubauz and [14] by Buchegger and Boudec. Hu *et. al.* [7] propose Ariadne, a new secure routing protocol.

Hu *et. al.* [16] define the wormhole attack and propose a solution using packet leashes. Their solution needs extra hardware to provide geographic information and all nodes in the network need their clocks to be synchronized. Also, it needs accurate prediction of packet sending time and receiving time. These requirements are not feasible for current common-used hardware and software.

Multi-path routing protocols in which multiple paths are returned in each route discovery can reduce the impact of the wormhole attack because the traffic is not focused on one route any more [15][17]. But still some precious resources are wasted in the permitted range. If these attacks are performed frequently, the loss can be remarkable.

In [11](Awerbuch *et. al.*), a confirmation for each data packet is required from the receiving node. By limiting the number of missed data packets, the route with poor quality will result in an investigation. It can detect the existence of wormhole only after the degradation has been detected and it still needs to find another route to go over from the beginning. Our proposal can prevent wormholes from being formed.

4   Our Proposal

4.1   Overview

It is important to note that the authentication of neighbors in AODV is weak. The neighboring mechanism in AODV is that each node adds a new neighbor to its neighbor list whenever it hears a "reliable enough" signal no matter *who* sends the signal. A malicious node can send this signal by simply recording packets, spoofing the MAC address (or not, if there is no MAC address mapping mechanism set up) to impersonate other nodes.

The weak neighbor authentication gives room to perform the above two replay attacks. The basic idea of our solution is to measure the Round Trip Time (RTT) between two nodes to decide if they are true neighbors. For the RREQ Flooding attack (Fig.1), when B receives an RREQ, if it could check that this packet could not be originated from direct neighbor nodes by looking at the value of RTT to the node, it would discard the request.

For the wormhole attack (Fig.2), the situation is more complicated. The wormhole attack has three scenarios if proper authentication mechanism has been used to discriminate between outsiders and insiders; thus outsiders cannot participate in the operation of the network because they do not have legitimate identifications: (1) M1 and M2 are both outsiders; (2) M1 is a colluding insider and M2 is an outsider or vice versa; (3) M1 and M2 are both insiders. In scenario (1), since M1 and M2 are all transparent nodes the RREQ received by Z can only come from A. In scenario (2), if M1 is an outsider and M2 selects not to hide (of course it will not hide because in that way Z would receive RREQ from A and would know the existence of wormhole from RTT), it is also possible for Z to receive RREQ forwarded by M2 or by M1 in the second case. If they exchange keys, then the case will be the same as in following scenario. In scenario (3), Z may receive from A, M1 or M2. If M1 and M2 are all outsiders(1) then by measuring RTT to A, node Z would know the existence of a wormhole. For scenario (2)(3), node Z would receive RTT replies from any node as long as they appear valid. Except those approaches detecting wormhole after QoS(Quality of Service) degrades, current approaches cannot detect this type of attack yet.

The proposal involves sending a verification message to un-trusted neighbors for which a node receives RREQ for the first time. An un-trusted neighbor is a neighbor that is not assured to be within transmission range. From the beginning of forming the network, each neighbor is set as trusted because we assume that there is no replay attack in that moment due to the spontaneous nature of the ad-hoc network. From then on, nodes move around across different nodes' transmission range. Upon receiving a RREQ from an un-trusted node, or "neighbor", a node will send a verification message and wait for the reply. Only after the node has approved the neighbor from its RTT value, the RREQ is forwarded continuously. It seems that this will increase the RREQ propagation delay for several folds, but our analysis will show this is not the case because the RREQ is broadcast and the path going through the trusted nodes have less resistance while verifications are being made on other paths.

Because RTT is such a variable value depending on node capability and traffic load, we measure the local average RTT as the threshold. A new RTT to an un-trusted neighbor will be compared with this threshold for accepting or rejecting. Basically, the threshold at a node equals the average RTT to its all trusted neighbors.

In order to attack successfully, the RREQ replay attack must be applied on the nodes far away from the originator because nodes around the originator have the freshest information. And the wormhole attack can form more serious harm when applied on larger range because this makes attackers have more control on the traffic. This large straddle necessitates at least two attackers; otherwise, one attacker will have to use powerful signals which will be heard by a large group of the nodes. By comparing neighbor lists among them to find the abnormal common neighbor, we would be able to detect this attack. The involvement of two attackers increases the possibility of detecting replays by comparing RTT times. Even though these attackers have powerful equipment, MAC delays depend only on local traffic. We assume that two attackers will make RTT remarkablely different between attacked scenarios and normal situation.

4.2   Verification Procedure

The verification procedure used to measure RTT between two nodes sending/receiving RREQ is illustrated in Fig.3. We assume that we have an efficient way to distribute a secret between each pair of nodes such that A and B share a key Kab. The random number is generated uniquely for each verification procedure to prevent replayed verification reply (VEF_REP). IPa and IPb are IP addresses of A and B. They are added to distinguish between the direction of A-B and B-A, otherwise anyone hearing VEF_REQ message can replay it back to forge a VEF_REP. Even though IP addresses are public, without Kab, nodes other than B cannot forge VEF_REP's to A. Each packet must be signed or encrypted depending on its efficiency because we need a mechanism for A and B to authenticate each other.

4.3   Threshold

We base the calculation of RTT threshold value on the Hello message exchange in AODV protocol. According to the protocol, nodes are required to broadcast

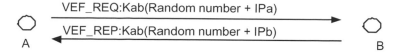

Fig. 3. Measuring RTT

Hello messages periodically to assure neighbors that the links between them are still alive. We use a slightly modified Hello message (RTT_REQ) by including a flag to request for an immediate reply (RTT_REP). By sending such a special Hello after every n Hello's, each node should get the RTT to each trusted neighbor by subtracting the receiving time of RTT_REP by the sending time of RTT_REQ. Trusted neighbors are those neighbors that have passed verification. At the beginning all neighbors are set as trusted because we assume there is no replay due to the spontaneous nature of the ad-hoc network. The RTT threshold is calculated by averaging the RTT's to all trusted neighbors and adding a margin depending on the stringency of security. RTT_REP messages also need to be slightly different to be distinguished from other common Hello's. To reduce overhead, RTT Hello's do not need to be encrypted or signed because neither confidentiality nor integrity is required (it is only within one-hop range). But a random number or time-stamp should be added to each RTT_REQ to be embedded in the RTT_REP to prevent fake replies.

For the nodes that currently do not have any route to maintain and thus are not broadcasting Hello's, RTT Hello's are required to maintain RTT thresholds for themselves. Since RTT Hello's have much longer interval, the overhead will be minimum.

5   Analysis

In this section, we give a mathematical analysis of the overhead caused by neighbor verification procedures. The overhead caused by RTT Hello is ignored since they are just Hello packets that need immediate responses.

Suppose in a network of N nodes, each node has neighbor change rate X, i.e., after a specific period of time, X percent new neighbors need to be verified. Also, we suppose during this period of time M RREQ's are processed.

We can deem X as the probability of the verification being launched in one-hop range. If the average length of routes in the network is Y hops, then according to Binomial distribution the probability of the verification being held in only one hop is:

$$P_1 = \binom{Y}{1} X^1 (1-X)^{(Y-1)} \tag{1}$$

The probability of i hops being verified is:

$$P_i = \binom{Y}{i} X^i (1-X)^{(Y-i)} \tag{2}$$

where $0 \leq i \leq Y$. Suppose the time consumed at each hop is t if no verification is needed, $3t$ if the verification is applied (one RREQ(t) and one RTT($2t$)), then the time of finding a route with i hops being verified is $Ti = 3t*i+t*(Y-i)$. The expectation E_i of the time consumed in finding a route will be $\sum_{i=0}^{i=Y} P_i * T_i$. For example, with $X = 10\%$ and $Y = 6$ we can have following table:

Table 1. Verification Probabilities

i	0	1	2	3	4	5	6
P_i	0.531	0.354	0.010	0.015	0.001	0.000	0.000
T_i	6t	8t	10t	12t	14t	16t	18t

From the table above, we can calculate $E_i = 7.2t$. Such that when neighbor verification probability is 10%, the average extra time for finding a new route would be $(7.203t - 6t)/6t = 20\%$. The fact that this overhead will be shared among the processing of M RREQ's makes it not significant.

6   Discussion and Conclusions

Adding more mechanisms always needs more protection. The question is: will this mechanism pose new security risks to the routing protocol? Can verification requests (VEF_REQ) be replayed to some other area in the network? We argue that this is not possible because each verification request needs the knowledge of the shared key between two nodes. The verification packets between two nodes cannot be applied to other pairs.

Can RTT_REP packets be forged by illegitimate nodes to fool the node to calculate wrong RTT threshold? This will not occur because if we have authentic trustable neighbors from initial stage of the network and each round of calculation of RTT threshold is based on the replies from trusted neighbors, the authentication of RTT_REPs will be guaranteed and the impact from a particular node will be balanced by other nodes after RTT averaging.

Can RTT_REQs be replayed to some other area in the network? When receiving a RTT_REQ replayed by an illegitimate node, the node may add the illegitimate node to its neighbor list and send back a RTT_REP. Since the cost of sending a RTT_REP is just trivial this will not impact the sending node much. The replayed RTT_REP will be discarded simply because the source node is not in the trusted neighbor list. The nodes in trusted neighbor list are added after the verification process.

What if a neighbor moves away while the node still has not sent another RTT_REQ since there is an interval? In this case, regular connectivity maintenance will discover the leaving of the neighbor and purge it out of its neighbor list.

In summary, we proposed a solution to prevent two important types of replay attacks, namely, the wormhole attack and the *RREQ flooding* attack, on the AODV routing protocol. Our technique is based on strengthening the neighbor authentication mechanism by a simple extension to the protocol. Analysis of the technique indicates that it achieves security with little overhead. Our future work entails simulation studies of the proposal, modelling other types of attacks on the AODV protocol and incorporation of the proposed technique in the AODV protocol.

References

1. C.-K. Toh, *Ad Hoc Mobile Wireless Networks : Protocols and Systems*, Prentice Hall, 2002.
2. C. E. Perkins, E. M. Belding-Royer and S. R. Das, "Ad hoc On-Demand Distance Vector (AODV) Routing," draft-ietf-manet-aodv-13.txt, Feb. 2003.
3. M.G. Zapata, "Secure Ad Hoc On-Demand Distance Vector(SAODV) Routing," draft-guerrero-manet-saodv-00.txt, Aug. 2001.
4. M. G.,Zapata and N.Asokan, "Securing Ad Hoc Routing Protocols," *Proc. ACM Workshop on Wireless Security(WiSe'02)*, Atlanta, Georgia, USA, Sep. 2002, pp.1-10.
5. P. Papadimitratos and Z. J. Haas, "Secure Routing for Mobile Ad Hoc Networks", *Proc. the SCS Communication Networks and Distributed Systems Modeling and Simulation Conference(CNDS 2002)*, San Antonio, TX, Jan. 2002.
6. Y. Hu, D. B. Johnson and A. Perrig, "SEAD: Secure Efficient Distance Vector Routing for Mobile Wireless Ad Hoc Networks," *Proc. the 4th IEEE Workshop on Mobile Computing Systems & Applications (WMCSA 2002)*, Calicoon, NY, Jun. 2002, pp. 3-13.
7. Y.Hu, A. Perrig and D. B. Johnson, "Ariadne: A Secure On-Demand Routing Protocol for Ad Hoc Networks," *Proc. the Eighth ACM International Conference on Mobile Computing and Networking (MobiCom 2002)*, Atlanta, GA, Sep. 2002, pp. 12-23.
8. S.Marti,T.J.Giuli,K.Lai and M.Baker, "Mitigating Routing Misbehavior in Mobile Ad Hoc Networks," *Proc. the Sixth annual ACM/IEEE International Conference on Mobile Computing and Networking*, 2000, pp. 255-265.
9. P. Papadimitratos and Z. J. Haas, "Secure Link State Routing for Mobile Ad Hoc Networks," *Proc. the IEEE Workshop on Security and Assurance in Ad Hoc*, Orlando, FL, Jan. 2003.
10. S. Yi, P. Naldurg and R. Kravets, "Security-Aware Ad hoc Routing for Wireless Networks," *Proc. ACM Mobihoc*, 2001.
11. B. Awerbuch, D. Holmer, C. Nita-Rotaru and H. Rubens, "An On-Demand Secure Routing Protocol Resilient to Byzantine Failures," *ACM Workshop on Wireless Security(Wise'02)*, Atlanta, Georgia, Sep.2002.
12. B. Dahill, B. Levine, E. Royer and C. Shields, "A Secure Routing Protocol For Ad Hoc Networks," *Technical Report UM-CS-2001-037*, University of Massachusetts, Department of Computer Science, Aug. 2001.
13. L. Buttyan and J. Hubauz, "Nuglets: a Virtual Currency to Stimulate Cooperation in Self-Organized Mobile Ad Hoc Networks," *Technical Report DSC/2001/001*, Swiss Federal Institute of Technology-Lausanne, Department of Communication Systems, Jan. 2001.
14. S. Buchegger and J. L. Boudec, "The Selfish Node: Increasing Routing Security for Mobile Ad Hoc Networks," *IBM Research Report RR 3354*, May 2001.
15. R. Ramanujan and A. Ahamad, "Techniques For Intrusion-Resistant Ad Hoc Routing Algorithms (TIARA)," *Proc. MILCOM 2000*, Los Angeles, Oct. 2000.

16. Y. Hu, A. Perrig and D. B. Johnson, "Packet Leashes: A Defense Against Worm-hole Attacks in Wireless Ad Hoc Networks," *Technical Report TR01-384*, Rice University Department of Computer Science, Dec. 2001.
17. Y. Yi, S. Lee, W. Su and M. Gerla, "On-Demand Multicast Routing Protocol (ODMRP) for Ad Hoc Networks," draft-ietf-manet-odmrp-04.txt, Feb. 2003.
18. H. Deng, W. Li, and Dharma P. Agrawal, "Routing Security in Ad Hoc Networks", *IEEE Communications Magazine, Special Topics on Security in Telecommunication Networks,* Vol. 40, No. 10, October 2002, pp. 70-75.

Resisting Malicious Packet Dropping
in Wireless Ad Hoc Networks

Mike Just[1], Evangelos Kranakis[2,*], and Tao Wan[2,**]

[1] Treasury Board of Canada, Secretariat, 2745 Iris St., Ottawa, ON, K1A 0R5, Canada
[2] School of Computer Science, Carleton University, Ottawa, ON, K1S 5B6, Canada

Abstract. Most of the routing protocols in wireless ad hoc networks, such as DSR, assume nodes are trustworthy and cooperative. This assumption renders wireless ad hoc networks vulnerable to various types of Denial of Service (DoS) attacks. We present a distributed probing technique to detect and mitigate one type of DoS attacks, namely malicious packet dropping, in wireless ad hoc networks. A malicious node can promise to forward packets but in fact fails to do so. In our distributed probing technique, every node in the network will probe the other nodes periodically to detect if any of them fail to perform the forwarding function. Subsequently, node state information can be utilized by the routing protocol to bypass those malicious nodes. Our experiments show that in a moderately changing network, the probing technique can detect most of the malicious nodes with a relatively low false positive rate. The packet delivery rate in the network can also be increased accordingly.

Keywords: Security, Denial of Service (DoS), Wireless Ad Hoc Networks, Distributed Probing, Secure Routing Protocols.

1 Introduction

A wireless or mobile ad hoc network (MANET) is formed by a group of wireless nodes which agree to forward packets for each other. One assumption made by most ad hoc routing protocols [16, 21] is that every node is trustworthy and cooperative. In other words, if a node claims it can reach another node by a certain path or distance, the claim is trusted. If a node reports a link break, the link will no longer be used. Although such an assumption can simplify the design and implementation of ad hoc routing protocols, it does make ad hoc networks vulnerable to various types of denial of service (DoS) attacks, which are discussed in detail in Section 2. One class of DoS attacks is malicious packet dropping. A malicious node can silently drop some or all of the data packets sent to it for further forwarding even when no congestion occurs.

Malicious packet dropping attack presents a new threat to wireless ad hoc networks since they lack physical protection and strong access control mechanism. An adversary

* Research supported in part by NSERC (Natural Sciences and Engineering Research Council of Canada) and MITACS (Mathematics of Information Technology and Complex Systems) grants.
** Research supported in part by OCIPEP (Office of Critical Infrastructure Protection and Emergency Preparedness) Research Fellowship.

S. Pierre, M. Barbeau, and E. Kranakis (Eds.): ADHOC-NOW 2003, LNCS 2865, pp. 151–163, 2003.

can easily join the network or capture a mobile node and then starts to disrupt network communication by silently dropping packets. It is also a threat to the Internet since the various software vulnerabilities would allow attackers to gain remote control of routers on the Internet. If malicious packet dropping attack is used along with other attacking techniques, such as shorter distance fraud, it can create more powerful attacks (i.e., *black hole* [12]) which may completely disrupt network communication.

Current network protocols do not have the capability to detect the malicious packet dropping attack. Network congestion control mechanisms do not apply here since packets are not dropped due to congestion. Link layer acknowledgment, such as IEEE 802.11 MAC protocol [1], can detect link layer break, but cannot detect forwarding level break. Although upper layer acknowledgment, such as TCP ACK, allows for detecting end-to-end communication break, it can be inefficient and it does not indicate the node at which the communication breaks. Moreover such mechanism is not available in connection-less transport layer protocols, such as UDP. Therefore, it is important to develop mechanisms to render networks the robustness for resisting the malicious packet dropping attack.

In this paper, we present a proactive distributed *probing* technique to detect and mitigate the malicious packet dropping attack. In our approach, every node proactively monitors the forwarding behavior of other nodes. Suppose node A wants to know if node B performs its forwarding functions, it will send a probe message to a node one hop away from node B, let us say to node C. C is supposed to respond to the probe message by sending back an acknowledgment to A. If A can receive the acknowledgment within a certain time period, it acts as a confirmation that node B forwarded the probe message to C. With the assumption that a probe message is indistinguishable from a normal data packet, A knows that B will forward all the other packets.

Our experiments demonstrate that in a moderately changing network, the probing technique can detect most of the malicious nodes with a relatively low false positive rate. The packet delivery rate in the network can also be increased if the detected malicious nodes are bypassed from network communication. We argue that the probing technique is of practical significance since it can be implemented in the application layer and does not require the modification of underlying routing protocols.

The remainder of the paper is organized as follows. In Section 2, we analyze the DoS attacks against a network infrastructure and review the corresponding prevention mechanisms. In Section 3, we define frequently used notation and terminology. In Section 4, we present our solution for monitoring wireless ad hoc networks. In section 5, we describe the implementation and simulation of our solution. We conclude the paper in the last section.

2   DoS Attacks on Routing Infrastructure

Wireless ad hoc networks are vulnerable to various types of DoS attacks, such as signal injection, battery drain, among others. This paper focuses on the DoS attacks on its routing infrastructure. Based on the types of traffic transmitted in a network, we can classify these DoS attacks into two categories: *DoS attacks on routing traffic* and *DoS attacks on data traffic*. Such classification is also applicable to the Internet.

2.1 DoS Attacks on Routing Traffic

An attacker can launch DoS attacks against a network by disseminating false routing information so that established routes for data traffic transmission are undesirable or invalid. There are at least three possible consequences. *Firstly*, data traffic may be captured in a *black hole* [13] and never leave out. For example, in a distance vector routing protocol, an attacker can attract data traffic by advertising shorter distance and then drop the attracted traffic. *Secondly*, data traffic may not flow through routing paths fairly and some of them are dropped due to network congestion. For example, an attacker can avoid some traffic or redirect traffic to other nodes by advertising carefully crafted routing update messages. *Thirdly*, an attacker may disseminate arbitrary routing information to mislead other routers to create invalid paths in their routing table. As a result, data traffic flowing through those paths will eventually be dropped due to network unreachability or life time expiration (i.e., in presence of routing loops).

2.2 DoS Attacks on Data Traffic

An attacker can launch two types of DoS attacks on data traffic. *First*, it can inject a significant amount of data traffic into the network to clog the network. If there is no protection mechanism in place for provisioning data traffic, legitimate user packets will be dropped along with malicious ones as the result of congestion control. In the worst case, the network could be completely shutdown.

Second, if a malicious user manages to join a network or compromise a legitimate router, it can silently drop some or all of the data packets transmitted to it for further forwarding. We call it the *malicious packet dropping attack*. Malicious packet dropping attack is a serious threat to the routing infrastructure of both MANET and the Internet since it is easy to launch and difficult to detect. To launch the attack, an attacker needs to gain the control of at least one router in the target network. The router used to launch the attack can be a specialized router or a computer running routing software. To gain access to a specialized router, an attacker can explore the software vulnerability of a router (e.g., buffer overflow) or explore the weakness of logon authentication process (i.e., weak password). Many routers run vulnerable software and open the vulnerability to the world. For example, a survey [17] on 471 Internet routers shows that majority of them run SSH, Telnet or HTTP and 17% of them accept connections from arbitrary IP addresses. An attacker can also explore the vulnerabilities of routing protocols to join the network with his own computer or a compromised inside machine. This is possible due to the fact that most routing protocols only deploy very weak authentication mechanisms, such as plain text passwords.

2.3 Preventing DoS Attacks on Routing Traffic

Significant work has been done to secure routing protocols against DoS attacks on routing traffic. Most of them apply cryptographic techniques (asymmetric or symmetric) to authenticating routing traffic.

Asymmetric cryptographic techniques, such as public-key based digital signatures, can be used to sign routing messages [24–26] to prevent external intruders from joining the network or malicious insiders from spoofing or modifying routing messages. The disadvantages are: 1) They are quite inefficient since both the signature generation and verification process involve the execution of computationally expensive functions. 2) They cannot prevent internal attacks.

Given the inefficiency of digital signature mechanisms, some researchers [7, 27] proposed to use symmetric cryptographic primitives (i.e., one-way hash chains, one-time signatures, authentication tree, etc.) for authenticating routing messages. Unfortunately, these approaches still do not prevent attacks from compromised internal routers. Hu, Johnson, and Perrig [13, 14] take the step further in securing distance vector routing protocols by forcing a node to increase metrics when forwarding routing update messages. Therefore, their approaches can prevent compromised nodes from claiming shorter distances. The disadvantage is that a malicious node can avoid traffic by claiming longer distances.

2.4   Preventing DoS Attacks on Data Traffic

It has been hypothesized that a network with QoS support can well resist DoS attacks since malicious packets will be dropped in the first place when facing network congestion. Other researchers proposed mechanisms [3, 6] to trace back to the origin of the malicious packets which cause the network congestion and drop them in the routers where they first enter into the victim network. *Ingress/Egress filtering* can also be helpful if IP spoofing is utilized in the attack.

Several approaches have been proposed to prevent DoS attacks on data forwarding level. Perlman [22] proposed hop-by-hop packet acknowledgment to detect packet dropping in a network. The disadvantage is that it will generate significantly high routing overhead. Cheung et al [8] proposed a probing method for defeating denial of service attacks in a fixed routing infrastructure using neighborhood probing. It requires a testing router to have a private address which allows it to generate a packet destined to itself but goes through the tested router. This requirement is not practical in MANETs. A distributed monitoring approach is proposed in [4] for detecting disruptive routers. The protocol is based on the principle that any packets sent to a router and not destined to it are supposed to leave that router. This principal is not applicable to MANET due to their changing network topology.

Marti et al [19] proposed and implemented two protocols for detecting and mitigating misbehaving nodes in wireless ad hoc networks by *overhearing* neighborhood transmissions. Their method is very effective for detecting misbehaviors in one-hop away. To monitor the behavior of nodes two or more hops away, one node has to trust and rely on the information from other nodes, which introduces the vulnerability that good nodes may be bypassed by malicious or incorrect accusation.

Buchegger and Le Boudec [5] developed the CONFIDANT protocol for encouraging node cooperation in dynamic ad-hoc networks. Each node monitors the behavior and maintains the reputation of its neighbors. The reputation information may be shared among friends. A trust management approach similar to Pretty GOOD Privacy (PGP) is used to validate received reputation information. Nodes with bad reputation may be isolated from the network. As a result, nodes are forced to be cooperative for their own interest. Our proposed probing technique can be used as one of the monitoring techniques in the CONFIDANT protocol.

Awerbuch et al [2] proposed a secure routing protocol for resisting byzantine failures in a wireless ad hoc network. The protocol requires an ultimate destination to send an acknowledgment back to the sender for each of its successfully received packets. If the loss rate of acknowledgment packets exceeds the predefined threshold, which is set to be slightly above the normal packet loss rate, the route used for sending packets from the source to the destination is detected as faulty and a binary search probing technique is deployed to locate the faulty link. The disadvantages of this protocol are: 1) it may incur significant routing overhead; 2) a data packet with an inserted probe list can be distinguished from those without probe lists, although the probe list is onion encrypted and cannot be tampered en route. Our proposed probing technique differs in that it can be implemented above the network layer (e.g., based on UDP), and the end-to-end encryption of IP payload using pair-wise shared keys can prevent intermediate nodes from distinguishing probe messages from data packets.

Padmanabhan and Simon [20] proposed a secure traceroute to locate faulty routers in wired networks. In their approach, end hosts will monitor network performance. If an end-to-end performance degrade is detected by a host to a destination, a *complaint* bit is set in all the subsequent traffic to that destination. The complaining host itself or the router sitting closest to the complaining host may start the troubleshooting process if it observes enough complaints. It first sends a secure traceroute packet to the next hop, which can be derived from its routing table. The router receiving the secure traceroute packet is expected to send a response back which also includes a next hop address. This process repeats until a faulty router is located (no response received from it) or every router on the path to the ultimate destination proves healthy. Our approach is different from the secure traceroute in that 1) our approach is proposed for MANET using source routing protocols (e.g., DSR), the secure traceroute is mainly used in wired networks. 2) our approach does not require modification to existing routing infrastructures, the secure traceroute may need to modify IP layer in order to monitor performance problem; 3) our approach utilizes redundant routing information for diagnosis, the secure traceroute does not.

Malicious nodes silently dropping packets exhibit the same behavior as selfish nodes, which may choose to drop packets for the sake of saving its own constraint resources, such as battery or CPU cycle. Selfishness and its threat to the network performance have been well studied by Roughgarden [23]. Incentive mechanisms have been proposed to encourage selfish nodes to be cooperative and to forward packets for others. Unfortunately, incentive mechanisms don't work for malicious users since they never play by rules. Our proposed probing scheme can be used to detect and mitigate selfishness problem.

3 Definitions and Assumptions

3.1 Node States

We classify the states of a node as follows. A node is *GOOD* if it responds to probe messages for itself and forwards other probe messages along their source routes. A node is *BAD* if it responds to probe messages destined to itself but fails in forwarding probe messages for others. A benign link failure may also be detected as *BAD* behavior if it is not cleared by other mechanisms (e.g., route error in DSR). A node is considered *DOWN* if 1) it is a neighbor node to the probing node and it doesn't respond to probe messages; or 2) it is not a neighbor node and it doesn't respond to probe messages through *all* the known paths. A node is considered at the *UNKNOWN* state if on all known paths from the probing node to the node, there exists at least one node in BAD or DOWN state.

3.2   Assumptions

Probe messages are indistinguishable from normal packets. One limitation of the probing technique is that it can be easily defeated if probe messages can be distinguished from normal data packets. For example, a malicious node may forward probe messages, but drop all the other data packets, thereby avoiding detection. This assumption can be realized using end-to-end encryption of IP payload by pair-wise shared keys. Since a malicious node can understand only the IP header, it does not have the information of upper layer protocols, such as TCP/UDP port numbers. By implementing the probing technique above the network layer (e.g., based UDP), an adversary will not be able to distinguish a probe message from a other data packet (e.g., HTTP or SMTP packet). Some other options are: 1) piggybacking a probe message on a normal data packet which requires acknowledgment, such as TCP SYN. The disadvantage is that such data packets may not be available during the time of probing. 2) assuming that an adversary cannot modify the forwarding software of the compromised router. Therefore, the adversary can only make decisions based on IP addresses, which does not allow for distinguishing a probe message from a normal data packet.

Multi-hop source routing protocols. The probing technique assumes a multi-hop source routing protocol since a probing node needs to specify the source route by which a probe message takes to get to the destination. This assumption is practical since some routing protocols, such as Dynamical Source Routing (DSR) [16], are multi-hop source routing protocols.

Bi-directional communication links. We assume that all communication links are bi-directional. This assumption is practical in some wireless networks, such as IEEE 802.11 [1], where all links have to be bi-directional for link layer acknowledgment to work.

4   The Distributed Probing Scheme

In order to monitor the behavior of mobile nodes by the probing technique, we need to decide which node should probe and how far it should probe. Given a network with n nodes, there are several interesting possibilities: 1) there is only one probing node and it probes all the other nodes; 2) there are k probing nodes ($1 < k < n$) and each probes over a distance of r ($1 < r < \infty$); 3) there are n probing nodes and each probes over a distance of infinity. The last approach is preferred in a MANET since it can detect selective packet dropping (i.e., based on IP addresses). Another advantage is that one node does not need to rely upon the information from other nodes to detect malicious ones. The disadvantage is that it may generate significant network overhead. The network overhead can be reduced if probe messages piggyback normal packets. To simplify the problem, we divided the probing technique into three algorithms: 1) the probing path selection algorithm; 2) the probing algorithm; and 3) the diagnosis algorithm. These are described below.

4.1   Probing Path Selection Algorithm

The probing paths are selected solely from the routing cache maintained by a mobile node. There are usually many redundant paths in the routing cache. Although probing over each of them may allow for validating all the known paths, it will also produce significant network overhead. The ideal strategy shall select a minimum number of paths but allows for monitoring the forwarding behavior of as many nodes in the routing cache as possible. The probing path selection algorithm returns a set of paths with the following properties.

1) For any two paths p_i and p_j, $p_i \nsubseteq p_j$. Since probing over a path can always disclose the forwarding behavior of those nodes in any of its subsets, any path which is a subset of another path will be eliminated.

2) For any two paths p_i and p_j, if the second farthest node in p_i is an intermediate node of p_j, the farthest node of p_i will be removed. For example, given two paths $p_1 = A \to B \to C \to D$ and $p_2 = A \to E \to F \to C \to G \to H$, node D will be removed from p_1. With D in p_1, A can monitor the forwarding function of C. Since such monitoring can be achieved by probing over p_2, there is no need to keep D in p_1. C will still be kept in p_1, since A needs to monitor B by sending a probe message to C.

3) The length of any path (in terms of number of hops) is greater than 1. Since we are interested in monitoring the forwarding function of mobile nodes, probing over a one hop path offers no information. A probe message is sent to a neighbor node only when a node subsequent to it doesn't respond to the probe message and the probing node needs to know if the neighbor node is BAD or has moved out of its transmission range.

4.2   The Probing Algorithm

With a set of selected probing paths, the probing algorithm will probe over each of them. Given a probing path, there are at least two ways of probing. One way is to probe from the farthest node to the nearest. The other way is to probe from the nearest node to the farthest. Each has its own advantages and disadvantages. Probing from far to near is better if the probing path is *GOOD* since it takes only one probe message and proves the goodness of all the intermediate nodes. But it may take more probe messages if a *BAD* node is located near the probing node. This method can be applied to a network where we have the confidence that the majority of the nodes in the network are *GOOD*. The advantage of probing from near to far is that it generates smaller number of probing messages to detect a *BAD* node located near the probing node. Another advantage is that we have the prior knowledge of the states of all the intermediate nodes along the path to the probed node except its immediate predecessor node. The disadvantage is the an intelligent attacker may be able to avoid detection by forwarding all packets (including probe messages destined to the downstream nodes) for a certain period of time immediately after receiving a probe message for itself. A received probe message therefore serves as a signature to an attacker that a diagnosis process is ongoing, and it would start to behave normally for a short period of time. Other search strategy (e.g., binary search) can also be deployed to reduce network overhead.

In this paper, we present the algorithm for the first method, probing from the farthest nodes to the nearest, since it is stronger than the other alternatives in detecting malicious nodes. For a probing path, the probing node sends a probe message to the farthest node. If an acknowledgment message is received within a certain period of time, all the intermediate nodes are shown to be GOOD. Otherwise, a probe message is sent to the second farthest node. This process is repeated until one node responds to the probe message or the nearest node (a neighbor node) is probed and it is not responsive. In the latter case, we know that the neighbor node in the probed path either is DOWN or has moved out to another location. Since the neighbor node is not responsive, there is nothing we can do to monitor the rest nodes in the path. Therefore, probing over this path is stopped. If an intermediate node is responsive but a node subsequent to it is not, it is possible: 1) the intermediate node failed forwarding the probe message to the next node; 2) the link between the two nodes is broken by location change; 3) the unresponsive node is incapable of responding to the probe message. The diagnosis algorithm will then be called to decide which one is the case.

4.3   The Diagnosis Algorithm

After the probing node detects a node (v_i) is responsive but the subsequent node (v_{i+1}) is unresponsive, it calls the diagnosis algorithm to determine if the link $v_i \leftrightarrow v_{i+1}$ is broken at the link level or forwarding level.

The probing node first searches the routing cache for another path to v_{i+1}. If such a path exists, it will probe v_{i+1} through this path. If v_{i+1} is still unresponsive, it searches the routing cache for another path. This process repeats until 1) there is a route (p) through which node v_{i+1} is responsive, or 2) the routing cache is exhausted.

In case 1, the diagnosis algorithm appends v_i to the path p and sends a probe message to v_i over p. If an acknowledgment is received from v_i, v_i is diagnosed as BAD since the link $v_{i+1} \rightarrow v_i$ is good but link $v_i \rightarrow v_{i+1}$ is not. Based on the assumption that any link is bidirectional, $v_i \leftrightarrow v_{i+1}$ should be good at link level. Therefore, it is broken at forwarding level. If v_i is unresponsive over the new path, the link $v_i \leftrightarrow v_{i+1}$ is diagnosed as broken in link layer. It is also possible that both v_i and v_{i+1} are BAD. Since there is no sufficient information available to distinguish this situation from the link layer break, we treat this situation as link layer break. It causes false negatives.

In case 2, node v_{i+1} may have moved out from its previous location and a new path to v_{i+1} is not discovered by the probing node yet. It is also possible that node v_{i+1} has moved out from the network or is DOWN. Although a route discovery may be able to disclose further information, it is also very expensive. Therefore, the diagnosis algorithm simply treats node v as being DOWN.

When a node is detected as BAD, the routing cache is updated by removing all nodes subsequent to the bad node. When a link is detected as broken, the routing cache is also informed and the link is truncated from all the paths. When the routing cache adds a route to the cache, it looks up the node state table and truncates the route accordingly if there is any BAD node in the path.

5   Simulations

We study the detection rate of the probing technique and and its impact on network performance using the *NS-2* network simulator [18] with the wireless extension from Rice University. The simulation is performed on Sun Ultra 10 workstations running Solaris 5.7.

5.1   Simulation Environment

We implemented the probing technique in *NS-2* version 2.1b9a with wireless extension. The routing protocol we use is Dynamic Source Routing (DSR) and the routing cache is path cache with a primary and a secondary FIFO cache [11]. The probing technique is implemented as a part of DSR and the probe message is a new type of DSR packet.

We simulate a network with 670m x 670m space and 50 mobile nodes. The simulation time is 100 seconds. The mobile nodes move within the network space according to the *random waypoint mobility model* [15] with a maximum speed of 20m/s. The pause time is 50 seconds, which represents a network with moderately changing topology. The communication patterns we use are 10 constant bit rate (CBR) connections with a data rate of 4 packets per second. Those simulation parameters are widely used by the community. We chose them to make our simulation results comparable with others.

We randomly choose 0, 3, 5, 8, 10, 13, and 15 BAD nodes in each of the simulation. Security researchers like to assume the worst case, but it rarely happens in real life. Since it is realistic that the majority of nodes in a network should be GOOD, we simulate at most 15 BAD nodes, which represent 30 percent of the total number of nodes.

5.2   Metrics

We chose the following metrics for measuring the probing technique: 1) *Detection Rate*, the ratio of the number of detected BAD nodes and the total number of actual BAD nodes. 2) *False Positive Rate*, the ratio of number of GOOD nodes mistakenly detected as BAD and the total number of GOOD nodes. The combination of this metric and the detection rate tells us the overall performance of the probing technique. 3) *Packet Delivery Rate*, the ratio of total number of data packets received and the total number of data packets sent in application level. In our simulation, the data packets refer to the CBR traffic. 4) *Network Overhead*, the ratio of total number of routing related transmissions (including all DSR related traffic and probe messages) and the total number of packet transmissions (including both routing related transmissions and data transmissions). Each packet hop is counted as one transmission.

5.3   Simulation Results

We study the probing technique using the above defined metrics. The standard DSR (Standard_DSR) is used as a baseline to compare with the DSR with the extension of the probing technique (DSR_Probe). We run the simulation 5 times and all the graphs (Figure 1) are plotted from the data averaged from the 5 runs.

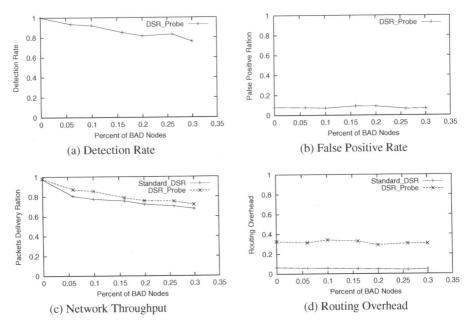

Fig. 1. Simulation Results

Detection Rate. Figure 1.a shows the detection rate. In the best case, 94% of the bad nodes can be detected. In the worst case, the detect rate is 76%. There are several reasons why a BAD node is not detected. First, the BAD node is not in any path in the routing cache each time when the probing technique starts to probe. Since the probing paths are selected solely based on the paths maintained by the routing cache, if a node is not contained in any path, its forwarding function will not be monitored. Second, there are two consecutive BAD nodes in a path, and the bad behavior of one is hidden by the other. The link between the two bad nodes is detected as link layer break, the the bad behavior is not detected. Although this affects the detection rate, it does not have impact on packet delivery rate since the link is removed from the routing cache in any way.

False Positive Rate. Figure 1.b shows the false positive rate. We can see from the graph that the highest false positive rate is below 9%, which is relatively low. The false positive is caused mainly by node movement since some link layer breaks are detected as forwarding level misbehavior. Therefore, it will decrease when the node motion becomes slower.

Packet Delivery Rate. The graph of packet delivery rate (Figure 1.c) has two curves and they represent the throughput of standard DSR and the DSR with the extension of the probing technique. The graph demonstrates that the DSR with the probing technique extension always performs better than the standard DSR. This is in line with our expectation since the bad nodes which failed in forwarding packets are removed from the routing cache. The result is that good paths are used for transmitting packets.

We can also see from the graphs that packet delivery rate sometimes is higher when there is a higher percentage of BAD nodes than when there is a lower percentage of BAD nodes. This is contrary to the common expectation. As explained in [19], the randomness of NS-2 results in this effect due to the fact that route replies may arrive at nodes in different orders in different runs. Therefore, a node may choose a path with BAD nodes in one run but choose a good path in another run.

Overhead. As shown in Figure 1.d, the routing overhead is increased significantly when the network topology changes faster or there is a high percentage of BAD nodes in the network. In both scenarios, a large number of probe messages have to be sent out to finalize the node states. The overhead can be reduced dramatically if probing messages piggyback normal data packets.

6 Conclusion

Wireless ad hoc networks are vulnerable to various types of DoS attacks. We presented a distributed probing technique to detect and migitate the malicious packet dropping attack in MANETs. We implemented the probing technique in NS-2 with wireless extensions. Our experiments show that in a moderately changing network, the probing technique can detect most of the malicious nodes with a relative low false positive rate. The packet delivery rate can also be increased if the node state information is shared with routing cache. We think the probing technique is of practical significance since it can be implemented independently from routing software and does not require modification to the existing infrastructure. The disadvantage of the probing technique is that it generates relatively high network routing overhead if probe messages do not piggyback data packets.

References

1. ANSI/IEEE std 802.11. Wireless LAN Medium Access Control (MAC) and Physical Layer (PHY) specification, 1999.
2. B. Awerbuch, D. Holmer, C. Nita-Rotaru, and H. Rubens. An On-Demand Secure Routing Protocol Resilient to Byzantine Failures. In *ACM Workshop on Wireless Security* (WiSe), September 2002.
3. S.M. Bellovin, M. Leech, and T. Taylor. ICMP Traceback Messages. Internet draft: draft-ietf-itrace-03.txt, January 2003.

4. K.A. Bradley, S. Cheung, N. Puketza, B. Mukherjee, and R.A. Olsson. Detecting Disruptive Routers: A Distributed Network Monitoring Approach. In *Proceedings of the IEEE Symposium on Research in Security and Privacy*, pages 115-124, May 1998.
5. S. Buchegger and J.Y. Le Boudec. Performance Analysis of the CONFIDANT Protocol (Cooperation Of Nodes - Fairness In Dynamic Ad-hoc NeTworks) In *Proceedings of the Third ACM International Symposium on Mobile Ad Hoc Networking and Computing*, (MobiHoc 2002), June 2002.
6. H. Burch and H. Cheswich. Tracing anonymous packets to their approximate source. In *Proceedings of USENIX LISA*. pages 319-327, New Orleans, LA, December 2002.
7. S. Cheung. An Efficient Message Authentication Scheme for Link State Routing. In *Proceedings of the 13th Annual Computer Security Applications Conference*, San Diego, California, USA, December 1997.
8. S. Cheung and K. Levitt. Protecting routing infrastructure from denial of service using cooperative intrusion detection. In *Proceedings of New Security Paradigms Workshop*, Great Langdale, Cumbria, UK, September 1997.
9. B.P. Crow, I.K. Widjaja, G. Jeong, and P.T. Sakai. IEEE 802.11 Wireless Local Area Networks. *IEEE Communications Magazine*, vol. 35, No. 9: pages 116-126, September 1997.
10. A. Habib, M. Hefeeda, and B. Bhargava. Detecting Service Violations and DoS Attacks. In *Proceedings of 2003 Internet Society Symposium on Network and Distributed System Security (NDSS'03)*, San Diego, California, USA. February 2003.
11. Y.C. Hu and D.B. Johnson. Caching Strategies in On-Demand Routing Protocols for Wireless Ad Hoc Networks. In *Proceedings of the Sixth Annual IEEE/ACM International Conference on Mobile Computing and Networking (MobiCom 2000)*, pages 231-242, August 2000.
12. Y.C. Hu, A. Perrig, and D.B. Johnson. Ariadne: A Secure On-Demand Routing Protocol for Ad Hoc Networks. In *Proceedings of the Eighth ACM International Conference on Mobile Computing and Networking (MobiCom 2002)*, September 23-28, 2002.
13. Y.C. Hu, D.B. Johnson, and A. Perrig. Secure Efficient Distance Vector Routing Protocol in Mobile wireless Ad Hoc Networks. In *Proceedings of the Fourth IEEE Workshop on Mobile Computing Systems and Applications (WMCSA 2002)*, June 2002.
14. Y.C. Hu, A. Perrig, and D.B. Johnson. Efficient Security Mechanisms for Routing Protocols. In *Proceedings of 2003 Internet Society Symposium on Network and Distributed System Security (NDSS'03)*, San Diego, California, USA. February 2003.
15. D. Johnson and D.A. Maltz. Dynamic Source Routing in Ad Hoc Wireless Networks. In *Mobile Computing*, chapter 5, pages 153-181. Kluwer Academic Publishers, 1996.
16. D. Johnson, D.A. Maltz, Y.C. Hu, and J.G. Jetcheva. The Dynamic Source Routing Protocol for Mobile Ad Hoc Networks (Internet-Draft). Mobile Ad-hoc Network (MANET) Working Group, IETF, February 2002.
17. G.M. Jones. The Case for Network Infrastructure Security. ;logon: The Magazine of USENIX and SAGE. pages 25-29, Volume 27, Number 6, December 2002.
18. K. Fall and K. Varadhan, editors. The *ns* Manual (formerly *ns* Notes and Documentation). April 14, 2002. http://www.isi.edu/nsnam/ns/doc/index.html
19. S. Marti, T.J. Giuli, K. Lai, and M. Baker. Mitigating Routing Misbehavior in Mobile Ad Hoc Networks. In *Proceedings of the Sixth Annual ACM/IEEE International Conference on Mobile Computing and Networking (MOBICOM 2000)*, August 2000.
20. V.N. Padmanabhan and D. R. Simon. Secure Traceroute to Detect Faulty or Malicious Routing. In *ACM SIGCOMM Workshop on Hot Topic in Networks* (HotNets-I), October 2002.
21. C.E. Perkins, E. M. Royer, and S.R. Das. Ad Hoc On Demand Distance Vector (AODV) Routing (Internet-Draft). June 2002.
22. R. Perlman. Network Layer Protocols with Byzantine Robustness. PhD thesis, Massachusetts Institute of Technology, August 1988.

23. T. Roughgarden. Selfish Routing. PhD thesis, Cornell University, May 2002.
24. B.R. Smith and J.J. Garcia-Luna-Aceves. Securing the Border Gateway Routing Protocol. In *Proceedings of Global Internet 1996*. London, UK. November 1996.
25. B.R. Smith, S. Murthy, and J.J. Garcia-Luna-Aceves. Securing Distance-Vector Routing Protocols. In *Proceedings of 1997 Internet Society Symposium on Network and Distributed System Security (NDSS'97)*, San Diego, California, USA. February 1997.
26. M.G. Zapata and N. Asokan. Securing Ad Hoc Routing Protocols. In *Proceedings of the ACM Workshop on Wireless Security (WiSe 2002)*, September 2002.
27. Kan Zhang. Efficient Protocols for Signing Routing Messages. In *Proceedings of 1997 Internet Society Symposium on Network and Distributed System Security (NDSS'98)*, San Diego, California, USA, March 1998.

A New Framework for Building Secure Collaborative Systems in True Ad Hoc Network

Hans-Peter Bischof, Alan Kaminsky, and Joseph Binder

Rochester Institute of Technology, 102
Lomb Mermorial Dr, Rochester, NY 14623
{hpb,ark,jsb7834}@cs.rit.edu

Abstract. Many-to-Many Invocation (M2MI) is a new paradigm for building secure collaborative systems that run in true ad hoc networks of fixed and mobile computing devices. M2MI is useful for building a broad range of systems, including service discovery frameworks; groupware for mobile ad hoc collaboration; systems involving networked devices (printers, cameras, sensors); and collaborative middleware systems. M2MI provides an object oriented method call abstraction based on broadcasting. An M2MI invocation means "every object out there that implements this interface, call this method." M2MI is layered on top of a new messaging protocol, the Many-to-Many Protocol (M2MP), which broadcasts messages to all nearby devices using the wireless network's inherent broadcast nature instead of routing messages from device to device. In an M2MI-based system, central servers are not required; network administration is not required; complicated, resource-consuming ad hoc routing protocols are not required; and system development and deployment are simplified.

Keywords: Collaborative systems, peer-to-peer systems, distributed objects, decentralized key management, ad hoc networking, server-less networking.

Introduction

This paper describes a new paradigm, Many-to-Many Invocation (M2MI), for building secure collaborative systems that run in true ad hoc networks of fixed and mobile computing devices. M2MI is useful for building a broad range of systems, including service discovery frameworks; groupware for mobile ad hoc collaboration.

We also address encryption and decryption of M2MI method invocations and a describe a decentralized key management in ad hoc networks.

M2MI provides an object oriented method call abstraction based on broadcasting. An M2MI-based application broadcasts method invocations, which are received and performed by many objects in many target devices simultaneously. An M2MI invocation means "Everyone out there that implements this interface, call this method." The calling application does not need to know the identities of the target devices ahead of time, does not need to explicitly discover the target devices, and does not need to set up individual connections to the target devices. The calling device simply broadcasts method invocations, and all objects in the proximal network that implement those methods will execute them.

S. Pierre, M. Barbeau, and E. Kranakis (Eds.): ADHOC-NOW 2003, LNCS 2865, pp. 164–174, 2003.

As a result, M2MI offers these key advantages over existing systems:

- M2MI-based systems do not require central servers; instead, applications run collectively on the proximal devices themselves.
- M2MI-based systems do not require network administration to assign addresses to devices, set up routing, and so on, since method invocations are broadcast to all nearby devices. Consequently,
- M2MI is well-suited for an ad hoc networking environment where central servers may not be available and devices may come and go unpredictably.
- M2MI-based systems allow to decrypt an encrypt method invocations using session keys [9].
- M2MI-based systems do not need complicated ad hoc routing protocols that consume memory, processing, and network bandwidth resources [10]. Consequently,
- M2MI is well-suited for small mobile devices with limited resources and battery life.
- M2MI simplifies system development in several ways. By using M2MI's high-level method call abstraction, developers avoid having to work with low-level network messages. Since M2MI does not need to discover target devices explicitly or set up individual connections, developers need not write the code to do all that.
- M2MI simplifies system deployment by eliminating the need for always-on application servers, lookup services, code-base servers, and so on; by eliminating the software that would otherwise have to be installed on all these servers; and by eliminating the need for network configuration.

M2MI's key technical innovations are these:

- M2MI layers an object oriented abstraction on top of broadcast messaging, letting the application developer work with high-level method calls instead of low-level network messages.
- M2MI uses dynamic proxy synthesis to create remote method invocation proxies (stubs and skeletons) automatically at run time - as opposed to existing remote method invocation systems, which compile the proxies, offline and which must deploy the proxies ahead of time.

This paper is organized as follows: the next chapter describes the target environment for M2MI based systems; the following chapter discusses the M2MI paradigm followed by a chapter showing how M2MI can be used to develop applications and service discovery frameworks. The last two chapters discuss a dynamic fault tolerant key management system.

Target Environmnet

M2MI's target domain is ad hoc collaborative systems: systems where multiple users with computing devices, as well as multiple standalone devices like printers, cameras, and sensors, all participate simultaneously (collaborative); and systems where the various devices come and go and so are not configured to know about each other ahead of time (ad hoc). Examples of ad hoc collaborative systems include:

- Applications that discover and use nearby networked services: a document printing application that finds printers wherever the user happens to be, or a surveillance application that displays images from nearby video cameras.
- Collaborative middleware systems like shared tuple spaces [1].
- Groupware applications: a chat session, a shared whiteboard, a group appointment scheduler, a file sharing application, or a multiplayer game.

In many such collaborative systems, every device needs to talk to every other device. Every person's chat messages are displayed on every person's device; every person's calendar on every person's device is queried and updated with the next meeting time. In contrast to applications like email or web browsing (one-to-one communication) or web-casting (one-to-many communication), the collaborative systems envisioned here exhibit many-to-many communication patterns. M2MI is designed especially to support applications with many-to-many communication patterns, although it also supports other communication patterns.

Devices come and go as the system is running, the devices do not know each other's identities beforehand; instead, the devices form ad hoc networks among themselves.

M2MI is intended for running collaborative systems without central servers. In a wireless ad hoc network of devices, relying on servers in a wired network is unattractive because the devices are not necessarily always in range of a wireless access point. Furthermore, relying on any one wireless device to act as a server is unattractive because devices may come and go without prior notification. Instead, all the devices - whichever ones happen to be present in the changing set of proximal devices - act in concert to run the system.

The M2MI Paradigm

Remote method invocation (RMI) [7] can be viewed as an object oriented abstraction of point-to-point communication: what looks like a method call is in fact a message sent and a response sent back. In the same way, M2MI can be viewed as an object oriented abstraction of broadcast communication. This section describes the M2MI paradigm at a conceptual level.

Handles

M2MI lets an application invoke a method declared in an interface. To do so, the application needs some kind of "reference" upon which to perform the invocation. In M2MI, a reference is called a handle, and there are three varieties, omnihandles, unihandles, and multihandles.

Omnihandles

An omnihandle for an interface stands for "every object out there that implements this interface." An application can ask the M2MI layer to create an omnihandle for a certain interface X, called the omnihandle's target interface. (A handle can implement more than one target interface if desired. An omnihandle for interface `Foo`; the omnihandle is named `allFoos` is created by code like this:

```
Foo allFoos = M2MI.getOmnihandle(Foo.class);
```

Once an omnihandle is created, calling method `doSomething` on the omnihandle for interface `AnInterface` means, "Every object out there that implements interface `AnInterface`, perform method `doSomething`." The method is actually performed by whichever objects implementing interface AnInterface exist at the time the method is invoked on the omnihandle. Thus, different objects could respond to an omnihandle invocationat different times. Three objects implementing interface `Foo`, objects A, B, and D, happen to be in existence at that time; so all three objects perform method y. Note that even though object D did not exist when the omnihandle `allFoos` was created, the method is nonetheless invoked on object D.

The target objects invoked by an M2MI method call need not reside in the same process as the calling object. The target objects can reside in other processes or other devices. As long as the target objects are in range to receive a broadcast from the calling object over the network, the M2MI layer will find the target objects and perform a remote method invocation on each one.

Exporting Objects

To receive invocations on a certain interface X, an application creates an object that implements interface X and exports the object to the M2MI layer. Thereafter, the M2MI layer will invoke that object's method Y whenever anyone calls method Y on an omnihandle for interface X. An object is exported with code like this:

```
M2MI.export(b, Foo.class);
```

`Foo.class` is the class of the target interface through which M2MI invocations will come to the object. We say the object is "exported as type Foo." M2MI also lets an object be exported as more than one target interface. Once exported, an object stays exported until explicitly unexported:

```
M2MI.unexport(b);
```

In other words, M2MI does not do distributed garbage collection (DGC). In many distributed collaborative applications, DGC is unwanted; an object that is exported by one device as part of a distributed application should remain exported even if there are no other devices invoking the object yet. In cases where DGC is needed, it can be provided by a leasing mechanism explicit in the interface.

Unihandles

A unihandle for an interface stands for "one particular object out there that implements this interface." An application can export an object and have the M2MI layer return a unihandle for that object. Unlike an omnihandle, a unihandle is bound to one particular object at the time the unihandle is created. A unihandle is created by code like this:

```
Foo b_Foo = M2MI.getUnihandle(b,Foo.class);
```

Once a unihandle is created, calling method Y on the unihandle means, "The particular object out there associated with this unihandle, perform method Y." When the statement b_Foo.y(); is executed, only object B performs the method. As with an omnihandle, the target object for a unihandle invocation need not reside in the same process or device as the calling object.

A unihandle can be detached from its object, after which the object can no longer be invoked via the unihandle:

```
b_Foo.detach();
```

Multihandles

A multihandle for an interface stands for "one particular set of objects out there that implement this interface." Unlike a unihandle which only refers to one object, a multihandle can refer to zero or more objects. But unlike an omnihandle which automatically refers to all objects that implement a certain target interface, a multihandle only refers to those objects that have been explicitly attached to the multihandle.

The multihandle is named someFoos, and it is attached to two objects, A and D. The multihandle is created and attached to the objects by code like this:

```
Foo someFoos = M2MI.getMultihandle(Foo.class);
someFoos.attach(a); someFoos.attach(d);
```

Once a multihandle is created, calling method Y on the multihandle means, "The particular object or objects out there associated with this multihandle, perform method Y." When the statement someFoos.y(); is executed, objects A and D perform the method, but not objects B or C. As with an omnihandle or unihandle, the target objects for a multihandle invocation need not reside in the same process or device as the calling object or each other.A multihandle can be created in one process and sent to another process, and the destination process can then attach its own objects to the multihandle.

An object can also be detached from a multihandle:

```
someFoos.detach(a);
```

M2MI-Based Systems

This section gives one examples showing how M2MI can be used to design a chat application and a print service discovery system. These examples show the elegance of ad hoc collaborative systems based on M2MI. Further examples can be found at [4].

Service Discovery – Printing

As an example of an M2MI-based system involving stand-alone devices providing services, consider printing. To print a document from a mobile device, the user must discover the nearby printers and print the document on one selected printer. Printer discovery is a two-step process: the user broadcasts a printer discovery request via an omnihandle invocation; then each printer sends its own unihandle back to the user via a unihandle invocation on the user. To print the document, the user does an invocation on the selected printer's unihandle.

Specifically, each printer has a print service object that implements this interface:

```
public interface PrintService {
    public void print(Document doc);
}
```

The printer exports its print service object to the M2MI layer and obtains a unihandle attached to the object. The printer is now prepared to process document printing requests. To discover printers, there are two print discovery interfaces:

```
public interface PrintDiscovery {
    public void request(PrintClient client);
}
public interface PrintClient {
    public void report(PrintService printer,
                       String name);
}
```

The client printing application exports a print client object implementing interface PrintClient to the M2MI layer and obtains a unihandle attached to the object. The application also obtains from the M2MI layer an omnihandle for interface PrintDiscovery. The application is now prepared to make print discovery requests and process print discovery reports.

Each printer exports a print discovery object implementing interface PrintDiscovery to the M2MI layer. The printer is now prepared to process print discovery requests and generate print discovery reports
The application first calls

```
printDiscovery.request(theClient);
```

on an omnihandle for interface PrintDiscovery, passing in the unihandle to its own print client object. Since it is invoked on an omnihandle, this call goes to all the printers. The application now waits for print discovery reports.
Each printer's request method calls

```
theClient.report(thePrinter,
                 "Printer Name");
```

The method is invoked on the print client unihandle passed in as an argument. The method call arguments are   the unihandle to the printer's print service object and the name of the printer. Since it is invoked on a unihandle, this call goes just to the requesting client printing application, not to any other print clients that may be present. After executing all the report invocations, the printing application knows the name of each available printer and has a unihandle for submitting jobs to each printer.

Finally, after asking the user to select one of the printers, the application calls:

```
c_Printer.print(theDocument);
```

where `c_Printer` is the selected printer's unihandle as previously passed to the report method. Since it is invoked on a unihandle, this call goes just to the selected printer, not the other printers. The printer proceeds to print the document passed to the print method.

Clearly, this invocation pattern of broadcast discovery request - discovery responses - service usage can apply to any service, not just printing. It is even possible to define a generic service discovery interface that can be used to find objects that implement any interface, the desired interface being specified as a parameter of the discovery method invocation.

M2MI Architecture

Our initial work with M2MI has focused on networked collaborative systems. In this environment of ad hoc networks of proximal mobile wireless devices, M2MI is layered on top of a new network protocol, M2MP. We have implemented initial versions of M2MP and M2MI in Java. Are detailed description of the design and architecture can be found at [4].

M2MI Security

Providing security within M2MI-based systems is an area of current development. We have identified these general security requirements:

- Confidentiality - Intruders who are not part of a collaborative system must not be able to understand the contents of the M2MI invocations.
- Participant authentication - Intruders who are not authorized to participate in a collaborative system must not be able to perform M2MI invocations in that system.
- Service authentication - Intruders must not be able to masquerade as legitimate participants in a collaborative system and accept M2MI invocations. For example, a client must be assured that a service claiming to be a certain printer really is the printer that is going to print the client's job and not some intruder.

While existing techniques for achieving confidentiality and authentication work well in an environment of fixed hosts, wired networks, these techniques will not work well in an environment of mobile devices, wireless networks, and no central servers.

A decentralized key management is necessary n order to achieve the security requirements.

Decentralized Keymanagement in Ad Hoc Networkd

State of the Art

Key management has been the thrust of several research initiatives in the ad hoc networking domain (e.g., [1, 6] et al). Each of these approaches seeks to establish a public key infrastructure within the constraints of ad hoc networks; each approach is discussed below.

"Securing Ad Hoc Networks" [10] was one of the first notable publications to propose a public key management service for ad hoc networks. The service itself encapsulates a public/private key pair K/k. The private key, k, is used to sign other nodes' public keys; the public key, K, is used to verify the signature. The service employs a (n, t+1) threshold scheme to distribute the private key and the digital signing process among n nodes. Each of the n nodes is denoted as a server node, as it has a special role in the signing service. Combiner nodes - which may be a subset of the server nodes or altogether different nodes - are also required to combine each server's partial signature. For example, to sign a certificate, each of the n server nodes must generate a partial signature using its share of the private (k_1, k_2, ... k_n) to compute a partial signature of the certificate. Once generated, each server node sends its partial signature to the combiner; the combiner then computes the entire signature. To its credit [10] was quite progressive at its inception, as its design is largely proactive and capable of handling a dynamic network state. Nonetheless, the service has remnants of its wired predecessor, namely, a trusted authority, and specialized server and combiner nodes. Although the threshold scheme employed allows $t < n$ servers to be compromised without sacrificing the service, its largely centralized approach encapsulates relatively few points of failure and attack.

"Providing Robust and Ubiquitous Security Support for Ad Hoc Networks [6] presents a natural extension to [1], wherein the signing service is distributed to any node n the network. For example, if a network member requires a certificate, it need only be in the proximity of *any* t+1 nodes. The service is otherwise similar to [6]. Despite the improved distribution, [6] still requires a trusted party at initialization. Further, because any node in the network may participate in the sharing, a malicious node may masquerade as t+1 bogus nodes and reconstruct the private key.

More recently, Hubaux et al have proposed a self organizing public key infrastructure in [1]. Unlike the previous two publications, [1] does not require a trusted authority or any specialized nodes; instead, each node issues its own certificates to other nodes. Each node maintains a limited repository of other nodes' certificates. When a node wishes to validate a certificate of another node, the nodes combine their certificate repositories; the validating node then examines the merged certificate repository for falsified certificates. If none are found, the certificate is accepted; otherwise it is rejected. The primary drawback of [1] is its initialization time. In long-lived ad hoc networks, such overhead may be admissable; it is likely to be prohibitive in more transient settings.

Although each of the above paradigms is effective in its own right, they are all based on a common assumption, namely, point-to-point communication. Public key infrastructures enable nodes with authentic public encryption keys that they may use to establish secure communication with one another. However, many ad hoc networks are collaborative, many-to-many environments. In these settings, public key cryptography is computationally intensive, as each group message must be encrypted *n-1* times. Group key management paradigms which provide a shared symmetric key that is shared among all group members, have been used throughout the wired networking domain to secure broadcast and many-to-many communication environments; however, very few attempts have been made to adapt group key management infrastructures to an ad hoc setting.

Dominant group key management paradigms include the well-known CLIQUES project [8], Kim et al [5], and several others. Each of these protocols is based on the generalized Diffie-Hellman problem, which requires every network member to contribute to the generation of the shared group key. Because they were developed for wired environments, many of these approaches require point-to-point and broadcast mediums, synchronous messaging, and static network topologies. Unfortunately, the wireless, amorphous, transient, many-to-many nature of ad hoc networks precludes many of the assumptions on which the above protocols were developed. We, therefore, introduce a new approach to key management that can effectively function within the constraints of an ad hoc network environment.

Looking Forward

The ad hoc network environment we envision is transient, dynamic in structure and membership, proximal, and broadcast-based. We also assume that network nodes wish to collaborate, that is, our primary goal is to ensure secure many-to-many communication. As a result, our paradigm is fully decentralized (i.e., it lacks server or otherwise specialized nodes), lightweight, and best-suited for small, spontaneous networks. The first protocol we present is not authenticated; the second is an extension of the first that includes authentication mechanisms.

The nucleus of our first protocol is a tuple-like entity, inspired by Gelerntner's tuples in [2], that is effectively a hash table shared among all members of the group. Each member of the group has an entry in the hash table, which includes that member's contribution to the group key.

The following atomic operations may be performed on the tuple:

- *take()* - removes the tuple from the space, such that no other group member may modify its contents.
- *read()* - reads the current contents of the tuple
- *write()* - writes the tuple into the space, overwriting the previous tuple

Although the tuple spaces are often implemented as a centrally-based service, the tuples used in this context are fully distributed: each member hosts its own entry in the tuple. Nodes may host more than one entry if replication is desired in the interest of availability.

Group Genesis

Group genesis requires two or more parties to be present.

1. Group members agree on a cyclic group, G, of order q, and a generator, α in G; each member then chooses a secret share, $N_i \in G$.
2. The first member, M_1, instantiates a Tuple Space and places a new tuple in the space. The tuple initially contains M_i's contribution and the current cardinal value. M_i then sends a broadcast message to the group stating that tuple has been created.

3. Upon receipt of the broadcast message, each member attempts to remove the tuple from the space in order to add its contribution. Because *take()* request will withdraw the tuple from the space; the other *take()* will block until the tuple is returned to the space. The member who receives the tuple then adds an entry in the tuple for itself and updates all existing intermediate values and the cardinal value. This step is repeated until $M_2 \ldots M_{n-1}$ have written their contributions into the tuple.

4. The last member of the group has special role in the key generation process. The last member is not pre-determined; it is simply the last member to send a *take()* request. M_n first performs a *take()* operation on the tuple. It then exponentiates each intermediate value in the tuple with its secret exponent, Sn, and adds in an intermediate value for itself. Unlike its predecessors, M_n does not update the cardinal value, as the final cardinal value is the group key. Instead, it writes the tuple back into the space with the previous cardinal value and the updated intermediate values. Mn then sends a broadcast message to the group, which informs them of the termination of the key generation phase.

Upon receipt of the broadcast message, each member *read()*s its intermediate value and uses it to compute the group key.

Member Addition – *join()*
A *join()* operation denotes the addition of a single group member. Semantics for *join()* entail a modification of the group key, such that the new member's share is included in the group key. The steps required for *join()* follow.

1. M_{n+1} *take()*s the tuple out of the space, adds its intermediate value, updates each existing intermediate values, and *write()*s the tuple back into the space.

2. M_{GC} performs a *take()* on the tuple, updates the cardinal value, *write()*s the tuple back into the space, and notifies all group members that the key generation is complete.

Following a *join()* operation, the new member becomes new group controller (i.e., $M_{n+1} = M_{GC}$).

By default, join does not ensure forward or backward secrecy. In many scenarios, this may be admissable; however, a simple extension to the join operation can ensure forward and backward secrecy. The revised protocol requires the existing group controller, M_n, factor its secret, S_n out of the existing cardinal and intermediate values, choose a new secret, S_n, and exponentiate each intermediate value with it.

Member Removal - *leave()*
Leave entails the removal of a group member's contribution to the group key, thereby prohibiting it from decrypting subsequent group messages. The following protocol assumes that the departure is voluntary. If the departure is not voluntary, the first step is clearly omitted, however, the excluded member is still unable to derive the group key.

1. The departing member, M_p, factors its contribution out of each entry in the tuple.

2. The group controller, M_{GC}, chooses a new secret S_{GC} and exponentiates each entry in the tuple with it.

Conclusion

We present a dynamic, fault--tolerant symmetric key management system. Unlike other key management paradigms, our approach does not require a specific order in which contributions are collected, nor does it rely on a trusted or centralized entity to combine the partial keys.

References

1. S. Capkun, L. Buttyan, and J. Hubaux. Self-organized public-key management for mobile ad hoc networks, 2002.
2. D. Gelernter. Generative communication in Linda. *ACM Transactions on Programming Languages and Systems*, 7(1):80-112, January 1985.
3. Internet Engineering Task Force. IP Routing for Wireless/Mobile Hosts (mobileip) Working Group. http://www.ietf.org/html.charters/mobileip-charter.html.
4. A. Kaminsky, Hans-Peter Bischof. Many-to-Many Invocation: A new object oriented paradigm for ad hoc collaborative systems. 17th Annual ACM Conference on *Object Oriented Programming Systems, Languages, and Applications (OOPSLA 2002)*, Onward! track, November 2002, to appear. Preprint:
   http://www.cs.rit.edu/~anhinga/publications/publications.shtml
5. Yongdae Kim, Adrian Perrig, and Gene Tsudik. Simple and fault-tolerant key agreement for dynamic collaborative groups. In Proceedings of the 7th ACM conference on Computer and communications security, pages 235244, 2000
6. H. Luo and S. Lu. Ubiquitous and robust authentication services for ad hoc wireless networks, 2000.
7. Michael Steiner, Gene Tsudik, and Michael Waidner. CLIQUES: A new approach to group key agreement. In Proceedings of the 18th International Conference on Distributed Computing Systems (ICDCS98), pages 380387, Amsterdam, 1998. IEEE Computer Society Press.
8. Jefferson S, Tuttle. Security in an Ad Hoc Network using Many-to-Many Invocation, http://www.cs.rit.edu/~jst1734
9. A. Wollrath, R. Riggs, and J. Waldo. A distributed object model for the Java system. Computing Systems, 9(4):265-290, Fall 1996.
10. S.-M. Yoo and Z.-H. Zhou. All-to-all communication in wireless ad hoc networks. In Proceedings of the 39th Annual *ACM Southeast Conferenc*e, pages 180-181, March 2001. http://webster.cs.uga.edu/~jam/acm-se/review/abstract/syoo.ps.
11. Lidong Zhou and Zygmunt J. Haas. Securing ad hoc networks. IEEE Network, 13(6):2430, 1999.

Computing 2-Hop Neighborhoods
in Ad Hoc Wireless Networks

Gruia Calinescu

Department of Computer Science, Illinois Institute of Technology, Chicago, IL 60616
calinesc@iit.edu

Abstract. We present efficient distributed algorithms for computing 2-hop neighborhoods in Ad Hoc Wireless Networks. The knowledge of the 2-hop neighborhood is assumed in many protocols and algorithms for routing, clustering, and distributed channel assignment, but no efficient distributed algorithms for computing the 2-hop neighborhoods were previously published.

The problem is nontrivial, as the graphs induced by ad-hoc wireless networks can be dense. We employ the broadcast nature of the wireless networks to obtain a distributed algorithm in which every node gains knowledge of its 2-hop neighborhood using a total of $O(n)$ messages, where n is the total number of nodes in the network, and each message has $O(\log n)$ bits, which we assume is enough to encode the ID and the geographic position of a node. Our algorithm operates in an asynchronous environment, and makes use of the geographic position of the nodes.

A more complicated algorithm achieves the same communication bounds when geographical positions are not available, but nodes are capable of evaluating the distance to neighboring nodes or the angle of signal arrival. We also discuss updating the knowledge of 2-hop neighborhoods when nodes join or leave the network.

1 Introduction

Wireless ad hoc networks can be flexibly and quickly deployed for many applications such as automated battlefield, search and rescue, and disaster relief. Unlike wired networks or cellular networks, no physical backbone infrastructure is installed in wireless ad hoc networks. A communication session is achieved either through a single-hop radio transmission if the communication parties are close enough, or through relaying by intermediate nodes otherwise.

In this paper, we assume that all nodes in a wireless ad hoc network are distributed in a two-dimensional plane and have an equal maximum transmission range of one unit. The topology of such wireless ad hoc network can be modeled as a *unit-disk graph, or UDG* (see [11] for many interesting properties of unit-disk graphs), a geometric graph in which there is a link between two nodes if and only if their distance is at most one.

The 1-hop neighborhood of a node v (denoted by $N_1(v)$) is simply the set of nodes adjacent to it in the UDG. We use $N_2(v)$ to denote the set of nodes

S. Pierre, M. Barbeau, and E. Kranakis (Eds.): ADHOC-NOW 2003, LNCS 2865, pp. 175–186, 2003.

of the UDG 2-hops away from v. The *2-hop neighborhood* of v is the bipartite graph with node set $N_1(v) \cup N_2(v)$ in which all the links of the UDG with one endpoint in $N_1(v)$ and the other endpoint in $N_2(v)$ are included.

Knowledge of the 2-hop neighborhoods is assumed in many distributed algorithm and protocols such as constructing structures [24,6], improved routing [20], broadcasting [9], and channel assignment [3]. The clusters used for channel control typically have diameter at most two [19]. The knowledge of the set of 2-hop neighbors is helpful in frequency assignment to avoid secondary interference. Also distributed algorithms for $L(2,1)$-Labeling ([12,8,10]) can use the information about 2-hop neighborhoods stored by every node. Knowledge of the 2-hop neighborhood can be used for efficient computation of multipoint relays, used for example in [14].

Our distributed algorithms operate in an asynchronous environment, and we use the number of messages as the measure of the efficiency of the algorithm. In our model a message can hold the ID of a node, the geographical position of a node, and $O(\log n)$ bits, where n is the total number of nodes in the network. Concentrating on the number and the length of the messages is justified by the limited resources available to wireless nodes. We assume nodes have $O(n)$ memory available.

In this model, computing the set of 1-hop neighbors with $O(n)$ messages is trivial: every node broadcasts a message announcing its ID. One can easily compute the 2-hop neighborhood with $O(n)$ messages of size $\Delta \log n$ each, where Δ is the maximum number of 1-hop neighnors. But we insist on messages of size $O(\log n)$ each, and therefore, as UDGs can be dense, computing the 2-hop neighborhood is not trivial.

The broadcast nature of the communication in ad hoc wireless networks is however very useful when computing local information. To our knowledge no distributed algorithm for computing 2-hop neighborhoods has been previously proposed and analyzed.

First we assume that each static wireless node knows its position information, either through a low-power Global Position System (GPS) receiver or through some other ways. Then to construct the 2-hop neighborhoods it is enough to know the IDs and positions of the 1-hop and 2-hop neighbors. With these assumptions, we present a simple distributed algorithm which allows every node to compute the positions of its 2-hop neighbors. The total number of $O(\log n)$-bit messages of the algorithm is $O(n)$.

Second, we assume that position information is not available, but every two adjacent nodes are capable of estimating their pairwise distance. Probing - lowering the transmission power over an interval of time - is one way which allows the computation of pairwise distances. A detailed discussion of location systems appears in [13]. Under this assumption, we present a distributed algorithm which allows every node to compute its 2-hop neighborhood. The total number of $O(\log n)$-bit messages of the algorithm is $O(n)$. The algorithm is based on triangulation and can be immediatly updated to work when the angle-of-arrival information is available (an assumption justified in [18] or [15]) instead of pairwise distances.

Our approach is based on the specific connected dominating set introduced by Alzoubi, Wan, and Frieder [2,21]. This connected dominating set is based on a maximal independent set (MIS), whose role in algorithms for unit-disk graphs was discovered by Marathe et. al [16]. An MIS is a dominating set: every node must have a 1-hop neighbor in the maximal independent set. In our algorithm, each node uses its adjacent node(s) in the MIS to broadcast over a larger area relevant information. Listening to the information about other nodes broadcast by the MIS nodes enables a node to compute its 2-hop neighborhood. There is a direct (without using a MIS) solution when node positions are available, but it is more complicated and requires synchronization in order to achieve $O(n)$ messages each of size $O(\log n)$ bits.

The example in Figure 1 shows that $\Theta(n/\log n)$ time is sometimes necessary for computing 2-hop neighborhoods (assuming one "step" allows the transmission of $O(\log n)$ bits), as the center node has to transmit $\Theta(n)$ bits to show the existance (or non-existance) of each node on one side to the nodes on the other side. This justifies our concentration on communication complexity, and not time complexity. And while our algorithms use heavily the nodes in the connected dominating set, the same example shows that overloading certain nodes is sometimes unavoidable.

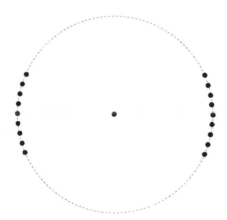

Fig. 1. The center node of this disk of radius 1 must send $\Theta(n)$ bits to allow the correct computation of the 2-hop neighborhoods

We also describe a straightforward procedure of updating the 2-hop neighborhoods when nodes join or leave the network. When leaving the network, the communication cost is $O(\log n)$ bits. When joining the network, the number of messages is bounded by a small constant times the number of nodes in the 2-hop neighborhood of the new node.

The paper is organized as follows. The next section clarifies the notation and explains the properties of the connected dominating set our algorithms use.

Section 3 describe the algorithm for the situation when geographic position is available. Section 4 describes the generalization of the algorithm to the situation when only pairwise distance in between adjacent nodes is available. Section 5 describes the recomputation of 2-hop neighborhoods due to changes in the network configuration. We conclude with Section 6.

2   Preliminaries

In this paper by broadcast we understand local broadcast - a packet send by a node, and received by every other node within the transmission range.

Recently [2,21] introduced a virtual backbone of the network, and our algorithms make heavy used of this virtual backbone. The next subsection quickly reproduces their construction, and lists the important properties of the virtual backbone.

2.1   The Virtual Backbone

The virtual backbone is a connected dominating set in the UDG. It is based on a maximal independent set (MIS), and we call the nodes in the maximal independent set *MIS nodes*. MIS nodes cannot be 1 hop away; if two MIS nodes are two or three hops away, we call them *virtually-adjacent*. One or two *connector* nodes are used to establish a path corresponding to a pair of virtually-adjacent MIS nodes. A node can participate as a connector for several pairs of virtually-adjacent MIS nodes. Only the links in between a connector node and the MIS nodes it connects, or in between two connector nodes which together establish the path corresponding to a pair of virtually-adjacent MIS nodes are added to the virtual backbone.

In [2,21] it is shown how the virtual backbone (including adding the connector nodes) can be constructed distributely with $O(n)$ messages, where the message size is $O(\log n)$ bits. They also show how to maintain the virtual backbone when the topology of the network changes.

Wan et. al. [2,21] proved that the virtual backbone is connected. Using an area argument, [2,21] proved that within three hops of an MIS node there could be at most 47 MIS nodes, and therefore the maximum degree of the virtual backbone is bounded by a constant we call Δ. Please refer to Figure 2 for intuition on the virtual backbone described above.

It was first proved in [16] that the size of any maximal independent set is at most five times the minimum dominating set in the UDG, as in fact for any node x can have at most five neighbors in an MIS. Alzoubi et al. [2,21] noticed that their virtual backbone is also within a constant the size of the minimum connected dominating set.

In addition, it is immediate that the virtual backbone of [2,21], together with links from every node to an MIS node adjacent to it, is a hop-spanner. Precisely, for every path in the UDG, there is a path on the virtual backbone with at most three times as many links from an MIS node adjacent to the origin of the path

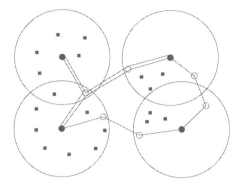

Fig. 2. An illustration of the virtual backbone of Alzoubi, Wan, and Frieder. The solid round nodes are the MIS node, which form a dominating set. Two virtually-adjacent MIS nodes are connected by paths of length at most three through connector nodes - the empty tiny circles in the figure. Nodes not in the virtual backbone are small solid squares in the figure

to an MIS node adjacent to the destination of the path. This fact was noticed by Alzoubi [1], and by Wang and Li [22], which also planarize the virtual backbone while keeping all its attractive properties.

3   Geographic Position Available

In this section we describe the distributed algorithm which allows every node to construct the list of its 2-hop neighbors, assuming every node knows its geographical position. With this information, every node can also easily compute the links between its 1-hop and 2-hop neighbors. Our algorithm is described in the simplest version, and we do not try to optimize the constant hidden in the O notation.

We start from the moment the virtual backbone is already constructed, and every node knows the ID and the position of its neighbors. The idea of the algorithm is for every node to efficiently announce its ID and position to a subset of nodes which includes its 2-hop neighbors.

The responsibility for announcing the ID and position of a node v is taken by the MIS nodes adjacent to v. Each such MIS node assembles a packet containing: $< ID, position, counter >$, with the ID and position of v, and a counter variable being set to 2. The MIS node then broadcasts the packet.

A connector node is used to establish a link in between several pairs of virtually-adjacent MIS nodes, and will not retransmit packets which do not travel in between these pairs of MIS nodes. The connector node will rebroadcast packets with nonzero *counter* originated by one of the nodes in a pair of virtually-adjacent MIS nodes, thus making sure the packet advances towards the other MIS node in the pair. Recall that the path in between a pair of virtually-adjacent MIS nodes has one or two connector nodes.

When receiving a packet of type $< ID, position, counter >$, an MIS node checks whether this is the first message with this ID, and if yes decreases the *counter* variable and rebroadcasts the packet.

A node listens to the packets broadcasted by all the adjacent MIS nodes (here it is convenient to assume a MIS is adjacent to itself), and, using its internal list of 1-hop neighbors, checks if the node announced in the packet is a 2-hop neighbor or not - thus constructing the list of 2-hop neighbors.

Theorem 1. *When finished, the algorithm described above correctly computes the 2-hop neighborhood for every node in the network, and uses $O(n)$ messages of size $O(\log n)$ each.*

Proof. The fact that the virtual backbone is a bounded-degree hop-spanner essentially implies the correctness of the algorithm. The precise argument is as follows. Assume nodes v and u share a neighbor x, and let $\bar{v}$, $\bar{u}$, and $\bar{x}$ be nodes in MIS which are adjacent to v, u, and x. Then $\bar{v}$ creates a packet with the ID and position of v, and with its counter set to 2. As $\bar{v}$ and $\bar{x}$ are virtually-adjacent, $\bar{x}$ will receive the packet and retransmit it with counter set to 1. As $\bar{x}$ and $\bar{u}$ are virtually-adjacent, $\bar{u}$ will also broadcast the packet, and therefore u finds out the ID and position of v.

Regarding the number of messages, we count the packets announcing the ID and position of x. Such packets are being sent by S_1, the MIS nodes adjacent to x, and we recall that $|S_1| \leq 5$. They are also sent by S_2, the MIS nodes virtually-adjacent to S_1, by S_3, the MIS nodes virtually-adjacent to S_2, and by the connector nodes in between pairs of virtually-adjacent MIS nodes inside $S_1 \cup S_2$, and by the connector nodes in between virtually-adjacent MIS nodes of S_2 and S_3. Thus the total number of nodes retransmitting packets announcing ID and position of x is $O(\Delta^2)$. As Δ, the maximum degree of the virtual backbone is constant, the total number of messages is $O(n)$.   □

We remark that with the counter of a packet being initially set to k (and decreased by one whenever a MIS node retransmits), the same argument as above implies that with $O(\Delta^k)$ messages every node can compute its k-hop neighborhoods.

4   Pairwise Distances Available

In this section we assume that neighboring nodes can compute their pairwise distance, but are not aware of their precise geographical position.

Our approach is based on the virtual backbone used before and *rigid pieces*, which we define as subgraphs containing one MIS node and a subset of its neighbors such that a system of coordinates can be locally established and in which the position of every node of the rigid piece is completely defined. A theory of geometric rigidity is well established [23]. We need only simple properties which are easily proved below.

First we describe the distributed algorithm for computing the rigid pieces. Before the actual construction, every node announces all the MIS nodes to which it is adjacent, and records the information transmitted by all its neighbors.

Every MIS node v constructs one after the other the rigid pieces in which it participates, and ensures these pieces are disjoint with the exception of v. Each such piece will have an ID, composed of the ID of the unique MIS which is in the piece and an integer in between 1 and 18. Once a node is assigned to a piece together with v, it announces in a broadcast message the ID of the rigid piece and its coordinates with respect to the rigid piece.

Let us describe the construction of one such rigid piece. The MIS node v always has coordinates $(0,0)$ with respect to the rigid piece. If all nodes adjacent to v are in a rigid piece with v, the procedure stops. Otherwise, v selects a neighbor x which is not in a rigid piece with v, and asks x to announce its participation in the rigid piece and its coordinates with respect to the rigid piece: $(\|xv\|, 0)$. Every node y adjacent to both v and x and not yet in some other rigid piece with v, computes its coordinates with respect to v and x based on the length of the sides of the triangle xyv. Actually, while the first coordinate of y is unique, the second one is not: only its absolute value can be computed exactly. If the angle $\widehat{yvx}$ is bigger than $\pi/3$, y will not participate in the rigid piece. If the second coordinate of y is 0, then y participates in the piece and announces its participation and its unique coordinates with respect to the rigid piece. If the angle $\widehat{yvx}$ is at most $\pi/3$ and the second coordinate of y is nonzero, y announces it is willing to participate in the piece. Node v will pick only one such y (assuming it exists), and announce that both of y's coordinates with respect to the rigid piece will be positive. See Figure 3 for intuition. At this moment y announces its participation in the rigid piece and its coordinates with respect to the rigid piece. Every node z adjacent to v, x, and y, and not yet in some other rigid piece with v, computes its unique coordinates with respect to the rigid piece, and announces its participation in the rigid piece and its coordinates.

The following theorem enumerates the important properties of the distributed algorithm described above.

Theorem 2. *Every non-MIS node is a member of at most five rigid pieces. Every MIS node is a member of at most 18 rigid pieces. Computing the nodes of a rigid piece and the coordinates with respect to the rigid piece of every node can be done with a number of messages bounded by a constant times the number of nodes adjacent to the MIS node in the piece. The total number of messages (each having $O(\log n)$ bits) until every node announces every piece in which it participates, together with its coordinates with respect to the rigid piece, is $O(n)$.*

Proof. Once we prove that a MIS node constructs at most 18 rigid pieces, the remaining assertions of the theorem follow from the description of the algorithm.

Let k be the number of rigid pieces constructed and let x_i be the first nodes selected by v when constructing the ith piece. Let y_i be the node picked by v as the first node of the rigid piece with nonzero second coordinate with respect to the ith rigid piece, if such a node exists. If y_i exists, define R_i be the sector of the unit disk centered at v consisting of the points z with angles $\widehat{zvx_i}$ and $\widehat{zvy_i}$ at most $\pi/3$. If y_i does not exists, let R_i be the sector of the unit disk centered at v

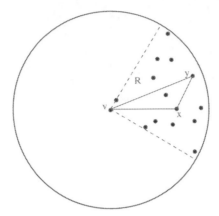

Fig. 3. The unnamed nodes in the figure can join the rigid piece started by v, x, and y. In the system of coordinates used, v is the origin, x has second coordinate 0, and y has the second coordinate positive. Notice that every node in the sector of the disk $R = R_i$ can join the rigid piece and that R covers at least 1/6 of the unit disk centered at v

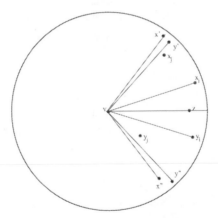

Fig. 4. There could be at most three sectors R_k which contain the point z: the first given by x_i and y_i (which are on opposite sides of the line vz) and then two sectors given by x' and y' (both above the line vz) and by $x"$ and $y"$ (both under the line vz). As shown in proof of Theorem 2, a fourth sector such as the one given by x_j and y_j cannot exists

consisting of the points z with angles $\widehat{zvx_i}$ at most $\pi/3$. Figure 3 again provides intuition.

We claim that any point z belongs to at most three sectors R_i, $1 \le i \le k$. See Figure 4 for intuition. Indeed, if there are $i < j$ with z in both R_i and R_j and x_i, y_i (if y_i exists), x_j and y_j (if y_j exists) are all on the same side of vz, then $x_j \in R_i$ as the angle $\widehat{x_ivx_j}$ is at most $\pi/3$ and $\widehat{y_ivx_j}$, if y_i exists, is also at most $\pi/3$. Therefore x_j should have entered rigid component i, a contradiction.

If both y_i and y_j exist, x_i and y_i are on different sides of vz, and x_j and y_j are also on different sides of vz, then we obtain a contradiction as follow. Assuming $i < j$ and x_i and x_j are on the same side of vz (the case when x_i and y_j are on the same side of vz is symmetric), we have that the angle $\widehat{y_i v x_j}$ is bigger than $\pi/3$, as otherwise x_j should have been taken in rigid piece i. Since the angle $\widehat{x_i v y_i}$ is at most $\pi/3$, we conclude that x_i is inside the angle $\widehat{z v x_j}$ and a symmetric argument yields that y_i is inside the angle $\widehat{z v y_j}$. Since $\widehat{y_i v x_j}$ is bigger than $\pi/3$, we conclude that $\widehat{x_j v y_j}$ is also bigger than $\pi/3$, a contradiction.

Thus only three sectors R_i can contain z: two with x_i on one side of vz and y_i nonexistent or on the same side as x_i, and one i with x_i and y_i on different sides of vz.

As any R_i covers at least $1/6$ of the unit disk centered at v, and any point belongs to at most three sectors, we conclude that there are at most 18 such sectors. This finishes the proof of Theorem 2. □

At this moment the rigid pieces are constructed, every node has announced its participation in the rigid pieces together with its coordinates with respect to that piece. We now describe the second phase of the distributed algorithm, in which every node v gives enough information to every of its 2-hop neighbors y to determine the fact that y is a 2-hop neighbor of v and which is the set of common neighbors. More precisely, for every rigid piece which intersects the 1-hop neighborhood of v, y will have enough information to compute which of its neighbors from the rigid piece are adjacent to v.

Every node records all the information passed by its neighbors. In the 1-hop neighborhood of a node v, there can be at most a constant number of rigid pieces, as the number of MIS nodes in the 1-hop and 2-hop neighborhood of v is bounded by 25: if we draw a disk of radius $1/2$ around every MIS node, we obtain disjoint disks of area $\pi/4$ which are included in the disk centered at v with radius $5/2$, whose area is $25\pi/4$ (this argument is also used in [2]).

Separately for every rigid piece which intersects the 1-hop neighborhood of v (including the rigid pieces containing v), v determines how many neighbors it has in the rigid piece. If v has at most two neighbors in the rigid piece, it ask all its neighbors in the rigid piece to announce they are neighbors with v.

If v has three non-coliniar neighbors in common with the rigid piece, using the distance to these three points, v can compute its coordinates with respect to the rigid piece. Then v asks its neighbors in MIS to announce its position with respect to the rigid piece. This is done exactly as the announcements in Section 3, with packets containing $< nodeID, pieceID, coordinates, counter >$. Any node receiving such a message evaluates whether it has neighbors in the rigid piece, and if yes the node computes the set of its neighbors from the rigid piece which are adjacent to v, based on their coordinates with respect to the rigid piece.

If v has three or more neighbors in common with a rigid piece, but they are coliniar, v cannot exactly compute its coordinates with respect to the rigid piece, but has exactly two possible value for its coordinates. Then v asks its neighbors in MIS to announce both positions with a packet containing

$< nodeID, pieceID, coordinates_1, coordinates_2, counter >$. Any node y receiving such a message evaluates if it has neighbors in the rigid piece, and if yes, as above, it can compute two sets S_1 and S_2 of nodes in the rigid piece which could be the common neighbors with v - assuming v has coordinates $coordinates_1$ or $coordinates_2$ with respect to the rigid piece. At least one of S_1 and S_2 is a set of colinear points, and if both are, they coincide. That set of colinear points is the set of neighbors common to y and v in the rigid piece.

All cases are taken care of and we conclude:

Theorem 3. *There is a distributed algorithm which, under the assumption that every node can estimate the distance to every adjacent node, computes for every node v the set of its 2-hop neighbors $N_2(v)$ and the links in between $N_1(v)$ and $N_2(v)$ with a total of $O(n)$ messages each of size $O(\log n)$ bits.*

5   Updating the 2-Hop Neighborhoods

In this section we discuss the message complexity of updating the 2-hop neighborhoods due to changes in network topology. We do not address updating the virtual backbone as this was done in [2]. The proposed protocol is straightforward and does not use the virtual backbone. We assume geographical knowledge is available in this section.

Before leaving the network, a node u uses its knowledge to let its 2-hop neighborhs know the fact it is leaving as described below. First the node u computes a maximal independent set (MIS) in the graph induced by its 2-hop neighborhs. Then u computes at most one "connector" node for each MIS node. As before, MIS is a dominating set, and using an area argument, has constant size. Node u prepares an $< ID, position, leaving, relay >$ message, with its own ID and position, the fact that it is leaving the network, and the full list of relay nodes. Each node, after receiving such a message, make a note that u is leaving and updates its 2-hop neighborhood accordingly, and, if it finds itself in the list of relay nodes, rebroadcast the message once.

When a node v joins the network, it will broadcasts its ID and position. Every existing node which receives this message will rebroadcast the ID and position of v. Every node y receiving such a message, will update its stored 2-hop neighborhood to reflect the presence of v. If y is adjacent to v, it will broadcast its ID and position. If y is a 2-hop neighbor of v, it selects a common neighbor x and asks x to relay to v the position and ID of y. The total bit complexity of message is $O(q \log n)$, where q is the size of the 2-hop neighborhood of v, and it cannot be improved by more than a constant factor since v must find out the IDs of the nodes in its 2-hop neighborhood.

6   Conclusions

The virtual backbone of Alzoubi, Wan, and Frieder [2,21] can be constructed without any geographical knowledge: their algorithm "operates" directly on the unit-disk graph. We need at least the distance in between any pair of adjacent nodes. Same arguments using rigid pieces apply when a node is able to compute the angle in between the segments to adjacent nodes. However, without any geographical knowledge we do not know whether it is possible to compute 2-hop neighborhoods with $O(n)$ messages each having $O(\log n)$ bits. This observation raises the interesting question whether there are any (meaningful) problems which have higher communication complexity on unit-disk graphs than on embedded (nodes aware of their geographical position) unit-disk graphs. Note that it is NP-Hard to recognize unit-disk graphs [7].

However, it follows from standard algebraic geometry results (page 542 of [17] or improved bounds in [4]) that the number of labeled unit-disk graphs of n nodes is between $2^{c_1 n \log n}$ and $2^{c_2 n \log n}$, for constants c_1 and c_2 and therefore a protocol with a total $O(n \log n)$ bits communication complexity is possible. An $O(n \log n)$ bits communication complexity would follow from a solution to an open problem in algebraic geometry [5]. It is worth mentioning that algebraic geometry solutions seem to have huge running time and space complexity.

Our model does not account for messages lost because of interference. It would be desirable to design synchronous distributed algorithms with low message complexity and low time complexity in a model where messages are lost either due to signal interference or due to node overloading.

Acknowledgments

The author thanks Peng-Jun Wan and Xiang-Yang Li, who inspired this paper by presenting their results. The author thanks Sougata Basu, Adrian Dumitrescu, and Peter Sanders for insight in the issue of extending the results to case when geographical knowledge is not available.

References

1. Khaled M. Alzoubi, "Distributed Algorithms for Connected Dominating Set in Wireless Ad Hoc Networks", Illinois Institute of Technology, 2002.
2. Khaled M. Alzoubi, Peng-Jun Wan and Ophir Frieder, "Message-Optimal Connected Dominating Sets in Mobile Ad Hoc Networks", in *ACM MOBIHOC '02*.
3. L. Bao and J. J. Garcia-Luna-Aceves, "Channel Access Scheduling in Ad Hoc Networks with Unidirectional Links", *5th International Workshop on Discrete Algorithms and Methods for Mobility*, 2001, Pages 9–18.
4. S. Basu "Different bounds on the different Betti numbers of semi-algebraic sets", to appear in *Discrete and Computational Geometry*. Available at http://www.math.gatech.edu/~saugata/.
5. S. Basu, R. Pollack, and M. F. Roy, Algorithms in Real Algebraic Geometry, Springer-Verlag, 2003.
6. V. Bharghavan and B. Das, "Routing in Ad Hoc Networks Using Minimum Connected Dominating Sets", *International Conference on Communications'97*, Montreal, Canada. June 1997.

7. Heinz Breu and David G. Kirkpatrick. "Unit disk graph recognition is NP-hard", *Computational Geometry. Theory and Applications*, 9:3–24, 1998.
8. T. Calamoneri and R. Petreschi, "L(2,1)-Labeling of Planar Graphs", *5th International Workshop on Discrete Algorithms and Methods for Mobility*, 2001, Pages 28–33.
9. G. Calinescu, I. Mandoiu, P.-J. Wan, and A. Zelikovsky, "Selecting Forwarding Neighbors in Wireless Ad Hoc Networks", *5th International Workshop on Discrete Algorithms and Methods for Mobility*, 2001, Pages 34–43.
10. I. Chlamtac and S. Pinter, "Distributed Nodes Organization Algorithm for Channel Access in a Multihop Dynamic Radio Network", *IEEE Trans. on Computers*, 36(6):728–737, 1987.
11. B. N. Clark, C. J. Colbourn, and D. S. Johnson, "Unit Disk Graphs", *Discrete Mathematics*, 86:165–177, 1990.
12. J. R. Griggs and R. K. Yeh, "Labeling Graphs with a Condition at Distance 2", *SIAM J. Disc. Math*, 5:586–595, 1992.
13. J. Hightower and G Borriello, "Location Systems for Ubiquitous Computing", *IEEE Computer*, vol. 34(8), 2001, pp 57–66.
14. P. Jacquet, A. Laouiti, P. Minet, and L. Viennot, "Performance analysis of OLSR multipoint relay flooding in two ad hoc wireless network models", in *RSRCP*, Special issue on Mobility and Internet, 2001.
15. K. Krizman, T. Bieda, and T. Rappaport, "Wireless position location: fundamentals, implementation strategies, and source of error", *Veh. Tech. Conf*, 1997, 919–923.
16. M. V. Marathe, H. Breu, H. B. Hunt III, S. S. Ravi and D. J. Rosenkrantz, "Simple Heuristics for Unit Disk Graphs", *Networks*, Vol. 25, 1995, pp. 59–68.
17. B. Mishra, "Computational Real Algebraic Geometry" in *Handbook of Discrete and Computational Geometry*, J. E. Goodman and J. O'Rourke (editors), CRC Press, 1997.
18. A. Nasipuri and K. Lim "A Directionality based Location Discovery Scheme for Wireless Sensor Networks", *WSNA 2002*.
19. S. Ramanathan and M. Steenstrup, "A survey of routing techniques for mobile communication networks", *ACM/Baltzer Mobile Networks and Applications*, 89–104, 1996.
20. I Stojmenovic and X. Lin, "Loop-free hybrid single-path/flooding routing algorithms with guaranteed delivery for wireless networks", *IEEE Transactions on Parallel and Distributed Systems*, 12:1023–1032, 2001.
21. Peng-Jun Wan, Khaled M. Alzoubi, and Ophir Frieder "Distributed Construction of Connected Dominating Set in Wireless Ad Hoc Networks", in *IEEE INFOCOM 2002*.
22. Yu Wang and Xiang-Yang Li, "Geometric Spanners for Wireless Ad Hoc Networks", in *ICDCS 2002*.
23. W. Whiteley "Rigidity and Scene Analysis", *Handbook of Discrete and Computational Geometry*, 893–916, ed. J. E. Goodman and J. O'Rourke, CRC Press, 1997.
24. J. Wu and H.L. Li, "On calculating connected dominating set for efficient routing in ad hoc wireless networks", *Proceedings of the 3rd ACM international workshop on Discrete algorithms and methods for mobile computing and communications*, 1999, Pages 7–14.

Topology Control Problems under Symmetric and Asymmetric Power Thresholds

Sven O. Krumke[1], Rui Liu[2], Errol L. Lloyd[2], Madhav V. Marathe[3],
Ram Ramanathan[4], and S.S. Ravi[5]

[1] Konrad-Zuse-Zentrum für Informationstechnik, Berlin (ZIB)
Takustraße 7, 14195 Berlin-Dahlem, Germany
krumke@zib.de
[2] University of Delaware, Newark, DE 19716
{ruliu,elloyd}@cis.udel.edu
[3] Los Alamos National Laboratory, MS M997, Los Alamos, NM 87545
marathe@lanl.gov
[4] Internetwork Research Department, BBN Technologies, Cambridge, MA 02138
ramanath@bbn.com
[5] University at Albany - SUNY, Albany, NY 12222
ravi@cs.albany.edu

Abstract. We consider topology control problems where the goal is to
assign transmission powers to the nodes of an ad hoc network so as to
induce graphs satisfying specific properties. The properties considered
are connectivity, bounded diameter and minimum node degree. The optimization objective is to minimize the total power assigned to nodes. As
these problems are NP-hard in general, our focus is on developing approximation algorithms with provable performance guarantees. We present
results under both symmetric and asymmetric power threshold models.

1 Introduction

It is well known that battery power is a precious resource in ad hoc networks.
Therefore, techniques for minimizing the energy consumed in ad hoc networks
have assumed importance. Topology control problems arise in that context. The
goal of such problems is to control the topology of networks through the assignment of suitable transmission powers to nodes. Formally, such problems are
specified by requiring the induced network to satisfy some graph theoretic properties while minimizing some function of the transmission powers assigned to
transceivers (nodes). Previous work in this area has considered properties such
as node and edge connectivity and optimization objectives such as minimizing
maximum power and minimizing total power. A summary of previous results in
this area is presented in Section 3.2.

In this paper, we study topology control problems for three graph properties,
namely connectedness, bounded diameter and minimum node degree (Precise
formulations of these problems are provided in Section 2.1.). Connectedness is
a basic requirement for any network. Ad hoc networks with small diameters are

S. Pierre, M. Barbeau, and E. Kranakis (Eds.): ADHOC-NOW 2003, LNCS 2865, pp. 187–198, 2003.
© Springer-Verlag Berlin Heidelberg 2003

desirable in practice since the diameter of a network determines the maximum end-to-end delay for message delivery. Networks in which the degree of each node is at or above a certain threshold value are useful from a reliability perspective. In such networks, the failure of a small number of nodes or links is unlikely to disconnect the network. For all of these properties, the problem of minimizing the maximum power can be solved efficiently; this follows directly from a general result presented in [10]. So, we consider topology control problems for these properties under the objective of minimizing total power. These problems are **NP**-complete in general. The focus of this paper is therefore on developing approximation algorithms with proven performance guarantees.

Previous work on topology control has assumed the *symmetric* power threshold model. In that model, the minimum transmission power (also called the **power threshold**) needed for a node x to reach a node y is assumed to be equal to the minimum transmission power needed for y to reach x. In practice, power threshold values for two nodes x and y may be asymmetric because of two reasons. First, the ambient noise levels of the regions containing the two nodes may be different. Secondly, one of the nodes may be equipped with a directional antenna [12] while the other node may have only an omnidirectional antenna. Motivated by these considerations, we study topology control problems under the *asymmetric* power threshold model. Our results show (as one would expect) that problems do become "harder" under the asymmetric power threshold model. In particular, we show that under the asymmetric power threshold model, the problem of obtaining a connected graph while minimizing the total power cannot be approximated to within a factor of $\Omega(\log n)$, where n is the number of nodes, unless $\mathbf{P} = \mathbf{NP}$. We also present an approximation algorithm with a performance guarantee of $O(\log n)$ for the problem. Under the symmetric power threshold model, it is known that this problem is **NP**-hard but can be approximated to within a constant factor [4,8].

2   Problems Considered

2.1   Model and Problem Formulation

We are given a set V of transceivers (nodes). For each ordered pair (u, v) of transceivers, we are given a **transmission power threshold**, denoted by $p(u, v)$, with the following significance: A signal transmitted by the transceiver u can be received by v only when the transmission power of u is at least $p(u, v)$. It is assumed that $p(u, v) > 0$ for all nodes u and v.

We study topology control problems under both *symmetric* and *asymmetric* power threshold models. Under the symmetric power threshold model, for each pair of transceivers u and v, $p(u, v) = p(v, u)$. The asymmetric threshold model is more general. Under this model, there may be some pairs of transceivers u and v such that $p(u, v) \neq p(v, u)$.

A power assignment is a function $f : V \to \mathbb{R}^+$ that specifies a nonnegative power value $f(v)$ to each transceiver $v \in V$. Two models for graphs induced by power assignments have been considered in the literature. In this paper we utilize the undirected graph model, in which the induced graph $G_f(V, E_f)$ has the undirected edge $\{u, v\}$ if and only if $f(u) \geq p(u, v)$ and $f(v) \geq p(v, u)$. For a power assignment f, the maximum power assigned to any node is given by $\max\{f(v) : v \in V\}$; the total power assigned to all nodes is given by $\sum_{v \in V} f(v)$.

Following [10], we denote each topology control problem by a triple of the form $\langle \mathbb{M}, \mathbb{P}, \mathbb{O} \rangle$. In such a specification, $\mathbb{M} \in \{\text{DIR}, \text{UNDIR}\}$ represents the graph model, $\mathbb{P}$ represents the desired graph property and $\mathbb{O}$ represents the minimization objective. In general, $\mathbb{O} \in \{\text{MAXP}, \text{TOTALP}\}$ (abbreviations of Max Power and Total Power respectively). However, for all the problems considered in this paper, $\mathbb{O} = \text{TOTALP}$.

Using this notation, we now define the main problems studied in this paper.

1. In the $\langle \text{UNDIR}, \text{DIAMETER}, \text{TOTALP} \rangle$ problem, we are given a set V of transceivers, the power threshold values $p(u, v)$ for each pair (u, v) of transceivers and a diameter[1] bound D. The goal is to compute a power assignment f such that the undirected graph G_f induced by f has diameter at most D, and the total power assigned is a minimum among all power assignments that induce graphs satisfying the diameter constraint.

2. In the $\langle \text{UNDIR}, \text{DEG LB}, \text{TOTALP} \rangle$ problem, we are given a set V of transceivers, the power threshold values $p(u, v)$ for each pair $(u, v) \in V$ and an integer Δ, where $2 \leq \Delta \leq |V| - 1$. The goal is to compute a power assignment f such that the undirected graph G_f induced by f is connected, the degree of each node in G_f is at least Δ, and the total power assigned is a minimum among all power assignments that induce connected graphs satisfying the degree constraint.

3. In the $\langle \text{UNDIR}, \text{CONNECTED}, \text{TOTALP} \rangle$ problem, we are given a set V of transceivers and the power threshold values $p(u, v)$ for each pair (u, v) of transceivers. The goal is to compute a power assignment f such that the undirected graph G_f induced by f is connected and the total power assigned is a minimum among all power assignments that induce connected graphs.

We study the $\langle \text{UNDIR}, \text{DIAMETER}, \text{TOTALP} \rangle$ and $\langle \text{UNDIR}, \text{DEG LB}, \text{TOTALP} \rangle$ problems under the symmetric power threshold model. The $\langle \text{UNDIR}, \text{CONNECTED}, \text{TOTALP} \rangle$ problem has been studied previously under the symmetric power threshold model [4,8]. We study it under the asymmetric threshold model (Section 5). Due to space limitations, we discuss only the results for $\langle \text{UNDIR}, \text{DIAMETER}, \text{TOTALP} \rangle$ and $\langle \text{UNDIR}, \text{CONNECTED}, \text{TOTALP} \rangle$ problems in the remainder of this paper.

The following graph theoretic definition is used throughout this paper.

Definition 1. *Let $G(V, E)$ be an undirected graph. An **edge subgraph** $G'(V, E')$ of G uses the same set V of nodes and a subset E' of the edge set E.*

---

[1] The **diameter** of G, denoted by $\text{DIA}(G)$, is the maximum over the lengths of shortest paths between all pairs of nodes in G.

2.2 Bicriteria Approximation

Our results for the diameter problem use the **bicriteria approximation** framework developed in [11] for dealing with computationally intractable optimization problems involving two objectives. We recall the relevant definitions and notation.

Definition 2. *Suppose a problem Π with two minimization objectives A and B is posed in the following manner: Given a budget constraint on objective A, find a solution which minimizes the value of objective B among all solutions satisfying the budget constraint. An (α, β)-**approximation algorithm** for problem Π is a polynomial time algorithm that provides for every instance of Π a solution satisfying the following two conditions.*

1. *The solution violates the budget constraint on objective A by a factor of at most α.*
2. *The value of objective B in the solution is within a factor of at most β of the minimum possible value satisfying the budget constraint.*

We note that $\langle$UNDIR, DIAMETER, TOTALP$\rangle$ is an example of an optimization problem with two objectives. In this problem, diameter of the induced graph and total power serve as the budgeted objective (with budget D) and the minimization objective respectively. Thus, an (α, β)-approximation algorithm for the problem provides a solution where the induced graph has diameter at most αD, and the total power assigned is within a factor β of the minimum total power needed to induce a graph with diameter at most D.

To obtain bicriteria approximation algorithms for the $\langle$UNDIR, DIAMETER, TOTALP$\rangle$ problem, we rely on known approximation results for another problem, called **Minimum Cost Tree with a Diameter Constraint** (MCTDC), also involving two minimization objectives. A formal definition of this problem is as follows.

Minimum Cost Tree with a Diameter Constraint (MCTDC)
Instance: A connected undirected graph $G(V, E)$, a nonnegative weight $w(e)$ for each edge $e \in E$, an integer $\delta \le n - 1$.

Requirement: Find an edge subgraph $T(V, E_T)$ of G such that $T(V, E_T)$ is a tree, $\overline{\text{DIA}(T)} \le \delta$ and the total weight of the edges in E_T is the smallest among all the trees satisfying the diameter constraint.

MCTDC is known to be NP-hard [11]. Bicriteria approximations for this problem have been presented in [2,9,11]. These results are used in Section 4.

3 Summary of Results and Related Work

3.1 Summary of Results

The following are the main results presented in this paper. For all the problems, n denotes the number of transceivers in the problem instance.

1. We show that if the diameter constraint cannot be violated, the ⟨UNDIR, DI-AMETER, TOTALP⟩ problem cannot be approximated to within an $\Omega(\log n)$ factor unless $\mathbf{P} = \mathbf{NP}$. This result holds even when the diameter bound $D = 2$. (Note that the problem is trivial when $D = 1$.)

2. We show that using any (α, β)-approximation algorithm for the MCTDC problem, one can devise a $(2\alpha, 2(1 - 1/n)\beta)$-approximation algorithm for ⟨UNDIR, DIAMETER, TOTALP⟩ problem. This result is based on a general framework presented in [10] for approximating the total power objective. Utilizing this general framework and known bicriteria approximations for the MCTDC problem, we obtain several bicriteria approximation algorithms for the ⟨UNDIR, DIAMETER, TOTALP⟩ problem. (See Section 4.2.)

3. For every fixed integer $\Delta \geq 2$, we show that the ⟨UNDIR, DEG LB, TOTALP⟩ problem is NP-complete. Also, we present an approximation algorithm with a performance guarantee of $2(\Delta+1)(1-1/n)$ for the problem. This algorithm produces a power assignment that induces a connected graph in which each node has degree at least Δ. The performance guarantee is with respect to the optimal total power value. (Details regarding these results will be included in a complete version of this paper.)

4. While the above results are under the symmetric power threshold model, we consider the ⟨UNDIR, CONNECTED, TOTALP⟩ problem under the asymmetric power threshold model. We show that the problem cannot be approximated to within an $\Omega(\log n)$ factor unless $\mathbf{P} = \mathbf{NP}$. We also present an $O(\log n)$ approximation algorithm for the problem.

3.2 Related Work

Reference [10] provides a general approach that leads to an approximation framework for minimizing total power. Using that framework, two new approximation algorithms for ⟨UNDIR, 2-NODE CONNECTED, TOTALP⟩ and ⟨UNDIR, 2-EDGE CONNECTED, TOTALP⟩ with an asymptotic approximation ratio of 8 are presented in [10]. Both of the approximation ratios are improved to 4 in [6]. Reference [3] shows that the ⟨DIR, STRONGLY CONNECTED, TOTALP⟩ problem is NP-complete and presents a 2-approximation algorithm for the problem. Calinescu et al. [4] improve the approximation ratio to $(1+\ln 2)$. The approximation ratio is further improved to $5/3$ in a journal submission based on [4].

4 Results for Diameter Problems

4.1 Lower Bound on Approximation

The following theorem can be proven using an approximation preserving reduction from the MINIMUM SET COVER (MSC) problem. The proof is omitted due to space constraint.

1. From the given problem instance, construct the undirected complete edge weighted graph $G_c(V, E_c)$, where the weight of each edge $\{u, v\}$ in E_c is equal to the power threshold value $p(u, v)$.
2. Use any approximation algorithm $\mathcal{A}$ for the MCTDC problem on graph $G_c(V, E_c)$ with diameter bound $2D$, and obtain a spanning tree $T(V, E_T)$ of G_c.
3. For each node (transceiver) u, assign a power value $f(u)$ equal to the weight of the largest edge incident on u in T.

Fig. 1. Outline of Heuristic GEN-DIAMETER-TOTAL-POWER

Theorem 1. *Let n denote the number of nodes in an instance of the ⟨UNDIR, DIAMETER, TOTALP⟩ problem. There is a constant δ_1, $0 < \delta_1 < 1$, such that there is no $\delta_1 \ln n$ approximation for the problem, unless $\boldsymbol{P} = \boldsymbol{NP}$. Moreover, this result holds even for instances in which the diameter bound $D = 2$.*
□

4.2   Bicriteria Approximations for Diameter and Power

Description of the General Algorithm. Recall that the specification for the ⟨UNDIR, DIAMETER, TOTALP⟩ problem includes a bound D on the diameter of the induced graph and that the goal is to minimize total power. Let n denote the total number of transceivers specified in the ⟨UNDIR, DIAMETER, TOTALP⟩ problem instance. Our general approximation algorithm for ⟨UNDIR, DIAMETER, TOTALP⟩, shown in Figure 1, is derived from the general outline presented in [10] for developing approximation algorithms under the total power objective.

In Step 2, we may use any approximation algorithm $\mathcal{A}$ for the MCTDC problem. As long as $\mathcal{A}$ runs in polynomial time, our heuristic also runs in polynomial time. The performance guarantee provided by the heuristic is a function of the performance guarantee provided by Algorithm $\mathcal{A}$.

Performance Guarantee of the General Heuristic. The solution produced by Heuristic GEN-DIAMETER-TOTAL-POWER is approximate in terms of both diameter and total power. So, we cannot directly apply the bound from [10] on the performance of the general approach to derive the performance guarantee provided by the heuristic. Our analysis uses a simple property of spanning trees generated by breadth-first-search (BFS).

Throughout Section 4.2, we use the following notation. Let I denote the given instance of the ⟨UNDIR, DIAMETER, TOTALP⟩ problem with n transceivers and diameter bound $D \geq 1$. Let f^* denote an optimal power assignment such that the graph G_{f^*} induced by f^* has diameter at most D, and let $OPT(I) = \sum_{v \in V} f^*(v)$. Let f denote the power assignment produced by the heuristic and let G_f denote the graph induced by f. Let $DTP(I) = \sum_{v \in V} f(v)$, the total power assigned by the heuristic for the instance I. The goal of this subsection is to prove the following result.

Theorem 2. *Suppose Algorithm $\mathcal{A}$ used in Step 2 of Heuristic* GEN-DIAMETER-TOTAL-POWER *is an (α, β)-approximation algorithm for the* MCTDC *problem. For any instance I of the $\langle$UNDIR, DIAMETER, TOTALP$\rangle$ problem, Heuristic* GEN-DIAMETER-TOTAL-POWER *produces a power assignment f satisfying the following two properties.*

1. $\text{DIA}(G_f) \leq 2\,\alpha\,D$.
2. $DTP(I) \leq 2\,\beta\,(1 - 1/n)\,OPT(I)$.

Our proof of Theorem 2 uses a few lemmas proved below. We begin with a simple lemma about spanning trees generated by carrying out BFS on a connected graph. The proof of this lemma is omitted.

Lemma 1. *Let G be a connected graph with diameter δ. Let T be any spanning tree for G generated by BFS. Then* $\text{DIA}(T) \leq 2\,\delta$. $\qquad\qquad\square$

The next lemma indicates why in Step 2 of Heuristic GEN-DIAMETER-TOTAL-POWER, we use the diameter bound of $2D$.

Lemma 2. *Consider the complete graph $G_c(V, E_c)$ constructed in Step 1 of Heuristic* GEN-DIAMETER-TOTAL-POWER. *There is a spanning tree $T_1(V, E_{T_1})$ of G_c satisfying the following two properties.*

(a) $\text{DIA}(T_1) \leq 2D$.

(b) *Let $W(E_{T_1}) = \displaystyle\sum_{\{x,y\} \in E_{T_1}} p(x, y)$ denote the total edge weight of T_1. Then,*
$$W(E_{T_1}) \leq (1 - 1/n)\,OPT(I).$$

Proof:
Part (a): Consider the graph G_{f^*} induced by the optimal power assignment f^*. Note that $\text{DIA}(G_{f^*}) \leq D$. Let v be node such that $f^*(v)$ has the largest value among all the nodes in V. Let $T_1(V, E_{T_1})$ be a spanning tree of G_{f^*} generated by carrying out a BFS on G_{f^*} with v as the root. Then, from Lemma 1, we have $\text{DIA}(T_1) \leq 2D$.

Part (b): Consider another assignment w of weights to the edges of T_1 as indicated below. Consider each edge $\{x, y\}$ in T_1, where y is the parent of x. Let $w(x, y) = f^*(x)$. Thus, the power value assigned by the optimal solution to each node except the root becomes the weight of exactly one edge of T_1. The power value $f^*(v)$ of the root is not assigned to any edge. Therefore,

$$\sum_{\{x,y\} \in E_{T_1}} w(x, y) \;=\; OPT(I) - f^*(v).$$

Since v has the maximum power value under f^* among all the nodes, we have $f^*(v) \geq OPT(I)/n$. Therefore,

$$\sum_{\{x,y\} \in E_{T_1}} w(x, y) \;\leq\; (1 - 1/n)\,OPT(I).$$

The following claim relates the weight $w(x,y)$ to the power threshold value $p(x,y)$. We omit the proof of this claim.

Claim. For each edge $\{x,y\} \in E_{T_1}$, $w(x,y) \geq p(x,y)$. □

As a simple consequence of the above claim, we have

$$W(E_{T_1}) \leq \sum_{\{x,y\}\in E_{T_1}} w(x,y) \leq (1-1/n)\,OPT(I),$$

and this completes the proof of Part (b) of the lemma. □

The next lemma, which follows from Lemma 2, uses the performance guarantee provided by the approximation algorithm $\mathcal{A}$ used in Step 2 of the heuristic.

Lemma 3. *Let $T(V, E_T)$ denote the tree produced by $\mathcal{A}$ at the end of Step 2 of Heuristic* GEN-DIAMETER-TOTAL-POWER. *Let $W(E_T) = \sum_{\{x,y\}\in E_T} p(x,y)$ denote the total weight of the edges in T. Let (α,β) denote the performance guarantee provided by $\mathcal{A}$ for the* MCTDC *problem. Then,*

(a) $\mathrm{DIA}(T) \leq 2\alpha D$.
(b) $W(E_T) \leq \beta(1-1/n)\,OPT(I)$. □

We are now ready to prove Theorem 2.
Proof of Theorem 2: Consider the spanning tree $T(V, E_T)$ produced in Step 2 of the heuristic. We will first show that every edge $\{x,y\} \in E_T$ is also in $G_f(V, E_f)$, the graph induced by the power assignment constructed in Step 3 of the heuristic. To see this, notice that $f(x)$ is the largest weight of an edge incident on x in T. Thus, $f(x) \geq p(x,y)$. Similarly, $f(y) \geq p(x,y)$. Thus, every edge in E_T is also in E_f. Since $\mathrm{DIA}(T) \leq 2\alpha D$, and addition of edges cannot increase the diameter, it follows that $\mathrm{DIA}(G_f) \leq 2\alpha D$.

To bound $DTP(I)$, we note from Lemma 3 that $W(E_T) \leq \beta(1 - 1/n)\,OPT(I)$. In the power assignment constructed in Step 3, the weight of any edge can be assigned to at most two nodes (namely, the end points of that edge). Thus, the total power assigned to all the nodes is at most $2W(E_T)$. In other words, $DTP(I) \leq 2\beta(1-1/n)\,OPT(I)$, and this completes the proof of Theorem 2. □

Obtaining Approximation Algorithms from Theorem 2. We now briefly indicate how several bicriteria approximation algorithms for the ⟨UNDIR, DIAMETER, TOTALP⟩ problem can be obtained using GEN-DIAMETER-TOTAL-POWER in conjunction with known bicriteria approximation results for the MCTDC problem.

1. For any fixed $\epsilon > 0$, a $(2\lceil\log_2 n\rceil, (1+\epsilon)\lceil\log_2 n\rceil)$-approximation algorithm is presented in [11] for the MCTDC problem. Using this algorithm and setting $\epsilon < 1/n$, we can obtain a $(4\lceil\log_2 n\rceil, 2\lceil\log_2 n\rceil)$-approximation algorithm for the ⟨UNDIR, DIAMETER, TOTALP⟩ problem.

2. For any *fixed* $D \geq 1$, a $(1, O(D \log n))$-approximation algorithm for the MCTDC problem is presented in [2]. Thus, for any fixed $D \geq 1$, we can obtain a $(2, O(D \log n))$-approximation algorithm for the ⟨UNDIR, DIAMETER, TOTALP⟩ problem.

3. For any D and any fixed $\epsilon > 0$, a $(1, O(n^\epsilon \log n))$-approximation algorithm for the MCTDC problem is presented in [9]. Thus, for this case, we can obtain a $(2, O(n^\epsilon \log n))$-approximation algorithm for the ⟨UNDIR, DIAMETER, TOTALP⟩ problem.

The above results are for inducing a bounded diameter graph over all the nodes. We can also obtain an approximation algorithm for the Steiner version of the ⟨UNDIR, DIAMETER, TOTALP⟩ problem where only a specified subset of the nodes (called the **terminals**) need to be connected together into a graph of bounded diameter. Letting η denote the number of terminals, reference [11] presents an $(O(\log \eta), O(\log \eta))$-approximation algorithm for the Steiner version of the MCTDC problem. Using this approximation algorithm in Step 2 of Figure 1, we obtain an $(O(\log \eta), O(\log \eta))$-approximation algorithm for the Steiner version of the ⟨UNDIR, DIAMETER, TOTALP⟩ problem.

5   Asymmetric Power Threshold Model Results

In this section, we consider the ⟨UNDIR, CONNECTED, TOTALP⟩ problem under the asymmetric threshold model. We begin with a lower bound on the approximability of the problem. This lower bound result can be proven in a manner similar to that of Theorem 1.

Theorem 3. *Let n denote the number of transceivers in an instance of the ⟨UNDIR, CONNECTED, TOTALP⟩ problem. There is a constant δ, $0 < \delta < 1$, such that there is no $\delta \ln n$ approximation for the problem, unless $\boldsymbol{P} = \boldsymbol{NP}$.* □

In the remainder of this section, we show this nonapproximability result is tight to within a constant factor by presenting an approximation algorithm with a performance guarantee of $O(\log n)$. These results should be contrasted with the fact that under the symmetric power threshold model, there are constant factor approximation algorithms for the problem [4,8].

The main idea behind the approximation algorithm is to reduce the problem to the computation of a connected dominating set for a graph with node weights. A simple observation allows us to restrict the class of solutions to the ⟨UNDIR, CONNECTED, TOTALP⟩ problem. Consider a node v_i and let $\gamma_i^1 \leq \gamma_i^2 \leq \ldots \leq \gamma_i^n$ denote the n power threshold values in nondecreasing order from v_i to the n nodes of the system. We may assume without loss of generality that in any solution, the power value assigned to v_i is one of $\gamma_i^1, \gamma_i^2, \ldots, \gamma_i^n$. This is because of the following:

(a) The power assigned to v_i cannot be less than γ_i^1, since in such a case, v_i cannot be adjacent to any other node in the induced graph.

1. Let $\alpha = 6 \ln n \sum_{i=1}^{n} \gamma_i^n$.
2. From the given instance of ⟨UNDIR, CONNECTED, TOTALP⟩, construct a graph $G_1(V_1, E_1)$ as follows.
   (a) For each transceiver v_i $(1 \leq i \leq n)$ in the problem instance, create a set $g_i = \{u_i^0, u_i^1, \ldots, u_i^n\}$ of $n+1$ nodes. Let the weight $w(u_i^0) = \alpha$. For $1 \leq j \leq n$, let the weight $w(u_i^j) = \gamma_i^j$. The node set V_1 is given by $g_1 \cup g_2 \cup \ldots \cup g_n$.
   (b) For each i, connect the nodes in g_i together as an $(n+1)$-clique. (Nodes u_i^0, $1 \leq i \leq n$, are not involved in any edges other than these clique edges.)
   (c) For any pair of nodes u_i^j and u_k^l, where $1 \leq j, l \leq n$, if $\gamma_i^j \geq p(v_i, v_k)$ and $\gamma_k^l \geq p(v_k, v_i)$, then add the edge $\{u_i^j, u_k^l\}$ to E_1. The edge set E_1 consists of the clique edges added in Step 2(b) and the edges added in Step 2(c).
3. Use the algorithm of [7] using a small value (say, 0.1) for ϵ to find a connected dominating set D_1 of approximately minimal weight for G_1.
4. If for some i, D_1 contains both u_i^j and u_i^k, where $j < k$, then delete u_i^j from D_1. Let D denote the resulting set after all such deletions have been carried out.
5. For each i, $1 \leq i \leq n$, if D contains u_i^j, then assign the power value γ_i^j to v_i.

Fig. 2. Approximation Algorithm for ⟨UNDIR, CONNECTED, TOTALP⟩

(b) A power value which is greater than γ_i^j but less than γ_i^{j+1} for some j can be replaced by γ_i^j without deleting any edges in the induced graph.
(c) Similarly, a power value greater than γ_i^n can be replaced by γ_i^n.

A solution in which for every node v_i, the power value assigned is one of γ_i^1, γ_i^2, ..., γ_i^n will be referred to as a **canonical solution**. Thus, we consider only canonical solutions in the remainder of this section.

Our approximation algorithm for the ⟨UNDIR, CONNECTED, TOTALP⟩ problem under the asymmetric threshold model is shown in Figure 2. The algorithm constructs a graph $G_1(V_1, E_1)$ from the given instance of ⟨UNDIR, CONNECTED, TOTALP⟩ and then invokes a known approximation algorithm for the *minimum weighted connected dominating set* problem [7]. We will prove the correctness and the performance guarantee of the algorithm through a series of lemmas.

For the remainder of this section, let I denote the given instance of the ⟨UNDIR, CONNECTED, TOTALP⟩ problem under the asymmetric threshold model, f^* denotes an optimal power assignment to the nodes for this instance and $OPT(I)$ denotes the total power assigned by the chosen optimal solution. As before, G_{f^*} denotes the graph induced by the optimal power assignment f^*.

For each transceiver v_i, the maximum power that can be assigned in a canonical solution is γ_i^n, $1 \leq i \leq n$. Thus, we have the following observation.

Observation 1 $OPT(I) \leq \sum_{i=1}^{n} \gamma_i^n$.   □

Our next two lemmas relate $OPT(I)$, the weight of an optimal connected dominating set for G_1 and the weight of a connected dominating set produced by the approximation algorithm. For space reasons, the proofs of the lemmas are omitted.

Lemma 4. *For the graph G_1 constructed in Step 2 of the algorithm, the weight of a minimum connected dominating set is at most $OPT(I)$.*   □

Lemma 5. *Consider the dominating set D_1 found in Step 3 of the algorithm (Figure 2).*

(a) *Let $W(D_1)$ denote the total weight of the nodes in D_1. Then, $W(D_1) < 6 \ln n\, OPT(I)$.*

(b) *D_1 does not contain any of the nodes u_1^0, u_2^0, ..., u_n^0.*

(c) *For every i, $1 \le i \le n$, D_1 contains at least one of the nodes from the set $g_i - \{u_i^0\}$.*   □

Lemma 6. *The set of nodes D computed in Step 4 of the algorithm is a connected dominating set of G_1. Further, for each i, $1 \le i \le n$, D contains exactly one vertex from the set $g_i - \{u_i^0\}$.*

Proof: From Part (c) of Lemma 5, the dominating set D_1 contains at least one node from each group g_i, $1 \le i \le n$. This property also holds for the set D since Step 5 eliminates a node u_i^j only when there is another node u_i^k from the same group g_i. In other words, D is also a dominating set for G_1.

Note also that for any two nodes u_i^j and u_i^k from the same group g_i, with $j < k$, the set of nodes to which u_i^j is adjacent is a subset of the corresponding set for u_i^k. Thus, the set of nodes remains a connected dominating set even after u_i^j is deleted. In other words, D is a connected dominating set for G_1.   □

It is easy to see that the approximation algorithm of Figure 2 runs in polynomial time. We now establish its correctness and performance guarantee.

Theorem 4. *The power assignment produced by the algorithm induces a connected graph. Further, the algorithm provides a performance guarantee of $O(\log n)$.*

Proof: From Lemma 6, the set D contains exactly one node from $g_i - \{u_i^0\}$, for each group g_i. Therefore, Step 5 of the algorithm assigns a power value to each transceiver. Since D is a connected dominating set and there is a one-to-one correspondence between D and the set of transceivers, the graph induced by the power assignment is also connected.

The total power assigned to all the nodes is equal to $W(D)$, the total weight of all the nodes in D. By Part (a) of Lemma 5, $W(D_1) < 6 \ln n\, OPT(I)$. Since $D \subseteq D_1$ and the node weights are nonnegative, it follows that $W(D) < 6 \ln n\, OPT(I)$. In other words, the approximation algorithm has a performance guarantee of $O(\log n)$.   □

6   Open Problems

Our work raises several open questions. First, it would be of interest to investigate whether there is a bicriteria approximation algorithm for the ⟨UNDIR, DIAMETER, TOTALP⟩ problem with a performance guarantee of $(O(1), O(\log n))$

for any given diameter value. A second problem is to improve the approximation ratio for the ⟨UNDIR, DEG LB, TOTALP⟩ problem. Finally, it would also be of interest to consider other topology control problems under the asymmetric threshold model.

References

1. D. M. Blough, M. Leoncini, G. Resta, and P. Santi, "On the Symmetric Range Assignment Problem in Wireless Ad Hoc Networks", *Proc. 2nd IFIP International Conference on Theoretical Computer Science*, Montreal, August 2002.
2. M. Charikar, C. Chekuri, T. Cheung, Z. Dai, A. Goel, S. Guha, and M. Li, "Approximation Algorithms for Directed Steiner Problems", *Journal of Algorithms*, Vol. 33, No. 1, 1999, pp. 73–91.
3. W. Chen and N. Huang, "The Strongly Connecting Problem on Multihop Packet Radio Networks", *IEEE Trans. Communication*, Vol. 37, No. 3, Mar. 1989.
4. G. Calinescu, I. Mandoiu, and A. Zelikovsky. "Symmetric Connectivity with Minimum Power Consumption in Radio Networks", *Proc. 2nd IFIP International Conference on Theoretical Computer Science*, Montreal, August 2002.
5. A. E. F. Clementi, P. Penna and R. Silvestri. "Hardness Results for the Power Range Assignment Problem in Packet Radio Networks", *Proc. Third International Workshop on Randomization and Approximation in Computer Science* (APPROX 1999), Lecture Notes in Computer Science Vol. 1671, Springer-Verlag, July 1999, pp. 195–208.
6. G. Calinescu and P-J Wan. "Symmetric High Connectivity with Minimum Total Power Consumption in Multihop Packet Radio Networks", Submitted for journal publication, 2003.
7. S. Guha and S. Khuller, "Improved Methods for Approximating Node Weighted Steiner Trees and Connected Dominating Sets", *Information and Computation*, Vol. 150, 1999, pp. 57–74.
8. L. M. Kirousis, E. Kranakis, D. Krizanc and A. Pelc, "Power Consumption in Packet Radio Networks", *Proc. 14th Annual Symposium on Theoretical Aspects of Computer Science* (STACS 97), Lecture Notes in Computer Science Vol. 1200, Springer-Verlag, Feb. 1997, pp. 363–374.
9. G. Kortsarz and D. Peleg, "Approximating the Weight of Shallow Light Trees", *Discrete Applied Mathematics*, Vol. 93, 1999, pp. 265–285. (Preliminary version appeared in *Proc. Eighth ACM-SIAM Symp. on Discrete Algorithms* (SODA'97), New Orleans, LA, Jan. 1977, pp. 103–110.)
10. E. L. Lloyd, R. Liu, M. V. Marathe, R. Ramanathan and S. S. Ravi, "Algorithmic Aspects of Topology Control Problems for Ad hoc Networks", *Proc. Third ACM International Symposium on Mobile Ad Hoc Networking and Computing* (MobiHoc'02), Laussane, Switzerland, June 2002.
11. M. V. Marathe, R. Ravi, R. Sundaram, S. S. Ravi, D. J. Rosenkrantz and H. B. Hunt III, "Bicriteria Network Design Problems", *Journal of Algorithms*, Vol. 28, No. 1, July 1998, pp. 142–171.
12. R. Ramanathan, "On the Performance of Ad Hoc Networks with Beamforming Antennas", *Proc. Second ACM International Symposium on Mobile Ad Hoc Networking and Computing* (MobiHoc'01), Long Beach, CA, Oct. 2001.
13. R. Ramanathan and R. Rosales-Hain, "Topology Control of Multihop Wireless Networks Using Transmit Power Adjustment", *Proc. INFOCOM 2000*.

IDEA: An Iterative-Deepening Algorithm for Energy-Efficient Querying in Ad Hoc Sensor Networks

Swapnil Patil

Department of Computer Science
State University of New York at Stony Brook
Stony Brook, NY 11790, USA
swapnil@cs.sunysb.edu

Abstract. The data-centric ad hoc sensor networks make efficient searching a crucial and challenging operation. Dynamic topology make flooding the most widely adopted solution at a cost of high bandwidth congestion leading to inefficient use of resources and low network lifetime. This paper presents IDEA, an efficient querying and searching technique for ad hoc sensor networks that reduces average energy consumption while maintaining the capacity and performance of the network. IDEA is based on iterative-deepening search which check-points the flooding of requests based on the results. This is further extended to a token-based approach called T-IDEA, which involves local decisions made by nodes to determine their participation in a virtual searching network. Results show that IDEA and T-IDEA significantly reduces the energy consumption compared to classical flooding approaches. Apart from that T-IDEA presents a highly distributed self-supervising topology formation which performs very well to increase the lifetime of the ad hoc sensor network.
. . .

1 Introduction

Recent advances in wireless and mobile networks have led to interest in building and deploying in ad-hoc and sensor networks. Due to their constrained environment and distributed operations, these networks provide an challenging research problems. Due to the distributed operations, most routing and searching operations in mobile ad-hoc and sensor networks are multi-hop. Thus, making flooding based approaches the most feasible. Many routing protocols [1,2,3] have been proposed using flooding or broadcast mechanisms.

In flooding, each node broadcasts the query to all its neighboring nodes. These nodes perform a local search for the query. If unsuccessful, these nodes broadcast the query to its neighbors and so on. Thus the query reaches the entire network, more than once for most of the nodes. But each node rejects query messages which have already been received once (based on their ID).

This paper introduces and evaluates new searching algorithm(s) in ad hoc sensor networks called *IDEA* and *T-IDEA*. These techniques achieve a significant gain over classical flooding based approach, and could easily scale to large networks.

S. Pierre, M. Barbeau, and E. Kranakis (Eds.): ADHOC-NOW 2003, LNCS 2865, pp. 199–210, 2003.
© Springer-Verlag Berlin Heidelberg 2003

IDEA uses the concept of *iterative deepening* by 'iteratively' sending the query messages to the increasing number of nodes until the query is answered (or not matched). This floods the query throughout the network only if the search does not yield a result at the certain 'check-points' i.e. end of each iteration. This technique is also applied to a self-organizing, token-based search, *T-IDEA* which adaptively takes into account the constraints of each node. Each sensor node identifies itself as 'available' or 'unavailable' for search operations, based on local decisions made towards overall network performance improvement.

The remainder of the paper is organized as follows. First, a brief summary of the problem is presented. Section 2 presents a brief description of the problem with the assumptions for that algorithm. Section 3 describes the *iterative deepening algorithm* in details, followed by an active token based version of the iterative deepening algorithm. Section 5 presents the experiments and the results of this protocol. Finally, Section 6 presents some related work and the following section concludes the paper.

2   Problem Overview

Information dissemination and gathering is one of the most important tasks from a mobile ad hoc network. Constraints like limited energy, dynamic network, low bandwidth etc. make it more challenging. Dynamic searching techniques are crucial for energy-efficient operations of mobile ad hoc networks.

This paper introduces a novel approach called *IDEA* to search in ad-hoc networks, using *iterative-deepening*, a known search technique to search over state space in artificial intelligence applications [14]. Iterative deepening searches are a combination of breadth-first searches with series of depth-first searches with increasing bounds of depth. This is further extended to use in conjunction with local information based on the interaction with neighboring nodes, called as token-based algorithm, *T-IDEA*. Nodes make local decisions, based on their energy constraints and importance in the search operation, to make themselves available for searching.

3   Iterative-Deepening

In applications where relevance of the search result is an important measure, iterative deepening is a better search technique than many classical algorithms. In the iterative deepening technique, multiple breadth-first searches are initiated with increasing depth limits, until the appropriate result has been found, or in case of routing destination has been reached, or as in the worst-case, the maximum depth limit D has been reached.

There are various motivations to support this search technique in sensor networks. First, the cost of querying at smaller depths is less than query-processing cost at larger depths. This is because of the fast growth of the number of nodes to be searched at growing depths. Secondly, this would also lead to reduced number of resources utilized for query processing in a constrained ad-hoc network. Compared to the traditional graph searching problems like BFS of depth d, this algorithm would asymptotically perform well by satisfying the query at a depth less than d. In deed, experiments show (Section 6) that for a fairly dense ad-hoc network with approximately 8 neighbors per node, with depth,d = 7, almost 70% of all queries can be satisfied at a depth less than d.

IDEA works on a network-wide rule called *i-Rule* which specifies the policy for depths of iteration for searching and the time-interval between successive iterations. Suppose the rule is, $i - Rule = \{i_1, i_2, i_3, T\}$, this means that the depth of search for the first iteration is i_1, the search depth for second iteration is i_2 and the third to depth i_3. Intuitively, if the search reaches the third-level iteration, then it has the same performance as BFS of depth i_3. The interval between these successive iterations known as the inter-iteration interval, T, which is required for the source (which initiates the query) to receive and analyze the response messages. One advantage of such a policy is that the rule could be based on any metric like the number of hops, time-to-search (like TTL) etc. But in case of ad hoc networks, the most common metric is the number of hops, considering that the communication cost is proportional to the number of hops. The algorithm details are explained further below.

For a i-Rule = {i,j,k,T}, a source node S initiates flooding (or BFS) till a depth i i.e. starting from the source the query is flooded to all nodes i hops away from the source. When the query reaches nodes that are i hops away, the query is *halted* and not flooded further. During this time the source S, may or may not receive the response messages to the query. After waiting for the *inter-iteration interval*, T, if it receives appropriate responses, then the source would stop querying. Else if the source does not receive appropriate results or no results, it initiates the second search iteration. It is important to note that, the definition of the term *'appropriate response'* or *'search results'* is application specific and is not addressed in this paper.

To initiate the second search iteration, source S will send the `Continue-j` message, indicating nodes to flood the message, j hops from the source. It should be noted that nodes which are i hops away from the source, have already processed the query and store the query with them (they are in *halted state*). Hence if these nodes (within i hops from the source) were to process the query, it would degrade the performance of the protocol by wasting energy of this constrained ad hoc network. Instead of re-processing the query, these nodes which are within i hops from the source would simply forward the received `continue-j` messages. Once the last node (i.e. i hops from the source) receives the message, these nodes re-send the halted query, and flood it to all the nodes which are 'j - i' hops from the present node (remember, j means j nodes from the source). One important point to note here is that, node will halt the query only for a time greater than the *inter-iteration interval*, T; before deleting its state.

It would always be a case that more than one query and consequently many `continue` messages are being flooded in the network. To match the queries with appropriate `continue` messages, every query is assigned a *unique identifier*, as in case of DSR [2]. The continue message will have the identifier for the corresponding query, thus nodes know which query is to be re-sent.

After flooding the query to nodes j hops away, the algorithm continues to the higher levels of depth in the subsequent iterations in the i-Rule. After the last iteration, in this case after k hops away, queries are not halted and the search is terminated.

4   *T-IDEA*: Token-Based Iterative-Deepening Algorithm

Iterative-deepening search is a good technique to keep a check on the flooding of the query by maintaining *'check-points'* i.e. iterations in the flooding the query. But it does not take into account the energy considerations of the mobile nodes, which is crucial to estimate the network lifetime for any protocol. A better algorithm should be adaptive to the energy constraints on the nodes in the ad hoc network, by involving nodes which have energy capacity to sustain as along as possible, thus having a minimal effect on the network lifetime.

T-IDEA implements this strategy by querying only a subset of the neighbors, thereby reducing the cost, but selecting neighbors which would converge to correct results. These neighbors identify themselves as participants in this search. We use a *token-based* scheme in which each node makes local decisions whether to take part in the search or not. Factors which are primarily functional in this token-based scheme are energy constraints, query processing capacity etc. The following sections describe the participation algorithm for nodes which participate in this search and the search in the participating nodes of ad hoc network.

4.1   *T-IDEA* Algorithm

To avoid involving highly constrained nodes or nodes who identify themselves as 'irrelevant' for particular search, this algorithm allows any source to query nodes which identify themselves as part of the search and are willing to accept and forward the queries.

Each node in the ad hoc network broadcasts participation token(s) to all of its neighbors. There are two types of tokens *q-tokens* and *t-tokens*. *q-token*(s) represent the number of query messages, q, for which the node can be the part of the search network and *t-token* represents the *search-TTL* for the node i.e. t time units. Thus, a source would send a query to only those nodes, from which it has a valid token. Once the node has replied/forwarded its q number of queries or has been a part for t time-units, it broadcasts a *zero-token message* sending its inability to take part in an any more active search query requests.

In aggregate, the *constained* nodes would declare their level of participation in the search operations, by means of number of queries or time-to-search. In case of *t-tokens*, sometimes the node receives and processes queries at a very high rate, leading to high energy drain of the ad hoc node capacity. So to reduce the query processing, the node can send a *q-token* before the expiration of the *t-token*, to control its energy drainage caused due to high query-processing rate. Similarly if a node has earlier sent a *q-token* to neighbors, but is receiving queries at a very low rate (thus main energy is drained by staying in active mode); in this case it can broadcast a *t-token* to declare a new token of its availability. Thus the generation of these tokens is primarily driven by the *energy constraints of respective nodes*. Section 5 describes the energy-based token generation model used during experiments.

One point of discussion in the token generation algorithm is the primary metric to be used. In the discussion so far energy has been the primary factor, and it would be the case for majority of applications. But suppose the network consists of mobile computers or handhelds which can be charged frequently, then energy might not be most important criteria. Another metric which plays an important role in searching is the connectivity and the ability to forward to nodes that lead to appropriate results. For example, nodes which have previously replied to the queries at a high frequency could be used to declare tokens of availability. Another criterion would be to use neighbors that have responded with messages with least number of hops. This is a very application specific metris, which is beyond the scope of this paper.

The *T-IDEA* search forwards the query message(s) to just a subset of its neighbors, based on the token-declaration made by the neighboring mobile nodes. By sending the query to a small subset of neighbor nodes, we will likely reduce the costs incurred by the nodes during query processing and forwarding. On the other hand, by selecting neighbors which wold produce many results, we can maintain quality of results to a large degree, even though fewer nodes are visited.

4.2 Resilient Search

This search algorithm is a controlled flooding approach which depends on nodes of the ad-hoc network identifying themselves to be a part of the searching network. One drawback of such a scheme, isthe increased probability of failure than flooding (where even if a node dies the query can propagated by some part of the network). This is possible in cases described earlier when a node experiences heavy energy drainage than predicted during the token declaration (due to increase in query processing rate).

This problem is alleviated by use of *keep-alive* messages. If a query is forwarded for certain number of hops, as declared in the i-Rule, we need to send explicit *keep-alive* messages. In case if the source node receives query response messages, they are considered to be implicit keep-alive messages. But if the source does not receive either of them it can re-issue the query.

4.3 Distributed Token Declaration Algorithm

A more distributed token declaration scheme for the nodes in the ad hoc network, would be to assign tokens to neighbors in proportion to their capacities. Based on the previous tokens received from your neighbors, the neighboring node which declares high capacity is assigned more tokens for querying the current node. This distributed token-algorithm is similar to fair queuing policies used for flow fairness in networks.

5   Simulation Model

Source node submits a search query to the network D times, spread over an interval, where maximum D = 7. The network assumes that each node can have a maximum of 8 neighbors and for experimental reasons the flooding has a limit of 7 hops from the source node.

The main aim of this algorithm is the *energy efficient operation*. The simulations have given varying weights to the energy consumed by different message. Some of the important messages in this algorithm are *Query Messages, Query-Response, Query-Result, Continue* with arbitrary maximum messages sizes of 100 bytes, 80 bytes, 70 bytes each and 40 bytes each. The energy required by each node is constant and is 1 micro-Joule per byte. These values are taken based on the approximate ratio of the packet sizes in packet networks which includes header, identifier and query-string (which is variable length). So in some cases it is possible that the packets might be of sizes smaller than the mentioned above. But the most important factor here is the energy required per byte, that determines the efficiency of the algorithm.

5.1   Aggregate Energy Consumption Model

Each query being propagated through the mobile network consumes energy. This metric gives us results about the energy consumption due the query processing.

Let,

E_q be the energy required to communicate the *query message*;

N_q be the number of nodes that process query k hops away

E_q be the number of nodes that process the query *again* k hops away i.e. redundant queries.

$$Total \quad Energy \quad per \quad Query \quad , TE_q = E_q \times (T_q + T_r) \qquad (1)$$

where,

$Total \ Query \ Messages, \ T_q = N_{(q,x)} + R_{(q,x)}$ and

$Total \ Response \ Messages, \ T_q = Resp_{(q,x)} + Rslt_{(q,x)}$

and

$N_{(q,x)}$ is the number of nodes that process query Q, k hops away from

$R_{(q,x)}$ is the number of redundant edges when the query Q is processed.

$Resp_{(q,x)}$ is the number of response messages received per query Q from a node x which is k hops from the source.

$Rslt_{(q,x)}$ is the number of results received per query Q.

This model is modified to accommodate the changes based on *IDEA* and *T-IDEA* methods

5.2   Energy Model for *IDEA*

In case of *IDEA*, the aggregate energy consumed per query must take into account the cost of sending continue messages, as well as the possibility that a query may terminate before reaching the maximum depth. Let R be the IDEA rule (the iterative depths) which would be used for evaluation. The rules are defined as a set of values, where the ith value, represents the number of hops a message is forwarded in the ith iteration (this is for $1 < i < N - 1$, for a N-item set, since the last value if the inter-iteration interval) or is halted. Let us assume d to be the item of the *IDEA* rule set i.e. we need to go d hops during the flooding approach.

So the energy consumption equation (1) would change to accommodate the *IDEA* algorithm. It would be a sum of the energy consumed to flood the query to n hops and receiving response messages from nodes at n hops. If the query ends before reaching nodes which are n hops away, then it will not be sent to nodes that are n hops away nor receive results from nodes which ar n hops away. This condition can be mathematically represented as a constant A, which is multiplied to the naive energy consumption relation achieved above in eqn (1).

```
A = 1 , if n is a rule in the IDEA-rule of iterations
            and query Q does not end at nodes n hops away
  = 0 , otherwise
```

Another important factor is the overhead caused by sending the continue messages. continue messages are sent to nodes n hops away if:
(a) depth k is in the *IDEA-rule*
(b) query is not satisfied within depth k before expiration
(c) depth k < D (max depth i.e. highest value of IDEA-rule elements)

5.3   Energy Model for *T-IDEA*

Energy consumption equation for token-based iterative deepening is similar to equation (1), for a classical flooding approach. The only change that is incorporated is the heuristic to decide whether a node would participate in the search operation in the ad hoc network, based on local decisions.

6   Results

6.1   Average Aggregate Energy Consumption

Figure 1 shows the cost of *each rule-item*, for differnt values of *inter-iteration interval, T,* in terms of average aggregate energy consumption. Along the x-axis, we vary the rule, *d* i.e. number of hops. Immediately obvious in these figures are the cost savings. IDEA-Rule R_1 at T = 8 units uses just about 10% of the aggregate bandwidth per query used by classical flooding, IDEA-Rule R_7, and just 41% of the aggregate processing cost per query.

To understand how such enormous savings are possible, we must understand the tradeoffs between the different rules and inter-iteration interval, T. Let us focus on energy consumption per query message, in Figure 1. First, notice that

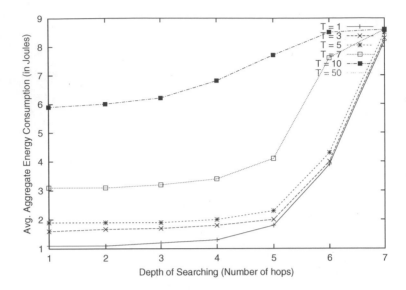

Fig. 1. Avg. Aggregate Energy Consumption for Iterative-Deepening Search

the average aggregate bandwidth for d = 7 (i.e. rule 7 or classical flooding) is the same, regardless of tt T. Since the *inter-iteration interval*, T is considered only between iterations, it does not affect *IDEA-Rule* $R_7 = \{7\}$, which has only a single iteration. Next, notice that as the number of hops for iteration, d, increases, the energy consumption of *IDEA-Rule* R_d increases as well. The larger d is, the more likely the rule will waste bandwidth by sending the query out to too many nodes i.e sending the query out to more nodes than necessary. Sending the query out to more nodes than necessary will generate more energy consumption for forwarding the query, and transferring response messages back to the source. Hence, as d increases, bandwidth consumption increases as well, giving *IDEA-Rule* R_7 or classical flooding gives the worst energy consumption performance.

Now, notice that as the inter-iteration interval, T, decreases, energy consumption per query usage increases. If T is small, then it is highly possible that source will assume that the query was not satisfied, leading to the overshooting effect as described earlier. For example, say T = 10 and d = 6, if a query Q can be satisfied at depth 6, but more time than T is required before certain number of results arrive at the client, then the client will only wait for T seconds, determine that the query is not satisfied, and initiate the next iteration at depth 7. In this case, the source overshoots the goal. The smaller T is, the more often the source will overshoot; hence, energy consumption usage increases as T decreases.

T-IDEA decreases the number of nodes which would take part in the query processing operation based on the *local decisions* and tokens created by each node. This helps in increasing the aggregate capacity of the network, since some nodes are *not* involved the energy consuming query processing process.

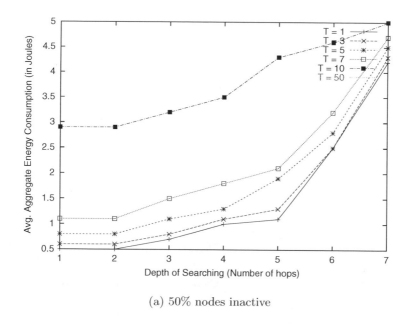

(a) 50% nodes inactive

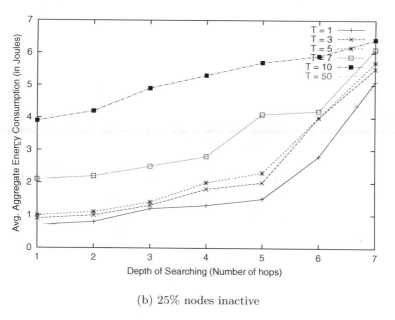

(b) 25% nodes inactive

Fig. 2. Avg. Aggregate Energy Consumption for Iterative-Deepening Search with *a fraction of inactive nodes.* (25% and 50%)

Figure 2, shows the variations in the avg. energy consumption for querying the ad-hoc sensor network using the *T-IDEA* algorithm.

Table 1. Probability of *correct* search results using IDEA for 4 and 8 neighbor nodes

No. of Search Results	Pr(4-neighbors per node)	Pr(8-neighbors per node)
50	0.698	0.756
100	0.597	0.673
150	0.534	0.572
200	0.458	0.502

6.2   Query Results *Correctness*

Performance of the iterative deepening technique and its rules, also depends on the value chosen for number of results expected, R. In addition, performance depends on the number of nodes that process each query, which in turn is determined locally by the number of neighbors (degree) that a node maintains. To see the effect of these two factors on the *IDEA* algorithm, the first simulation assumed 8 neighbors per node (approximately), then for 4 neighbors per node. Over each data set, we then ran analysis for four different values of search results, first with R = 50, then R = 100, R = 150 and R =200

Table 1 shows the performance of each variation in terms of query satisfaction probability, $Pr()$. It represents all iterative deepening policies as it is independent of the inter-iteration interval, T. As expected, when the definition of query satisfaction, Q_{def}, increases, the level of satisfaction decreases. However, it is encouraging to note that satisfaction does not drop very quickly as number of search results increase.

Also as expected, when the number of neighbors per node decreases, the probability that a query is satisfied also decreases, since fewer neighbors generally translates to fewer number of results. However, we find it interesting that satisfaction probability with 4 neighbors per node is not much lower than satisfaction probability with 8 neighbors per node. The reason being that with 8 neighbors per node, the source node usually receives more results than needed to satisfy the query. When the number of neighbors per node decreased to 4, the source received significantly fewer results, but in most cases it was able to produce the query results.

7   Related Work

Various research groups have worked on optimization of searching algorithms in ad-hoc sensor networks. Earlier research was focused on optimal flooding or broadcasting in ad hoc networks. [11] proposes a heuristics based algorithm, which decides to forward the packet based on various factors like probabilistic neighbor node selection. On the other hand, this paper gives a *deterministic* selection policy based on the tokens or the policy for iterative deepening. [8] uses the concept of elimination scheme based on the information of nodes to which the packet was broadcast from the transmitter. Node(s) do not broadcast to nodes which have already received the packet. This *'broadcast cover state'* is maintained by the transmitter after transmitting to the new nodes. [4] also uses the concept of elimination set based on the tuple of location co-ordinates and neighbor nodes for a particular node; it also uses a negative acknowledgement scheme for retransmissions. Algorithm in this paper does not keep the state of the broadcast either for the *IDEA* search or for the token-based, *T-IDEA* search, since it scales robustly to any change in the topology.

Many approaches use the topological information of the ad hoc network. [10] selects a set of neighbors called multipoint relays (MPR), based on the topological information such that each node covers the same network region, which the complete set of neighbors does. The computation of this minimal set is NP-Complete problem. The iterative-deepening search in this work does not exploit any topology information, but tries to exploit tokens received from the neighbor nodes (in case of *T-IDEA*).

Several works use the concept of dominating set to optimally broadcast the packets, which are different from the approach addressed by this paper which does an optimal flooding based on a policy(s). [9] proposed a distributed deterministic algorithm, which defined a set as dominating if all nodes in that graph are either the neighbor nodes belonging to the set or in the set of neighbors. Two rules are proposed to reduce the number of internal nodes.

Another genre of solutions was the cluster-based approach, where nodes organize themselves in clusters and nominate cluster heads to do the routing. [6], [12], [13] perform clustering for a hierarchical routing scheme, but not mainly for efficient flooding. This scheme depends on the complete neighbor information and incurs an overhead due to exchange of 'HELLO' messages. These classical clustering approaches focus on forming clusters but do not have an optimal connected set with least number of clusters. More recent work like [5] is based on self-pruning methods that makes local decisions on the forwarding status. This depends on small clustered topology creation which changes dynamically. However, this work forms a topology of nodes involved in the searching, but does not have clustering, it is completely granular to single node level.

8   Conclusion

This paper examined a novel approach for efficient searching in ad-hoc sensor networks. The results in section 7 show the trade-off(s) involved in the *IDEA* algorithm(s),described in this paper, compared to classical search and query algorithms. This paper includes several contributions. First, *IDEA* algorithm is probably the first algorithm (to the best of author's knowledge) to use the iterative-deepening algorithm to efficiently control the flooding in a ad-hoc sensor network. Second, the token-based algorithm *T-IDEA*, gives a self-stabilizing approach to dynamically adapt to the constraints of individual nodes in the network. Unlike previous approaches, this does not use any complex algorithm but is based on the local decisions, based on per-node characteristics, made by the nodes *individually* in the sensor network. Finally simulations show that the *IDEA* algorithm (s) perform better that the classical approaches in an energy-constrained network environment. *IDEA* and *T-IDEA* reduce the aggregate energy consumption of the network significantly, while maintaining the quality of results. This increases the average *lifetime* of the network with an increased performance and scalability.

Acknowledgements

I would like to thank my advisor Prof.Samir R. Das for his valuable suggestions and discussions during this research work. I am also grateful to my colleagues at the WINGS Lab in the Department of Computer Science at SUNY at Stony Brook.

References

1. C. Perkins, E. Royer, S. Das, Ad hoc on demand distance vector(AODV) routing, http://www.ietf.org/internet-drafts/draft-ieftmanet-aodv-03.txt, 1999.
2. D.B. Johnson and D.A. Maltz, Dynamic source routing in ad hoc wireless networks, Mobile Computing, Academic Publishers,1996. pp. 153-181.
3. S.-J. Lee, W. Su, and M. Gerla On-Demand Multicast Routing Protocol (ODMRP) for Ad Hoc Networks, Internet Draft, draftietf-manet-odmrp-02.txt, Jan. 2000
4. I. Stojmenovic, M. Seddigh, and J. Zunic, Internal node based broadcasting algorithms in wireless networks, in Proceedings of the HawaiiInt. Conf. on System Sciences, Jan. 2001.
5. B. Chen, K. H. Jamieson, and R. Morris An energy-efficient coordination algorithm for topology maintenance in Ad Hoc wireless networks, Mobicom, 2001
6. C.R. Lin and M. Gerla, Adaptive Clustering for Mobile Wireless Networks, IEEE Journal on Selected Areas in Communications, Vol. 15, No. 7, Sep. 1997, pp. 1265-1275.
7. I. Stojmenovic, M. Seddigh, and J. Zunic, Dominating sets and neighbor elimination-based broadcasting algorithms in wireless networks, IEEE Transactions on Parallel and Distributed Systems, vol. 12, no. 12, Dec.2001.
8. W. Peng and X.C. Lu, On the reduction of broadcast redundancy in mobile ad hoc networks, in Proceedings of the Annual Workshop on Mobile and Ad Hoc Networking and Computing (MobiHOC 2000), Boston, Massachusetts, USA, Aug 2000, pp. 129-130.
9. J. Wu and H. Li, A dominating-set-based routing scheme in ad hoc wireless networks, in Proceedings of the Third Int'l Workshop Discrete Algorithms and Methods for Mobile Computing and Communications (DIALM), Aug.1999, pp. 7-14.
10. A. Qayyum, L. Viennot, and A.Laouiti, Multipoint relaying for flooding broadcast messages in mobile wireless networks, in Proceedings of the 35th Annual Hawai International Conference on System Sciences (HICSS'02), Hawaii, 2002.
11. Sze-Yao Ni, Yu-Chee Tseng, Yuh-Shyan Chen, and Jang-Ping Sheu, The broadcast storm problem in a mobile ad hoc network, in Proceedings of the MobiCom'99 Seattle WA, Aug. 1999.
12. Gerla, M and Tasi, J., A Multi-cluster, mobile, multimedia radio network, ACM Baltzer Journal of Wireless Networks, Vol. 1 No.3, 1995
13. Krishna, P., Vaidya, N.H., Chatterjee, M., Pradhan, D.K., A cluster-based approach for routing in dynamic networks, Computer Communication Review, vol.27 (no.2), ACM, April 1997.
14. S. Russel and P. Norvig. Artificial Intelligence: A Modern Approach. Prentice-Hall 1995.

On the Interaction of Bandwidth Constraints and Energy Efficiency in All-Wireless Networks

Tommy Chu and Ioanis Nikolaidis

Computing Science Department
University of Alberta
Edmonton, Alberta T6G 2E8, Canada
{tommy,yannis}@cs.ualberta.ca

Abstract. The minimization of expended energy for unicast and broadcast communication between nodes in a wireless network has been studied mostly as a path optimization problem without particular regard for the traffic load demands. In this paper, we consider the call admission problem whereby given a traffic load (described as source-destination rate demands) the required expended energy is minimized. In addition, we explicitly model bandwidth capacity constraints. The capacity constraints reflect the fact that, from the perspective of a single node, traffic includes data that the nodes originate and forward, as well as traffic they receive and is of no interest to them. This last class of traffic is unavoidable due to the transmission radius of nearby stations. Under the assumption that the MAC protocol behaves in an ideal fashion, we consider two centralized algorithms that attempt to admit the given load and we remark on their relative performance, especially with respect to their energy consumption and blocking (connection rejection) rate.

1 Introduction

Apart from eliminating the need for a pre-existing infrastructure, a serious motivation behind the deployment of wireless networks is the ability to establish connectivity among, hopefully, unrestricted number of wireless nodes. Naturally, the extent to which this is doable has to do with the volume of carried traffic and the extent to which the radio spectrum is wisely reused among all participating nodes. While a sufficiently intense traffic load can render any network inadequate, the idea of proper reuse of the radio spectrum in light of competing nodes is not at all solved. In fact, there have been studies, such as the one in [1] suggesting that wireless ad hoc networks are, in principle, not scalable. That is, the foreseen benefit of spatial reuse of the radio spectrum produces diminishing returns. The rate at which bandwidth reuse increases with increasing number of nodes cannot catch up with the increasing load demands that the additional nodes place on the network.

The broadcast wireless advantage, furthermore in [2], illustrates the property of manipulate the transmission range. Increasing the transmission power can reach more neighbors while it decreases the spatial reuse and increase the

S. Pierre, M. Barbeau, and E. Kranakis (Eds.): ADHOC-NOW 2003, LNCS 2865, pp. 211–222, 2003.

energy demands. Clearly, in reality, not all nodes are expected to place the same load burden on a wireless network. One cannot discount the value of nodes that potentially forward traffic but do not introduce their own traffic (or lots of their own traffic). In this sense, even if the asymptotic diminishing returns of spatial reuse dominate, the scale of the network at which the introduced load dominates over space reuse may be sufficiently large for most applications. What becomes evident at this point is that the allocation of source-destination demands on actual paths in a wireless network, specifically in an ad hoc network, has not been widely studied as a function of the source-destination loads. The concept of a *load matrix* is certainly old and quite heavily used in circuit switching networks [3]. We will use it in the context of path establishment in a packet-switching wireless network. The idea is to assign all the source-destination paths in an ad hoc network, such that their global (over all nodes) energy demands are collectively minimized. How packet transmissions take place, at each inter-mediate hop, are left outside the scope of the paper. Suffice is to say that we consider the existence of an *ideal* MAC protocol which properly synchronizes the transmission of competing nodes without causing any undue reduction on the effective capacity. The particular problem of implemented MAC protocols that provide coordinated access to the medium in order to capitalize on spatial reuse is left outside the scope of the paper, and it is a research topic on its own right, e.g., [4,5].

In a wireless environment, each transmission is a local broadcast, that is, it is received by more than just the next hop along the path. As such, even though nodes outside the path from source to destination are uninterested in the specific packet (they will neither forward it, nor they are the intended destination), they will nevertheless "hear" it. Assuming the energy required to receive a packet is substantially small, at least compared to transmitting a packet, overhearing packets can result in a small energy penalty but, more importantly, it results in a congestion penalty. That is, the nodes that overhear the packet transmissions have to refrain from using the medium for their own purposes.

Consider a specific node, i, that is within the range of several other nodes. Let us assume that these other nodes transmit traffic of τ_i bits per second total. The specific node is liable for forwarding some of this received traffic, Θ_i bits per second (clearly $\Theta_i \leq \tau_i$). Finally, the node also originates some of the traffic (acts as a source) for a load of A_i bits per second. From the perspective of the node, the wireless medium provides a capacity of C bits per second. In order for the routes in the network to be feasible, the *capacity constraint* $\tau_i + \Theta_i + A_i \leq C$ must be satisfied *at each* nodes in the network. We note however, that any technique that attempts to minimize energy consumption, will attempt to minimize transmission radii. The result is that energy consumption reduction is expected to force τ_i to be reduced as well, except for the fraction that has to be absolutely (because of the topology) forwarded by node i which implies that the least traffic possible as seen from node i is $2\Theta + A_i \leq C$. Nevertheless, τ_i also reflects the fact that nodes are assigned a particular transmission range to ensure the connectivity of the network, or more specifically, the existence of paths from source to destination (if we care only about particular source-destination pairs and not about the connectivity of nodes that have no traffic to send or receive). Hence, too low a τ_i can result in a disconnected topology.

For the sake of exposition, we will consider the problem of finding paths between source-destination pairs in *static* wireless networks. We will be given a traffic matrix which indicates the traffic load between each source-destination pair. The load is asymmetric (that is, the load from A to B is not necessarily equal to the load from B to A) and is non-zero for all source-destination pairs (but can be arbitrarily small). Each source-destination pair will subsequently have to be routed using multiple intermediate hops. If no capacity constraints are present, a source-destination pair can pick the lowest energy cost from source to destination using a conventional shortest path algorithm. That is, the shortest path algorithm can be applied on a graph in which the costs stand for the distance between the nodes raised to the loss exponent. For a single source-destination pair, and if no other traffic or bandwidth constraints existed, that would be the optimum solution. We note that the unicast energy minimization problem is in fact a shortest path problem, compared to the multicast/broadcast energy minimization problem, which is known to be NP-hard [6].

Unfortunately, the general case of the problem (multiple source-destination demands, capacity constraints and wireless broadcast nature) is sufficiently complex to defy currently a simple answer. We note that the first two features (multiple source-destination demands, capacity constraints) render the problem a case of the *node-capacitated multi-commodity flow* class of problems. That is, each node has a capacity as a constraint while each node pair need to established a flow and deliver amount of traffic concurrently in a network. This problem is also shown to be NP hard [7], and many approximation algorithms have been proposed. Unfortunately, the literature on the topic does not consider the broadcast nature of the wireless medium, and thus the fact that τ_i is not just a function of the traffic *intended* to be received by i but also of the traffic of other nodes that are "near" i - i.e., it depends on the geometric features of the topology. It is thus not surprising that results on node-capacitated multi-commodity flow are specific to particular topologies, e.g., rings [8]. The general case appears to be an extremely complex case, even without the presence of mobility.

We point out that several versions of the basic problem exist. For example, transmission and reception by a node are lumped into one capacity constraint only. Clearly, separate channels can be used for transmitting and receiving, with separate fixed capacities. Secondly, we assume knowledge of the traffic loads between all source-destination. If the traffic load matrices present only knowledge of the average load.

2   Algorithms

Let us assume that we have been given the costs D[u][v] between any two nodes u and v. The costs are determined as the Euclidian distance between the points, raised to the power of the loss exponent. The following two simple heuristics aim at producing paths among all source-destination pairs subject to the per-node bandwidth constraints and with the objective of minimizing the power consumption.

The initial topology graph considered, G(V,E), is completely connected. Subsequently edges are removed to ensure that the capacity constraints are not violated. T is the set of source-destination demands, whereby $T_{i,j}$ is the demand (bits per second) from node i to node j. Ps is the set of all paths established with the intention of minimizing energy cost. Shortest_Path(G(V,E), D, source, destination, Path) is any straightforward implementation of the shortest path algorithm from source node to destination node on a graph whose connectivity is captured by G(V,E) and the edge costs are captured by D. The Shortest_Path() returns true if a path is found and false if it could not, because the source and destination are in two different connected components.

2.1   Successive Minimum Energy Paths

Successive_Minimum_Energy_Paths (Input: V, D, P, T; Output Ps)

```
 1: for all u ∈ V do
 2:    C[u] ←0
 3: end for
 4: G(V, E) ← completely connected graph of V nodes
 5: while T ≠ ∅ do
 6:    select T_{i,j} from T
 7:    R ← C
 8:    repeat
 9:       reconstruct ← false
10:       if (!Shortest_Path(G, P, i, j, Path)) then
11:          return (INFEASIBLE)
12:       else
13:          for all (u, v) ∈ Path do
14:             for all x ∈ V do
15:                if D[u][v] ≤ D[u][x] then
16:                   R[x] ← R[x] + T_{i,j}
17:                end if
18:                if R[x] > capacity then
19:                   reconstruct ← true
20:                end if
21:             end for
22:          end for
23:          if (!reconstruct) then
24:             C ← R
25:             Ps ← Ps ∪ Path
26:          else
27:             for all (u, v) ∈ Path do
28:                for all (x, y) ∈ E do
```

```
29:                    if (R[x] > capacity and D[x][y] > D[x][u] − D[u][v]) then
30:                        E = E\(x, y)
31:                    end if
32:                end for
33:            end for
34:        end if
35:    end if
36:    until (!reconstruct)
37:    T ← T\T_{i,j}
38: end while
39: return (FEASIBLE)
```

Psuedocode 1: Successive Minimum Energy Paths

In the first algorithm we create one (for each source-destination pair) minimum energy path at a time. Such path construction can employ the shortest path algorithm, e.g. the Bellman Ford algorithm, with the transmitting energy as the cost. We continue adding paths, until the capacity constraint of one node is exceeded. We will reduce the maximum transmission power level of each node whose transmission violated the capacity of the node in question. The reduction of the transmission power is captured by removing the corresponding edges and, subsequently, rerunning the shortest path. Hypothetically, it is still possible that the new path will influence the load of the removed edges, due to the nearby nodes relaying its traffic. Subsequently, we remove nodes as relays from consideration and re-run the shortest-path algorithm. The process continues until a path can be found that does not exceed the capacity constraints of any of the nodes it traverses through and of any of the nodes that overhear its transmission. Therefore, the order of path construction is critical to the outcome. We also note that certain traffic load matrices are simply infeasible they cannot be accommodated, because one or more source-destination paths cannot be established.

2.2 All Pairs Minimum Energy Paths

All_Pairs_Minimum_Energy_Paths(Input: G(V,E),D,P,T; Output Ps)

```
1: All_Pairs_Shortest_Paths(G, P, Ps)
2: for all u ∈ V do
3:     C[u] ← capacity
4: end for
5: for all T_{i,j} ∈ T do
6:     for all (u, v) ∈ Ps_{i,j} do
7:         for all x ∈ V do
8:             if (P[u][v] ≤ P[u][x]) then
9:                 C[x] ← C[x] − T_{i,j}
10:            end if
11:        end for
```

```
12:    end for
13: end for
14: while (true) do
15:    minNode ← −1
16:    minC ← inf
17:    for all x ∈ V do
18:       if (C[x] < 0 and C[x] < minC) then
19:          minNode ← x
20:          minC ← C[x]
21:       end if
22:    end for
23:    if (minNode = −1) then
24:       return (FEASIBLE)
25:    end if
26:    for all Ps_{i,j}inPs do
27:       for k = 1 to |Ps_{i,j}| − 2 do
28:          s ← Ps_{i,j}[k]
29:          via ← Ps_{i,j}[k + 1]
30:          d ← CurrentPath[k + 2]
31:          if (P[s][via] ≤ P[s][minNode] and
                 P[via][d] ≤ P[via][minNode]) then
32:             for all v ∈ V do
33:                if (D[via][v] ≤ D[via][d]) then
34:                   C[v]+ = T_{i,j}
35:                end if
36:                Ps_{i,j} ← Ps_{i,j} − via
37:             end for
38:          end if
39:       end for
40:    end for
41:    if C[minNode] < 0 then
42:       return (INFEASIBLE)
43:    end if
44: end while
```

Psuedocode 2: All Pairs Minimum Energy Paths

In the second heuristic, All_Pairs_Shortest_Paths(V,D,Ps) is an all pairs shortest path algorithm applied on a V vertex completely connected graph. The costs of the edges are provided in D and the resulting paths returned in Ps. Ps is a set of paths. With slight abuse of syntax, $Ps_{i,j}$ represents the path from i to j in set Ps, and $Ps_{ij}[k]$ is the kth vertex along the paths from i to j from set Ps. This algorithm is based on the idea of all-pairs shortest paths algorithm such as the Floyd-Warshall algorithm, which runs in $O(N^3)$ time. We examine the nodes where the bandwidth constraint is violated. We first find out a node such that its bandwidth violation is the most. Then, we will fix the transmission radii of the nodes in the vicinity of the node found in violation of the capacity constraint. If a path being overheard by the node with the capacity violation turns out to consist of a two-hop sub-path and two corresponding transmission radii cover that constrained node, we will reduce it to one-hop path if possible. The idea is taken from the triangulation relaxation of the shortest path construction. Note that when we compute the minimum energy path, the relaxation step by adding an edge to any existing path can reduce the cost of a specific node. To this end, we have computed the minimum energy path but we would like to satisfy the bandwidth constraint by sacrificing energy consumption. Hence, we will take the reverse procedure of the relaxation step. We will select the first two-hop sub-path with such property, and after we replace it by a single relay, we will remove the load from these two hops and recalculate the load for the new single transmission. This examination process will continue until no node handling or overhearing the traffic or no path re-construction has been made. This approach differs from the previous one in the process of repairing path. The previous algorithm considers the reconstruction of the entire path while this approach considers the subsection of a path that causes the bandwidth constraint violation. Again, it is totally within reason to end up with an infeasible configuration.

3   Simulation Study

The cost we computed captures the global energy consumption. After we construct the paths from all source-destination pairs, we compute the total transmission energy consumption. If a node is involved the communication of different source-destination pair, its transmission power may not be the same in these different paths. As a result, we consider the total transmission power Energy$[i][j]$ for a particular path, and using $T_{i,j}$ as the weight for each paths' transmission power, the average global energy consumption is equal to the sum of $T_{i,j}$ x Energy$[i][j]$. That is, furthering the concept of an ideal MAC, the cost calculation assumes that a node can vary its transmission power depending on the destination of the packet being handled each time. The simulations were conducted with randomly placed nodes within a 1500x500 rectangular area and without node mobility. The capacity of each node, C, was the unit of bandwidth, hence C=1 throughout the simulations. The traffic load matrix, $T_{i,j}$, was produced in a random fashion. Specifically, each element of $T_{i,j}$, is uniformly randomly generated to be a demand between 0 and L/2(N-1) (where L is a parameter controlling the relative load over all nodes and N is the number of nodes). The particular formula guarantees that the sum of traffic originating from source i is less than which guarantees that the load of another nodes due to forwarding the load of this source-destination pair is going to be less than 1 (i.e. $A_i = y : x = i \wedge j \in V \wedge y = \sum T_{x,j} \leq 1$). Note however the restriction of the traffic matrix and capacity generation is not sufficient to avoid infeasible solution scenarios. The relative load is a predefined parameter while the absolute load is defined as the

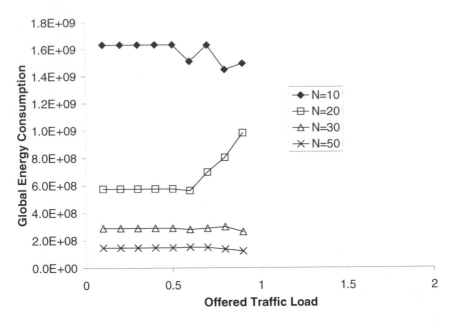

Fig. 1. The global energy consumption for Successive Minimum Energy Paths Algorithm selecting the source-destination paths to admit in arbitrary node number.

sum of all traffic load divide by the relative load $(\sum T_{i,j}/\text{L})$. The simulation are for N=10, 20, 30, and 50 nodes and, L=0.1 to 2 in 0.1 interval. The traffic load is between 0 and 2 because we would like to show the spatial reuse feature of ad hoc network (revealed when the load is larger than 1). The simulation results are shown in Fig. 1 and Fig. 2 for the first algorithm and in Fig. 3 and Fig. 4 for the second algorithm.

The results are expressed as a relation between global energy consumption and the average traffic load. Our intuition would suggest that the more the load, the more the nodes that end up violating their respective constraints, the longer the paths to avoid such congested nodes, hence, the more the energy required. This situation is partly what happens in Fig. 1. However, the graph shows that the energy consumption drops when the load is near 0.8. The figure is misleading in this respect because what is missing is the fact that several of the runs that correspond to the point at 0.8 and higher resulted in infeasible scenarios. Fig. 2 demonstrates the ratio of infeasible solution corresponding to the result generated from the successive minimum energy paths algorithm with unordered path construction. With 50 nodes, the number of infeasible solution is around 90 A more definite result from our simulations suggests that the global energy decreases as the number of nodes increases with the same simulation area. This is because and additional node provides an opportunity for paths to be split along a longer path where the sum of energy required over the entire path is lower than with fewer intermediate hops. Nevertheless, this cannot counter the fact that additional nodes produce a higher node density and a higher probability

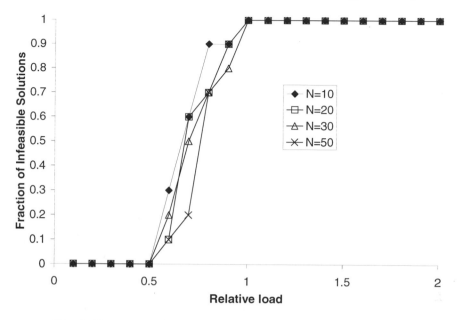

Fig. 2. Percentage of infeasible solution corresponding to Fig. 1.

that the transmissions will congest other nodes, and hence it restricts paths from being available. The unavailability of a path therefore has a direct impact on the ability of conserve energy by using it as a relay.

Applying the Successive Minimum Energy Paths construction, we find that it is highly unlikely to obtain the optimal solution (or even a solution when the traffic demand is high) for this multiple source-destination demands and energy consumption optimization problem. Our experiments were also run with different order for path construction, such as increasing traffic demands or decreasing traffic demands for the source-destination pairs, but the results are similar to the ones reported here. The Successive Minimum Energy Paths algorithm constructs one path at a time using the shortest path algorithm using the energy cost matrix. Each path construction achieves the minimum energy consumption requirement of a particular path. However, in the Successive Minimum Energy Paths algorithm, the bandwidth reservation is performed by ignoring the already allocated source-destination pairs.

Furthermore, we observe that in order to minimize the energy consumption, the path construction process will likely end up with a longer path. Assume S is the source node and D is the destination node. An additional node R can either act a relay transmission such that the path is from S to R and from R to D or sitting there overhearing it. The distance between a node pair is denoted $d_{i,j}$, and the energy usage is $(d_{i,j})^{\alpha}$, where α is corresponding to the loss exponent $(2 \leq \alpha \leq 4)$. Assume $d_{S,D}$ is equal to 5 units of distance; $d_{S,R}$ and $d_{S,D}$ are equal to 3 units. With the energy as the cost matrix, since $(d_{S,D})^{\alpha} \geq (d_{S,R})^{\alpha} + (d_{R,D})^{\alpha}$, R will not act as an relay, and therefore D will increase the

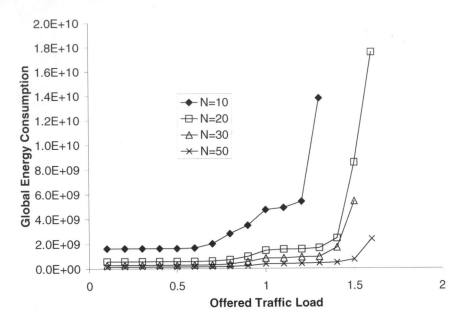

Fig. 3. Simulation results for All Pairs Minimum Energy Paths

overhearing traffic twice. Although the energy consumption is reduced, each node consumes capacity because of overhearing the traffic (S will overhear the relaying transmission of R, and R will receive and transmit, acting as relay, the traffic from S, thus consuming its capacity). As the loss exponent intensifies, this scenario occurs more often. Even if we attempt to reconstruct the particular constraint-violated path, there may not exist a route from the source to the destination. As a result, most of the overhearing traffic drains out the bandwidth capacity and results in the infeasible solution.

Our conclusion from the Successive Minimum Energy Paths construction is that using the shortest path algorithm with the transmission power as the cost matrix results in the minimum energy path construction for the particular source-destination pair. However, the overall traffic demands grow due to the use of relays and the capacity allocation becomes infeasible. The All Pairs Minimum Energy Paths Construction creates all minimum energy paths without the knowledge of the traffic demands. The resulting energy cost is subsequently compromised (increased) in order to "fix" the capacity violations, using the inverse of the triangle relaxation.

The simulation of All Pairs Minimum Energy paths uses the same input parameters, and the result are also expressed with global energy consumption and the probability of having infeasible solution shown in Fig. 3 and Fig. 4 respectively. A comparison of Fig. 2 and Fig. 4 suggests that All Pairs Minimum Energy provides a better potential for spatial reuse (the load exceeds 1 and still results in feasible configurations). In fact, when the traffic load is equal to 1.3 in the 50 nodes scenario, the number of infeasible solutions was 0. The energy

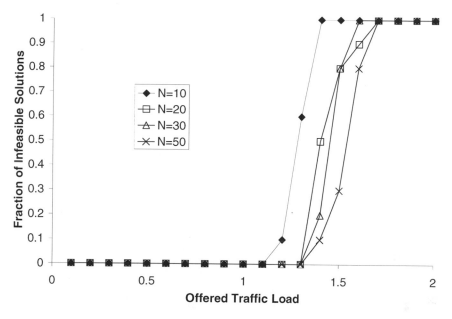

Fig. 4. Percentage of infeasible solutions corresponding to the Fig. 3.

consumption curve rises when the load is around than 0.6. A similar observation from the previous result is that the global energy consumption increases as the number of participating node decreases. Still, as the load increases, the global energy consumption curve rise abruptly capturing an, almost exponential, increase of energy consumption in order to produce a feasible solution. The first explanation is that the energy consumption rises quickly because we do not find two particular small amount energy transmissions to replace via the inverse-triangle relaxation. Another reason is, of course, that the load cannot be accommodated.

4   Conclusions

In this paper we considered the problem of energy minimization in a bandwidth constrained ad hoc wireless environment. The apparent tendency of energy minimization to reduce transmission radii would appear to be in agreement with spatial reuse (which benefits from transmission radius reduction as well). Unfortunately, in a bandwidth constrained setting, it becomes immediately obvious that minimization of energy results in longer paths, adding to the congestion and diminishing the benefits of spatial reuse. The observation applies to large network configurations, leaving potential for smaller network configurations (load-wise and node-wise) to still benefit by the choice of the right path construction heuristic. Two such heuristics are studied in this paper, and it appears that a scheme that attempts to reduce the load of relay nodes could result in improved performance for a modest sacrifice in energy consumption. The next step is to attack the problem from the viewpoint of optimization to appreciate how the introduction of each capacity constraint diverts the search towards the mimimal energy solution.

References

1. P. Gupta and P. Kumar, "Capacity of Wireless Networks," IEEE Transactions on Information Theory, vol.46, no 2, p.388-404, Mar 2000.
2. J. E. Wieselthier, G. Nguygen, and A. Ephremides, "On the Construction of Energy Efficient Broadcast and Multicast trees in Wireless Networks," In Proc. of IEEE INFOCOM 2000, Tel-Aviv, Israel, vol.2, p.585-594, Mar 2000.
3. A. Medina, N. Taft, K. Salamatian, S. Bhattacharyya, and C. Diot, "Traffic Matrix Estimation: Existing Techniques and New Directions," In Proc. of the 2002 conference on Applications, Technologies, Architectures, and Protocols for Computer Communications, Pittsburgh, Pennsylvania, USA, p.161-174, Aug 2002.
4. M. Marina, G. Kondylis, and U.C. Kozat, "RBRP: A Robust Broadcast Reservation Protocol for Mobile Ad Hoc Networks," In Proc. of IEEE/ICC-2001, Helsinki, Finland,vol.3, p.878-885, Jun 2001.
5. Z. Tang and J. Garcia-Luna-Aceves, "A Protocol for Topology-Dependent Transmission Scheduling," In Proc. of IEEE WCNC 1999, New Orleans, Las Angelas, USA, vol.3, p.1333-1337, Sep 1999.
6. M. Cagalj, J.-P. Hubaux, and C. Enz, "Minimum-Energy Broadcast in All-Wireless Networks: NP-Completeness and Distribution Issues," In Proc. of 8th Annual International Conference on Mobile Computing and Networking, Atlanta, Georgia, USA, p.172-182, Sep 2002.
7. V. Guruswami, S. Khanna, B. Shepherd, R. Rajaraman and M. Yannakakis, "Near-Optimal Hardness Results and Approximation Algorithms for Edge-Disjoint Paths and Related Problems," In Proc. of the 31st Annual ACM Symposium on Theory of Computing, Atlanta, Georgia, USA, p.19-28, May 1999.
8. V. Tandon, A. Frank, and Z. Vegh, "Node-capacitated Multicommodity Routing for a Ring," Math. of Operations Research, vol.27, no.2, p.372-383, May 2002.

Automated Meter Reading and SCADA Application for Wireless Sensor Network

Francisco Javier Molina, Julio Barbancho, and Joaquin Luque

Departamento de Tecnologia Electronica, University of Seville,
C/ Virgen de Africa, 7. Seville 41011, Spain
{fjmolina,jbarbancho,jluque}@us.es
Tel.: (+034) 954 55 28 35, Fax: (+034) 954 55 28 33

Abstract. Currently, there are many technologies available to automate public utilities services (water, gas and electricity). AMR, Automated Meter Reading, and SCADA, Supervisory Control and Data Acquisition, are the main functions that these technologies must support. In this paper, we propose a low cost network with a similar architecture to a static ad-hoc sensor network based on low power and unlicensed radio. Topological parameters for this network are analyzed to obtain optimal performances and to derive a pseudo-range criterion to create an application-specific spanning tree for polling optimization purposes. In application layer services, we analytically study different polling schemes.

Keywords: Automated Meter Reading Application, SCADA, Ad Hoc Networks, Spanning Tree Algorithm, Multihop Routing Protocol.

1 Introduction

Since the 70's, many technologies have been developed for Automatic Meter Reading functions (AMR) and Distribution Automation (DA) for utility applications (water, gas and electricity) [1,2]. Many studies show that solutions based on low power radio networks are viable and that they offer the best cost/performance ratio[3,4,5]. However, it is only in the late 90's that, radio and microcontroller technologies have allowed the development of smart sensor networks. We propose (in this paper) the use of ad-hoc network technologies to support this application because:

- ad-hoc protocols are best suited to low power systems,
- nodes can be located without pre-planning,
- and topology is more flexible, making management simpler.

Public utilities' management has many different aspects closely interrelated that must be coordinated within a corporative network: (e.g. Meter reading from customer meters, Distribution management, Economic dispatch...) These applications are often distributed throughout many computers. But, customer data polled from sensor networks, queries, remote control orders and network management messages must be processed by a unique computer named UC -

S. Pierre, M. Barbeau, and E. Kranakis (Eds.): ADHOC-NOW 2003, LNCS 2865, pp. 223–234, 2003.

Utility Controller. UC works as the master that controls many remote units (sensor nodes), like a well known architecture called SCADA, Supervisory Control and Data Acquisition. We have called this application ASCADA, Augmented SCADA, as it has to support typical SCADA functions and additional AMR services. Some of these are listed below:

a) **AMR functions.** Meter reading or checking an individual customer, a group-cluster of meters, and all meters (global reading). These queries could be simultaneous, and the execution time must be as short as possible to reduce the reading period. Nowadays, the on-site reading period, carried out by an operator, is about two months. The goal is to manage global readings daily or weekly.

b) **Telemetry functions.** These services obtain data from sensors, and they control some elements located at selected points of the distribution network (flow, power, state of valves or switches, etc.) Distribution sensing and automation will enhance supply services, reducing failure, alarm and response times. All this data must be polled periodically, and the completion time must be as short as possible to reduce the bandwidth load.

c) **Remote control orders.** Security and reliability are the main characteristics of these services. Minimization of transmission time is a general objective. In this case, it is quite important to reduce multiple hops and provide dynamic routing capabilities to enhance reliability. Further functions would be:
   - encrypted data and sender identification for secure operations,
   - order sequencing to avoid duplication,
   - receipt request for confirmation,
   - message transmission indicating the end of the command.

c) **Alarm transmission.** From distribution elements, nodes detect transmissions caused by an exceptional situation. Nodes from customer meters must not have this service to avoid network overload.

Some AMR services need to use a high percentage of network capacity. This fact will be present throughout the paper. The next section analyses topology characteristics more closely related to individual and overall polling. Section 3 presents different strategies to compute and optimize global polling and simulation results. Finally, we outline future work and alternative solutions that are currently being tested.

2   Topological Model and Properties

The IEEE working group SC-31 has proposed a set of topological models for AMR systems. Figure 1 shows the model based on a fixed radio network. Each element is conceptual and does not necessarily exist in the form shown. Devices in the topology could be combined or reduced to a null element. When they are present, elements A, B, C, D and X are intermediate devices. The topology can support multiple delivery points, as shown, separated by service boundaries.

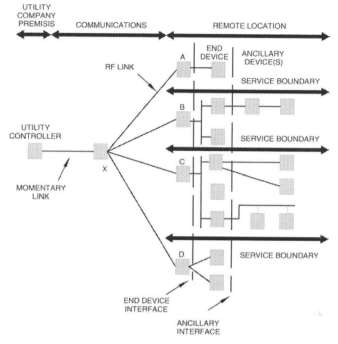

Fig. 1. Radio AMR topology model.

Table 1. Radio ranges on different scenarios and 100mW radiated power.

Band	Indoors	Outdoors (with obstacles)	Outdoors (with in line-sight)
433 MHz	<100 meters	<500 meters	1 - 2 Km
800 MHz	<50 meters	<300 meters	0.8 - 1.5 Km
900 MHz	<30 meters	<200 meters	< 800 meters

Sensors connected to End Device and data stored in memory may vary depending on location or application: customer buildings (meters) or distribution surveillance and control (flow meters, power meters, valves or switchers states).

For this project, only application, network, link and physical layers have been developed. Many of them were inspired by ad-hoc network protocols [6,7,8]. Application layer services include ASCADA functions for local and cluster meter reading, tampering warning, alarm warning and the remote control of actuators. The Network Layer supports a multihop routing protocol and network topology maintenance. Data Link Layer offers a medium access contention (CSMA CA), sequence and synchronization control. But the Physical Layer is probably one of the most important aspects. License exempt standards fix frequency bands and radiated power output for telemetry applications, so consequently radio range is also limited. Table 1 shows European ETS300-220 limits and bands for telemetry, and radio ranges for different locations.

The End Devices used include radio OEMs which are compliant with ETS 300-220. In European cities, most of them are located indoors (but shouldn't), so the radio range ends up being about 100 m (433 MHz band). The Radio Network consists of many End Devices forming a dense network where each node needs a multihop transmission to reach the Utility Controller node. There is no planning to select node locations, so the network topology has an arbitrary structure like an ad-hoc network, although the amount of nodes will be greater (thousands for medium size cities).

Currently, ad-hoc networks are classified into two categories:

1. Mobile Ad-hoc Networks -MANET.
2. Sensor Ad-hoc Networks.

Although the proposed network has some common aspects with sensor networks, it differs from both:

- Nodes have no mobility.
- Communication is usually between nodes and the UC.
- Power is not a main priority.
- Nodes are prone to failure.
- They are densely deployed within the range area.
- There are few topological changes (on very few occasions a node is added or eliminated and radio range changes rarely occur).

Ad-hoc networks do not have any special nodes. However, for an application layer, the Utility Controller will have true a special node. We will use this property for optimizing network performance. To measure the relationship between these network performances and topology, we define a simple parameter - Medium Number of Hops a node needs to reach the Utility Controller. is related to medium access time from UC to a node, and to global polling time, also. We have estimated in various scenarios. The first one is shown in figure 2, and it assumes the following conditions:

1. Network nodes are uniformly distributed across the city, so we can define a density parameter.
2. All devices are indoors, so the radio range will be short and the same for all of them.
3. UC is located at center of the net.

We will refer to this topology as SR - Short Range Topology. We have proven that depends mainly on geometric parameters, and it can be computed approximately when $R_{GC} \gg R_{SR}$ as it follows:

$$\overline{NH} \approx \frac{2}{3}H = \frac{2}{3}\frac{R_{GC}}{R_{SR}} \tag{1}$$

Where

- R_{GC} - *Global Radius*. It is the radius of a circle that covers all nodes in the city or a significant number of them.

UC- Utility Controller

Node

R_R- Radio Range

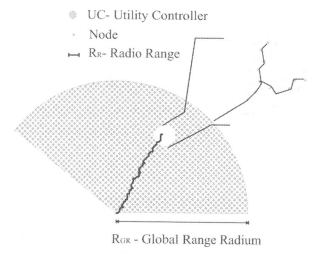

R_GR - Global Range Radium

Fig. 2. SR - Short Range Topology.

– R_{SR} - *Short Radio range.* It is the medium value of short radio range of communication equipment.

$\overline{NH}$ is closely related to network topology. It enables computing network performances for several protocols and topologies. Minimizing $\overline{NH}$, will greatly enhance the completion time for the global reading service. This optimization is very important because global reading would probably be the major load service. Other functions like order and alarm transmission will also be enhanced. As $\overline{NH}$ decreases, transmission reliability will grow and service execution time will reduced. One way to enhance previous SR topology is to use those nodes that have much longer range than R_{SR} as a bridge to reach UC reducing intermediate hops. For example, equipment located outdoors may have a range up to ten times greater. Long range nodes could act as a long range subnetwork, able to connect any city area with the UC through fewer intermediate hops. Network protocols must enable message flow between short range nodes and the closest long range nodes in order to continue through the long-range subnetwork. In this way, the network is divided into different clusters within a main node that belongs to the long range subnet (see figure 3). We will refer to this topology as SR-LR (Short Range - Long range) architecture.

In this case, as usual, the number of short range nodes is significantly greater than the number of large range nodes. It can be shown that $\overline{NH}$ will be:

$$\overline{NH}_{SR-LR} \approx \overline{NH}_{SR} + \overline{NH}_{LR} = \frac{2}{3}\frac{R_{LR}}{R_{SR}} + \frac{2}{3}\frac{R_{GC}}{R_{LR}} \qquad (2)$$

Note that $\overline{NH}_{SR}$ does not depend on geometric parameters. It depends on radio transmission characteristics. However, $\overline{NH}$ depends on city geometry. A medium number of hops resulting from SR-LR architecture is significantly less

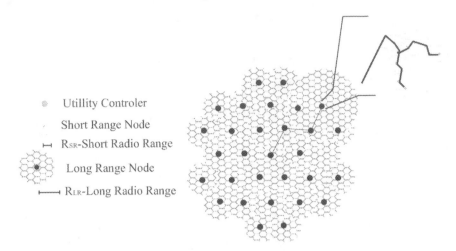

Utillity Controler

Short Range Node

R_{SR}-Short Radio Range

Long Range Node

R_{LR}-Long Radio Range

Fig. 3. SR-LR architecture.

than SR architecture. For example, in a medium-size, European city like Seville (Spain) 95% of the population is located inside of a circle of about $R_{GC} \approx 5Km$. Then:

$$\overline{NH}_{SR} \approx \frac{2}{3}\frac{5000}{1000} = 33.33 \quad \overline{NH}_{SR} \approx \frac{2}{3}\left(\frac{1000}{100} + \frac{5000}{1000}\right) = 10 \qquad (3)$$

$\overline{NH}$ is generally a geometric parameter that we have to relate to network or protocol characteristics. As is shown in figure 4, messages flow between the UC, and nodes. They can be described as a token moving through a spanning tree. Each node represents a hop, and its branches show some of the equipment inside its radio range. A first approximation of the number of transmissions may be derived from in a simple way, in terms of tokens moving in a spanning tree, under the following conditions:

1. There is only a token in the tree, so there are no collisions.
2. The token always jumps between parent and children without retransmission.
3. Tokens flow following minimum paths.

Then:

$$\overline{NT} = \overline{NH} \qquad (4)$$

These conditions draw an ideal scenario:

- where a tree grows,
- radially without loops,
- all the nodes are in service,
- and there are no collisions or interference.

A first approch for real scenarios can be derived, supposing that p is the error rate of transmissions between nodes. In this way retransmissions will be necessary. We assume also, that p is uniform within the radio range and the same for all nodes in the network. Then, we have proved that the maximum value for medium number of transmissions is:

$$\overline{NT}_{MAX} = \frac{1}{1-p} \cdot \overline{NH} \tag{5}$$

Again $\overline{NT}$ depends on $\overline{NH}$ directly. So, finding the short paths to the UC sold be the basis of topology management and performance optimization. We have developed a custom algorithm to find these paths, but any other would be possible [9,10,11,12,13,14] (most of these are designed for point to point communication). However, ASCADA defines the UC at the application layer as the main node. It is often present as a transmitter or receiver. Topologies described previously make use of these properties to minimize multiple hops and create preference paths to the UC (based on a minimum hops criteria). Moreover, sensor networks usually select paths based either on power or quality link criteria [15,16]. Because a minimum number of transmissions is the objective, our selected criteria quantify these transmissions as close as possible to reality. Let i, j be two neighbor nodes. We define as *pseudo-range from node i to the UC through j*:

$$\rho_{ij} = d_j + \overline{nt_{ij}} \qquad d_j = \min(\rho_{jx}) \qquad \forall x \tag{6}$$

Where d_{ij} is the minimum pseudo-range between j and UC, and $\overline{nt_{ij}}$ is the medium number of transmissions between i and j, then, equation 6 assures the location of a short path following a reasonable link quality along it, and loop free.

3   Augmented Scada Application Optimization

Nodes always know the path to root through the parent node, but any selected algorithm must allow the root node (UC) to have an approximate image of current tree topology. This way, packets from nodes to the root do not contain information about routing. Conversely, messages from root to nodes generally use an explicit routing scheme, so packets must contain information about the path. Packet header size must be optimized because radio frames should be as short as possible to reduce the transmission error ratio, effective bandwidth, etc.

We propose in the paper to optimize application services in such a way so that message traffic and routing header size are minimal. In the following, we summarize all these application services:

AMR services:

- *AMR_Read*: Root sends a read message to a node to read data or check a meter.
- *AMR_Poll*: Root sends a message to read data from all meters in a node.
- *AMR_Collect*: Root sends a message again, to read data from all nodes of a subtree.

Telemetry services:

- *TLM_Polling.* This service initiates a periodic polling to a subset of nodes, located at distribution points. It creates and updates a table called Image Table with data from sensors which include a timestamp. This table may be accessed by custom primitives (*TLM_Read*).
- *TLM_Read.* Return data from image table with integrity information (timestamp).

Remote control services:

- *RC_Send.* Root sends a message with orders to control remote device (valves, breakers...). Message must be sequenced to avoid duplication and encrypted for secure operation. It may be necessary to notify a receipt message, and an order completion message.
- *AMR_Collect* service may be the most complex because time minimization and network overhead reduction are quite difficult to optimize simultaneously. There are many studies for polling optimization [17,18], but most of them use a well defined topology (rectangular, hexagonal...). The proposed network is a random network, we only suppose that nodes are uniformly deployed either as short-range nodes or long-range nodes. Execution time and network overheads can be evaluated approximately from topology parameters (see previous section). These values may be used to compare how different polling schemes may optimize network performance.

From the medium number of transmissions, we can compute execution time in an ideal context (equation 4) or in a more real one (equation 5). We can consider the collect message like a token moving along the branches from root to nodes, and the answer as another token returning back from nodes. The first approach is a simple collecting schedule consisting of polling each node from the root and, individually, waiting for the answers. Nodes do not have an application layer in that case, and there is only a single token moving in the tree. Moreover, let us consider an ideal scenario with the following conditions:

1. There is no interference or collisions.
2. Waiting time to medium access is zero.
3. Protocol stack computing time is negligible.
4. Only one frame is necessary to transmit all data from a node.
5. All tokens have exactly the same size.

An approach value of execution time, for an overall collect order, may be derived from a medium number of transmissions (equations 2,3,4), as:

$$CT_1 = \left(\text{PTS} \cdot \text{N} \cdot \overline{NT} + \text{ATS} \cdot \text{N} \cdot \overline{NT}\right) \cdot \frac{\text{CharSize}}{\text{BaudRate}} \tag{7}$$

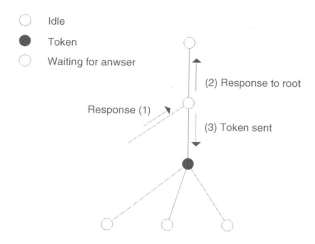

○ Idle

● Token

○ Waiting for anwser

(2) Response to root

Response (1)

(3) Token sent

Fig. 4. Polling scheme 2.

Where

- CT_1 - Collect Time for schedule 1.
- N - Network nodes.
- PTS - Polling Token Size.
- ATS- Answer Token Size.

Second collecting algorithm is shown in figure 4. An application layer in all the nodes receives the order. This node passes the token to one of its children, and waits for a response. When it is received, (figure 4(1)) the node sends data to the root (figure 4(2)), and it passes the collect order to the next child (figure 4(3)). Only when there are no more children to be polled, will the node answer with its own data. Only one token is being passed between parent and children, so no path header information is necessary, and the answer token is simultaneously flowing to the root. To prevent a lost token, a timeout period guarantees the token passes to the next child.

Using the previous scenario, we can compute an approximate collecting time value for that schedule. Neglecting initial transmissions and second order effects, the majority of transmissions are caused by answer token passing. Polling token passing occurs simultaneously and so does not have any significant effect. In this way:

$$CT_2 = N \cdot ATS \cdot \overline{NT} \cdot \frac{\text{CharSize}}{\text{BaudRate}} \tag{8}$$

The ratio between CT_2 and CT_1:

$$\frac{CT_2}{CT_1} = \frac{N \cdot ATS \cdot \overline{NT}}{N \cdot PTS \cdot \overline{NT} + N \cdot ATS \cdot \overline{NT}} = \frac{1}{1 + \dfrac{PTS}{ATS}} \tag{9}$$

Usual PTS and ATS values allow a relative decrement of about 70 per cent. A greater reduction can be reached by using multiple tokens. If collisions or

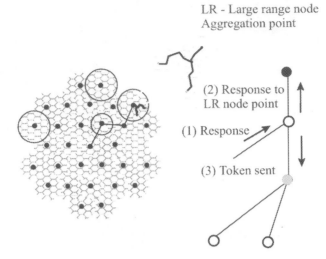

Fig. 5. Simultaneously collecting tokens: polling scheme 3.

interference are not considered, then collecting time decreases to a factor equal to the number of tokens. Obviously, collisions increase as the number of tokens increase. Many response tokens travel to the root node, creating an implosion problem [19] (which is greater near the root node). To avoid this, we propose a polling scheme based on Short range - Long Range topology. The root node must send polling tokens to each node in the LR subnet. Each LR node collects data from all their sub-trees, aggregating and storing data, without passing them to root. LR nodes only send data when the UC requests them. The node application protocol does not change significantly with respect to the second schedule, so this variant is fundamentally the same one for nodes. As it is shown in figure 5, polling tokens are passing simultaneously over different areas, so collision probability is low. Supposing that all the clusters have approximately the same number of nodes (long-range and short range densities are homogeneous), neglecting collisions and second order effects, then:

$$CT_3 = \left(\frac{N_{SR}}{N_{LR}} \cdot \text{ATS} \cdot \overline{NT}_{SR} + N \cdot \text{ATS} \cdot \overline{NT}_{LR}\right) \cdot \frac{\text{CharSize}}{\text{BaudRate}} \quad (10)$$

And supposing that $N_{SR} \approx N$, and applying (equation 2)

$$\frac{CT_3}{CT_1} = \frac{1}{1 + \dfrac{PTS}{ATS}} \cdot \frac{1}{1 + \dfrac{\overline{NT}_{SR}}{\overline{NT}_{LR}}} \quad (11)$$

Using topology parameters for Seville, this scheme may reach $CT_3 = CT_1/10$.

4 Future Works

For the proposed network, we are currently working on network simulation, self-configuration and optimization of ASCADA services performance. For the latter, we are studying two complementary polling schedules: Avalanche token passing and data catching/pre-collecting. To reduce interference and to enhance polling order diffusion, each node sends two or more tokens to the children and responses are aggregated to send a single response to the root. Nodes must be selected in such a way that interference probability would be the least possible. Data catching and pre-collecting schemes use data validation to update stored data. Nodes save data from their children with a timestamp, so when a polling token arrives, the node may use this data without passing it to them. Catching update strategies, as periodically or predictive pre-collecting, must be designed to minimize data age and optimize network performance.

Execution times and network overload optimization are the main objectives of algorithms presented in this paper. Reliability however, must also be an important characteristic for a SCADA system.

Currently, we are working on reliability enhancement. Services such as order messaging and alarm transmissions can be critically affected by local failures, interference or collisions. Reliability must be present in all protocol layers, but especially in application and network layers. Some characteristics previously outlined may raise reliability in application layers: order sequencing, encrypted messages, etc. At the network layer, we are researching a routing algorithm able to find an alternate path when an error has occurred. We are testing a modification of the Fish Eye Routing and other algorithms [20,21].

Acknowledgment

All of this work has been made possible thanks to the project being jointly financed by the Spanish Government (*MYCT-Ministerio de Ciencia y Tecnología*) and private companies, in particular EMASESA and ISOTROL.

References

1. A. Bond. The water industry (automatic meter reading). *IEE Colloquium on 'Low Power Radio and Metering'. IEE, London, UK*, (Digest No.1994/060), 1994.
2. Philips M. Adams, B. Trends towards standard communications for metering. *Ninth International Conference on Metering and Tariffs for Energy Supply. IEE, London, UK*, (Conf. Publ No.462), 1999.
3. A.M. Fox. The business case for radio based amr. *IEE Colloquium on 'Low Power Radio and Metering'. IEE, London, UK*, (Digest No.1994/060), 1999.
4. Radford D. Mak, S. Design considerations for implementation of large scale automatic meter reading systems. *IEEE Transactions on Power Delivery*, 10(Digest No.1994/060), Jan 1995.
5. F.J. Molina, M.G. Gordillo, J. Luque, and J. Barros. Radio network architecture for automatic meter reading. *Conférence Internationale des Grandes Réseaux Électiques, Krakow POLAND*, 1999.

6. J. Broch, D. A. Maltz, D. B. Johnson, Y.C. Hu, and J. Jetcheva. A performance comparison of multi-hop wireless ad hoc network routing protocols. *MOBICOM, Dallas, TX*, Aug 1998.
7. E. Royer and C.K. Toh. A review of current routing protocols for ad hoc mobile wireless networks. *IEEE Personal Communications*, 6, Apr 1999.
8. S. Das, C. Perkins, and E. Royer. Performance comparison of two on-demand routing protocols for ad hoc networks. *IEEE, INFOCOM 2000*, 2000.
9. Z. J. Wang and J. Crowcroft. Analysis of shortest-path routing algorithms in a dynamic network environment. *ACM SIGCOMM*, 1992.
10. J. Behrens and J. J. Garcia-Luna-Aceves. Hierarchical routing using link vectors. *IEEE, INFOCOM 1998*, 1998.
11. J. J. Garcia-Luna-Aceves and M. Spohn. Source-tree routing in wireless networks. *Proc. IEEE ICNP 99, 7th ntl. Conference on Network Protocols, Toronto, Canada*, Oct 1999.
12. D. B. Johnson and D. A. Maltz. *Dynamic Source Routing in Ad Hoc Wireless Networks. In Mobile Computing.* Kluwer Academic Publishers, 1996.
13. C. E. Perkins and E. M. Royer. Ad-hoc on demand distance vector routing. *WM-CSA'99, New Orleans, LA*, Feb 1999.
14. Couto and B Aguayo. Performance of multihop wireless networks: Shortest path is not enough. *Proceedings of the HotNets, Princeton, New Jersey*, Oct 2000.
15. S. Singh and C. S. Raghavendra. Power aware routing in mobile ad hoc networks. *Proceedings of MOBICOM*, 1998.
16. Chandrakasan A. Heinzelman, R. W. and H. Balakrishnan. Energy-eficient routingprotocols for wireless microsensor networks. *In Hawaii International Conference on System Sciences (HICSS '00)*, Jan 2000.
17. G.A. Cheston and S. T. Hedetniemi. Polling in tree networks. *Proc. Second West Coast Conf. Computing in Graph Theory*, 1983.
18. A. A. Rescigno. Optimal polling in communication networks. *Parallel and Distributed Systems, IEEE Transactions on*, 8, May 1997.
19. T. Imielinski and S. Goel. Querying and monitoring deeply networked collections of physical objects. *Proceedings of MobiDE'99, (Seattle, Washington)*, 8, Aug 1999.
20. Gerla M. Pei, G. and T. W. Chen. Fisheye state routing: A routing scheme for ad hoc wireless networks. *Proceedings of ICC 2000, New Orleans, LA*, 8, Jun 2000.
21. M. Spohn and J.J. Garcia-Luna-Aceves. Neighbourhood aware source routing. *Proc. of ACM Symposium on Mobile Ad Hoc Networking and Computing (Mobi-HOC '01)*, 2001.

Range Assignment for High Connectivity in Wireless Ad Hoc Networks

Gruia Calinescu and Peng-Jun Wan

Department of Computer Science, Illinois Institute of Technology,
Chicago, IL 60616
calinesc@iit.edu, wan@cs.iit.edu

Abstract. Depending on whether bidirectional links or unidirectional links are used for communications, the network topology under a given range assignment is either an undirected graph referred to as the symmetric topology, or a directed graph referred to as the asymmetric topology. The Min-Power Symmetric (resp., Asymmetric) k-Node Connectivity problem seeks a range assignment of minimum total power subject to the constraint the induced symmetric (resp. asymmetric) topology is k-connected. Similarly, the Min-Power Symmetric (resp., Asymmetric) k-Edge Connectivity problem seeks a range assignment of minimum total power subject to the constraint the induced symmetric (resp., asymmetric) topology is k-edge connected.
The Min-Power Symmetric Biconnectivity problem and the Min-Power Symmetric Edge-Biconnectivity problem has been studied by Lloyd et. al [21]. They show that range assignment based the approximation algorithm of Khuller and Raghavachari [17], which we refer to as **Algorithm KR**, has an approximation ratio of at most $2(2 - 2/n)(2 + 1/n)$ for Min-Power Symmetric Biconnectivity, and range assignment based on the approximation algorithm of Khuller and Vishkin [18], which we refer to as **Algorithm KV**, has an approximation ratio of at most $8(1 - 1/n)$ for Min-Power Symmetric Edge-Biconnectivity.
In this paper, we first establish the NP-hardness of Min-Power Symmetric (Edge-)Biconnectivity. Then we show that **Algorithm KR** has an approximation ratio of at most 4 for both Min-Power Symmetric Biconnectivity and Min-Power Asymmetric Biconnectivity, and **Algorithm KV** has an approximation ratio of at most $2k$ for both Min-Power Symmetric k-Edge Connectivity and Min-Power Asymmetric k-Edge Connectivity. We also propose a new simple constant-approximation algorithm for both Min-Power Symmetric Biconnectivity and Min-Power Asymmetric Biconnectivity. This new algorithm is best suited for distributed implementation.

1 Introduction

Recently, range assignment problems for wireless ad hoc networks have been studied extensively. In wireless ad hoc networks no wired backbone infrastructure is installed and communication sessions are achieved either through a single-hop transmission if the communication parties are close enough, or through

S. Pierre, M. Barbeau, and E. Kranakis (Eds.): ADHOC-NOW 2003, LNCS 2865, pp. 235–246, 2003.
© Springer-Verlag Berlin Heidelberg 2003

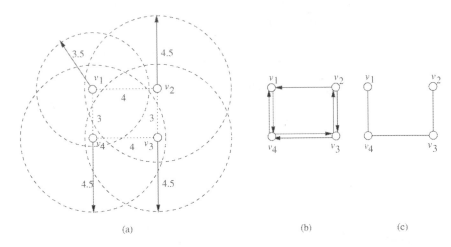

Fig. 1. The network topology: (a) the nodes and their transmission ranges, (b) the asymmetric topology, and (c) symmetric topology.

relaying by intermediate nodes otherwise. Omnidirectional antennas are used by all nodes to transmit and receive signals. Such antennas are attractive due to their broadcast nature. A single transmission by a node can be received by many nodes within its vicinity. We assume that every node can dynamically adjust its transmitting power based on the distance to the receiving node and the background noise. In the most common power-attenuation model [22], the signal power falls as $\frac{1}{d^\kappa}$ where d is the distance from the transmitter antenna and κ is a real *constant* between 2 and 5 dependent on the wireless environment. We assume that all receivers have the same threshold for signal detection, and normalize this threshold to one. With these assumptions, the power required to support a link between two nodes separated by a distance d is d^κ.

The network topology of a wireless ad hoc network, which consists of all possible one-hop communication links among the nodes, is determined by the transmission ranges of the nodes. Depending on whether *unidirectional* links or *bidirectional* links are used for communications, the network topology is represented by either a directed graph referred to as the *asymmetric topology*, or an undirected graph referred to as the *symmetric topology*. In the asymmetric topology, there is an arc from a node u to another node v if and only v is within the transmission range of u. In the symmetric topology, there is an edge between two nodes u and v if and only they are within the transmission ranges of each other. An example is depicted in Figure 1. Figure 1 (a) gives the positions and the transmission ranges of all nodes. The asymmetric topology and the symmetric topology are given in Figure 1 (a) and (b) respectively.

Connectivity is one of the most important properties of an wireless ad hoc network. By asymmetric k-node (resp., k-edge) connectivity we mean the asymmetric topology is k-node (resp., k-edge) (strongly) connected, and by symmetric k-node (resp., k-edge) connectivity we mean the symmetric topology is k-node

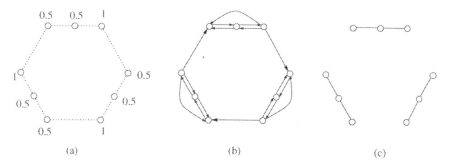

Fig. 2. Asymmetric topology may have higher connectivity than symmetric topology. (a). The nodes lie in a regular hexagon of side equal to one, and their transmission ranges are given beside the nodes. (b) The asymmetric topology is connected. (c). The symmetric topology is disconnected.

(resp., k-edge) connected. For $k = 1$, edge and node connectivity are identical to each other, and thus are simply referred to as connectivity. For $k = 2$, 2-node connectivity is simply referred to as biconnectivity, and 2-edge connectivity is simply referred to as edge-biconnectivity. With the same transmission ranges, the asymmetric connectivity is always not lower than the symmetric connectivity. If the transmission ranges are not identical, the asymmetric connectivity may be higher than the symmetric connectivity. Figure 2 shows an example in which the asymmetric topology is connected but the symmetric topology is disconnected. The network consists of nine nodes lying on a regular hexagon of side equal to one, with six nodes at the vertices of the hexagon and the other three nodes at the midpoints of three alternate sides of the hexagon. Three alternate nodes at the vertices have transmission range of one, and all others have the transmission range of one half. The asymmetric topology is connected, but the symmetric topology is not. On the other hand, if all nodes have the same transmission range, the asymmetric topology and the symmetric topology always have the same connectivity.

The requirement on the network connectivity (either asymmetric or asymmetric) imposes a constraint on the transmission ranges of all nodes. A crucial issue is how to find a range assignment of the smallest total power to meet a specified connectivity requirement. The Min-Power Symmetric (resp., Asymmetric) k-Node Connectivity problem seeks a range assignment of minimum total power subject to the constraint the induced symmetric (resp. asymmetric) topology is k-connected. Similarly, the Min-Power Symmetric (resp., Asymmetric) k-Edge Connectivity problem seeks a range assignment of minimum total power subject to the constraint the induced symmetric (resp., asymmetric) topology is k-edge connected. Clearly, the smallest total power for asymmetric k-node (resp., edge) connectivity is no more than the smallest total power for symmetric k-node (resp., edge) connectivity.

The study of the Min-Power Asymmetric Connectivity problem was started by Chen and Huang [5], who gave a 2-approximation algorithm based on minimum spanning tree. Further contributions were made in [19] and [8]. The related broadcast problem was studied in [27], [25], and [6]. The recent survey [9] presents the state of the art for these "asymmetric" problems. The Min-Power Symmetric Connectivity problem was proposed in [2] and [4]. Both papers claim that Min-Power Symmetric Connectivity is NP-Hard, and [4] presents a $(1 + \ln 2)$-approximation algorithm. In the journal submission of [4], this approximation ratio is improved to $5/3$.

The Min-Power Symmetric Biconnectivity problem has been first studied by Ramanathan and Rosales-Hain [23], which proposed one reasonable heuristic but without a proven approximation ratio. Lloyd et. al [21] studied both Min-Power Symmetric Biconnectivity and Min-Power Symmetric Edge-Biconnectivity. Among other results, they show that the range assignment based the approximation algorithm of Khuller and Raghavachari [17], which we refer to as **Algorithm KR**, has an approximation ratio of at most $2(2 - 2/n)(2 + 1/n)$ for Min-Power Symmetric Biconnectivity, and the range assignment based on the approximation algorithm of Khuller and Vishkin [18], which we refer to as **Algorithm KV**, has an approximation ratio of at most $8(1 - 1/n)$ for Min-Power Symmetric Edge-Biconnectivity.

In this paper, we present a reduction that establishes the NP-Hardness of both Min-Power Symmetric Two-Node Connectivity and Min-Power Symmetric Two-Edge-Connectivity. The NP-Hardness holds for plane instances, not only for arbitrary graph weights. We show that the range assignment based on the **Algorithm KR** has an approximation ratio of at most 4 for both Min-Power Symmetric Biconnectivity and Min-Power Asymmetric Biconnectivity. Specifically, we prove that the total power of this range assignment is less than four times the smallest power for asymmetric biconnectivity. We also show that the range assignment based on **Algorithm KV** has an approximation ratio of at most $2k$ for both Min-Power Symmetric k-Edge Connectivity and Min-Power Asymmetric k-Edge Connectivity. Specifically, we prove that the total power of this range assignment is less than $2k$ times the smallest power for asymmetric k-edge connectivity. As both algorithms are graph algorithms, the approximation ratios hold also if the nodes are in the three dimensional space, if the possible ranges come from a discrete set of values, if obstacles completely block the communication in between certain pairs of nodes, and if there is a maximum value on the ranges.

Although the range assignments based **Algorithm KR** and **Algorithm KV** have very complicated implementations and are not practical for wireless ad hoc networks. This motivates us to seek a trade-off between the approximation ratio and the implementation complexity. We propose a very simple range assignment which achieves both symmetric and asymmetric biconnectivity. The total power of this range assignment is less than 8 for $\kappa = 2$, or $3.2 \cdot 2^\kappa$ for $\kappa > 2$ times the smallest power for asymmetric connectivity for plane instances.

The remaining of this paper is organized as follows. Due to space limitations, we omit the reduction proving the NP-hardness of Min-Power Symmetric

(Edge-) Biconnectivity. In Section 2, we introduce related graph-theoretic results and some terms and notations. In Section 3 and Section 4, we derive tighter upper bounds on the approximation ratios of the range assignments based **Algorithm KR** and **Algorithm KV** respectively. In Section 5, we present the new algorithm, MST-Augmentation, and analyze its approximation ratio. Finally, in Section 6, we conclude the paper and report preliminary experimental results.

2   Preliminaries

A directed graph $D = (V, A)$ is said to be a *branching* (or arborescence) rooted at some vertex $s \in V$ if $|A| = |V| - 1$ and there is a path to s from any other vertex. In other words, branchings in directed graphs are a directed analog to spanning trees in undirected graphs.

Theorem 1 (Edmonds). *[11] Suppose that, given a directed graph $D = (V, A)$ and a specified vertex $s \in V$, there are k arc-disjoint paths to s from any other vertex of D. Then D has k arc-disjoint branchings rooted at s.*

Theorem 2 (Whitty). *[26] Suppose that, given a directed graph $D = (V, A)$ and a specified vertex $s \in V$, there are two internally vertex-disjoint paths to s from any other vertex of D. Then D has two arc-disjoint branchings rooted at s such that for any vertex $v \in V - s$ the two paths to s from v uniquely determined by the branchings are internally vertex-disjoint.*

Consider a directed graph $D = (V, A)$, a specified vertex $s \in V$, and a positive integer k. The cheapest subgraph of D that has k arc-disjoint paths to s from every other vertex, if there is any, must be the union of k arc-disjoint branchings rooted at s and can be found in polynomial time by the weighted matroid intersection algorithm due to Lawler [20] and Edmonds [12]. The fastest implementation of a weighted matroid intersection algorithm is given by Gabow [14]. Given a vertex $r \in V$, the cheapest subgraph of D that has k internally vertex-disjoint paths to r from every other vertex, if there is any, can also be found in polynomial time by an algorithm due to Frank and Tardos [13], or a faster algorithm due to Gabow [15].

We will also make use of a corollary of Menger's Theorem, the so-called Fan Lemma.

Theorem 3 (Fan Lemma). *[10] Suppose that D is a k-vertex connected directed graph and U is a proper subset of its vertices with $|U| = k$. Then for any vertex v not in U, there are k internally vertex-disjoint paths that link v to distinct vertices of U.*

The bidirected version of an undirected graph G is a directed graph obtained by replacing every edge of G with two oppositely oriented arcs. The undirected version of a directed graph D is an undirected graph obtained by ignoring the directions of the arcs of D.

From now on, we model the wireless ad hoc network by a weighted complete graph $G = (V, E, c)$ with $c(e) = \|e\|^{\kappa}$ where $\|e\|$ is the length of the edge e. Every range assignment is specified by a spanning graph H as follows. The transmission power of node v with respect to H, denoted by $p_H(v)$, is defined by $p_H(v) = \max_{u \in N_H(v)} c(vu)$. Clearly, the symmetric topology induced by this range assignment contains H as a subgraph, and the asymmetric topology induced by this assignment contains the bidirected version of H as a subgraph. Thus, the range assignment specified by H achieves at least the connectivity of H.

For any spanning subgraph H of G, we define the power cost of H as $p(H) = \sum_{v \in V(H)} p_H(v)$. Then $p(H)$ is exactly the total power of the range assignment induced by H. We also define the weight of H as $c(H) = \sum_{e \in E(H)} c(e)$.

The two parameters $p(H)$ and $c(H)$ are related by the following previously known lemma.

Lemma 1. *For any spanning subgraph H of G, $p(H) \leq 2c(H)$.*

Proof. Let H be a subgraph of G. Then,

$$p(H) = \sum_{v \in V} p_H(v) = \sum_{v \in V} \max_{u \in N_H(v)} c(vu)$$

$$\leq \sum_{v \in V} \sum_{u \in N_H(v)} c(vu) = 2 \sum_{e \in E(H)} c(e) = 2c(H).$$

For directed spanning subgraphs Q, we define similarly $p_Q(v) = \max_{vu \in Q} c(cu)$ for every vertex v, and $p(Q) = \sum_{v \in V} p_Q(v)$.

3   Algorithm KR for k-Edge Connectivity

Algorithm KR [17] constructs a k-edge connected spanning subgraph H as follows. For some node s, let D_s be the minimum-weight directed subgraph of the bidirected version of G in which there are k arc-disjoint paths to s from every other vertex in V. Let H be the undirected version of D_s for an arbitrary node s. Then, as shown in [17], H is k-edge connected.

Let *opt* be the power cost of an optimum range assignment for asymmetric k-edge connectivity. We have the following theorem.

Theorem 4. $p(H) \leq 2k \cdot opt$.

Proof. Consider Q, the directed graph given by the optimum range assignment. Q is strongly k-edge connected, and therefore by Theorem 1 Q contains k arc-disjoint branchings rooted at s: $T_1, T_2, \cdots, T_k$.

As $\cup_{i=1}^{k} T_i$ is a feasible solution solution for the directed subgraph computed by the algorithm, $c(D_s) \leq \sum_{i=1}^{k} c(T_i)$. For any vertex v and $1 \leq i \leq k$, denote by $a_i(v)$ the parent of v in $T_i(v)$. Given v, $p_Q(v) = \max_{vu \in Q} c(uv) \geq \max_{1 \leq i \leq k} c(va_i(v)) \geq \frac{1}{k} \sum_{1 \leq i \leq k} c(va_i(v))$, and therefore

$$opt = p(Q) \geq \frac{1}{k} \sum_{1 \leq i \leq k} c(T_i).$$

Using Lemma 1, we conclude: $p(H) \leq 2c(H) \leq 2c(D_s) \leq 2\sum_{i=1}^{k} c(T_i) \leq 2k \cdot opt$

Theorem 4 implies that the approximation ratio of **Algorithm KR** is at most $2k$.

4 Algorithm KV for Biconnectivity

Algorithm KV [18] constructs a 2-node connected spanning subgraph H as follows.

1. Let xy be the edge of G of minimum weight and s an vertex not in V. Construct weighted directed graph D as follows: Replace every edge of G with two oppositely-oriented arcs of the same weight and then add two arcs xs and ys of weight 0.
2. Let D' be the minimum-weighted subgraph of D in which there are two internally vertex-disjoint directed paths to s from every vertex in V. (D' can be obtained by using the algorithm of Frank and Tardos [13], or a faster algorithm by Gabow [15]).
3. Output the subgraph H of G which contains the edge xy and every edge of G with at least one of its two directed copies in D'.

As shown in [18], H is two-connected. Let opt be the power cost of an optimum range assignment for asymmetric 2-node connectivity. We have the following theorem.

Theorem 5. $p(H) \leq 4 \cdot opt.$

Proof. Consider Q, the directed graph given by the optimum range assignment, to which we add the arcs xs and ys of weight 0. Using Theorem 3 (Fan Lemma), for any vertex v other than x and y, Q has two internally vertex-disjoint directed paths that link v to x and y respectively. Therefore, in Q, every vertex v has two internally vertex-disjoint directed paths linking it to s. Using Theorem 2, Q has two arc-disjoint branchings rooted at s: A_1 and A_2 such that, for every vertex $v \in V$, the two paths in A_1 and A_2 from v to r are internally vertex-disjoint.
As $A_1 \cup A_2$ is a feasible solution for the directed subgraph we needed in step 2 , $c(D') \leq c(A_1)+c(A_2)$. For any vertex v and $1 \leq i \leq 2$, denote by $a_i(v)$ the parent of v in $A_i(v)$. Given v, $p_Q(v) = \max_{vu \in Q} c(uv) \geq (c(va_1(v)) + c(va_2(v)))/2$, and therefore $opt = P(Q) \geq (c(A_1) + c(A_2))/2$.
Using Lemma 1, we conclude:

$$p(H) \leq 2c(H) = 2c(D') \leq 2(c(A_1) + c(A_2)) \leq 4opt$$

Theorem 5 implies that the approximation ratio of **Algorithm KR** is at most 4.

5 Algorithm MST-Augmentation for Biconnectivity

In this section, we present a simple algorithm which produces a biconnected spanning graph H by augmenting an MST. The algorithm first finds an Euclidean MST T and initializes H to T. At any non-leaf node v of T, a local Euclidean MST T_v over all the neighbors of v in T is constructed and added to H. Thus the H is a union of the big MST T and many small MSTs. H is 2-connected, as it follows from the following argument. Only internal nodes of T can be articulation points; let u be such a node. Removing u from T creates a number of connected components of T, each having one vertex neighbor with u in T. But the neighbors of u in T remain connected by T_u, the local MST which does not include u.

We refer to this algorithm as **MST-Augmentation**. Besides being simple and very fast (as every vertex has constant degree in T, total running time is dominated by constructing T and is $O(n \log n)$), this algorithm is best suited to efficient distributed implementation. Another advantage of this algorithm is the independence of the path-loss exponent.

To bound the approximation ratio of **MST-Augmentation**, we introduce a geometric constant α defined below. Let o be the origin of the Euclidean plane. A set U of at least two points is called as a *star-set* if its Euclidean MST for $\{o\} \cup U$ is a star centered at o. The star is denoted by S_U. Note that each star-set contains at least two but at most six points. For any star-set U, let T_U be the minimum spanning tree of U. Then α is defined as the supreme of the ratio $c(T_U)/c(S_U)$ over all star-sets.

Lemma 2. *For any $\kappa \geq 2$, $2^{\kappa-1} \leq \alpha \leq 1.6 \cdot 2^{\kappa-1}$. If $\kappa = 2$, then $\alpha = 2$.*

Proof. The lower bound $2^{\kappa-1}$ is achieved by U consisting of two points u_1 and u_2 such that o is the midpoint of the line segment $u_1 u_2$. Next, we prove the upper bound $1.6 \cdot 2^{\kappa-1}$. Consider any star-set U. If U has exactly six points, then these points form a regular hexagon centered at o, and hence $c(T_U) = \frac{5}{6}c(S_U) < 1.6 \cdot 2^{\kappa-1}c(S_U)$. So we assume U has $m \leq 5$ points. For any two points u and w in U,

$$c(uw) = \|uw\|^{\kappa} \leq (\|ou\| + \|ow\|)^{\kappa} = 2^{\kappa} \left(\frac{\|ou\| + \|ow\|}{2} \right)^{\kappa}$$

$$\leq 2^{\kappa} \frac{\|ou\|^{\kappa} + \|ow\|^{\kappa}}{2} = 2^{\kappa-1} (c(ou) + c(ow)).$$

Thus, the total weight of the convex polygon formed by the points of U is at most $2^{\kappa}c(S_U)$. On the other hand, as removing the largest edge of the polygon creates a tree on U, $c(T_U)$ is at most $\left(1 - \frac{1}{m}\right)$ times the total weight of this polygon. Thus, $c(T_U) \leq \left(1 - \frac{1}{m}\right) \cdot 2^{\kappa}c(S_U) \leq \left(1 - \frac{1}{5}\right) \cdot 2^{\kappa}c(S_U) = 1.6 \cdot 2^{\kappa-1}c(S_U)$. The lemma thereby follows.

Now we assume $\kappa = 2$ and show that $\alpha = 2$. Since $\alpha \geq 2$, we only have to show that $\alpha \leq 2$. Consider a star-set $U = \{(a_i, a_i) : 1 \leq i \leq m\}$. Let K_U denote the complete graph over U. We first claim that $c(S_U) \geq \frac{1}{m}c(K_U)$. To see this, we make use of the following inequality. $\sum_{i=1}^{m} a_i^2 = \frac{\left(\sum_{i=1}^{m} a_i\right)^2 + \sum_{1 \leq i < j \leq m} (a_i - a_j)^2}{m} \geq \frac{\sum_{1 \leq i < j \leq m} (a_i - a_j)^2}{m}$. Thus, $c(S_U) = \sum_{i=1}^{m} \left(a_i^2 + b_i^2\right) \geq \frac{\sum_{1 \leq i < j \leq m} [(a_i - a_j)^2 + (b_i - b_j)^2]}{m} = \frac{1}{m} c(K_U)$.

Next, we claim that $c(T_U) \leq \frac{2}{m} c(K_U)$. This claim can be proved by a simple counting argument. Note that a complete graph of order m has m^{m-2} spanning trees, and each edge appears in

$$\frac{m^{m-2}(m-1)}{\frac{m(m-1)}{2}} = 2m^{m-3}$$

spanning trees (see, for example, Chapter 2 of [24]). The total weight of all spanning trees of K_U is thus $2m^{m-3}c(K_U)$. Hence, $c(T_U) \leq \frac{2m^{m-3}c(K_U)}{m^{m-2}} = \frac{2}{m} c(K_U)$. From the two previous claims, we have

$$\frac{c(T_U)}{c(S_U)} \leq \frac{\frac{2}{m} c(K_U)}{\frac{1}{m} c(K_U)} = 2.$$

So the lemma follows for $\kappa = 2$.

Now we are ready to represent the upper bound on $p(H)$ in terms of α and the power cost of an optimum range assignment for asymmetric connectivity which is denoted by opt.

Theorem 6. $p(H) < 4\alpha \cdot opt$.

The proof of this theorem consists of the following several lemmas. The next lemma is implicit in previous work and it follows immediatly from the fact that T is a minimum spanning trees and one argument used in the proof of Theorem 4.

Lemma 3. $c(T) < opt$.

Let E_1 be the set of all edges of T incident to leaves. Let E_2 be the set of all edges of the trees T_v for all non-leaf nodes v. Let H' be the graph $(V, E_1 \cup E_2)$. Then H' is a subgraph of H, and thus $p(H) \geq p(H')$. The next lemma states that the equality actually holds.

Lemma 4. *For every node v, $p_H(v) = p_{H'}(v)$, and consequently $p(H) = p(H')$.*

Proof. We prove the lemma by contradiction. Assume that $p_H(v) > p_{H'}(v)$ for some node v. Let $p_H(v) = c(uv)$. Then uv must be an edge of T and neither of u and v is a leaf. Since u is not a leaf, u has an neighbor w other than v such that vw is an edge in T_u. So vw is an edge of E_2. Since both uv and uw are edges of the MST, $\|uv\| \leq \|wv\|$, and thus $c(uv) \leq c(wv)$. Therefore,

$$p_H(v) = c(uv) \leq c(wv) \leq p_{H'}(v),$$

which is a contradiction.

The next lemma provides an upper bound in the total weight of H'.

Lemma 5. $c(H') \leq 2\alpha \cdot c(T)$.

Proof. From Lemma 2, we have $c(T_u) \leq \alpha \sum_{uv \in T} c(uv)$. Then

$$c(H') = c(E_1) + c(E_2) = \sum_{u \text{ leaf}} \sum_{vu \in T} c(uv) + \sum_{u \text{ internal}} c(T_u)$$

$$\leq \alpha \sum_{u \text{ leaf}} \sum_{vu \in T} c(uv) + \alpha \sum_{u \text{ internal}} \sum_{vu \in T} c(uv) = 2\alpha c(T),$$

as every edge of T appears exactly twice in the summation.

Now Theorem 6 follows immediately from Lemma 1, Lemma 3, Lemma 4, and Lemma 5:

$$p(H) = p(H') \leq 2c(H') < 4\alpha \cdot c(T) < 4\alpha \cdot opt.$$

Theorem 6 and Lemma 2 imply that the approximation ratio of **MST-Augmentation** is at most 8 for $\kappa = 2$ and at most $3.2 \cdot 2^\kappa$ for general κ.

6   Conclusion

We presented improved analysis for existing algorithms for Min-Power Symmetric Biconnectivity and Min-Power Symmetric k-Edge Connectivity, and showed the symmetric output of these algorithms is also a good approximation for Min-Power Asymmetric Biconnectivity and Min-Power Asymmetric k-Edge Connectivity, respectively. We showed that Min-Power Symmetric Biconnectivity and Min-Power Symmetric Edge-Biconnectivity is NP-Hard. We introduced the new algorithm **MST-Augmentation** and showed it also has constant approximation ratio.

We are aware of instances where the min-power asymmetric two-connected topology uses only 7/10 of the min-power symmetric two-connected topology. It would be interesting to find how small this ratio could be. By our analysis of the Min-Power Biconnectivity **Algorithm KR**, the ratio is at least 1/4, and in fact we can show the ratio is at least 1/3. By comparison, the ratio of min-power symmetric connected topology to min-power asymmetric connected topology is known to be at least 1/2, and this bound is tight (see for example the journal version of [4]).

Preliminary experimental results for Min-Power Symmetric Biconnectivity show that on random instances with 100 nodes, the following hold:

- "smart" local optimization algorithms improve by an average of 6% the Ramanathan and Rosales-Hain algorithm, with a maximum improvement of 18%. The Ramanathan and Rosales-Hain algorithm has a local optimization phase and on average uses 29% less power than **MST-Augmentation**.

- Our best heuristics have power 75% to 250% more than the cost of the minimum spanning tree (the only easily computable lower bound for the problems). The average power used is 110% more than the cost of the minimum spanning tree.
- For our best algorithms, the power required to ensure Symmetric Biconnectivity is on average 61.6% higher than the power required for Symmetric Connectivity. Our heuristics for Symmetric Connectivity are very good [1], but we still do not know the quality of the Symmetric Biconnectivity solutions our heuristics produce. Note that the minimum power for Symmetric Biconnectivity could be higher than the minimum power for Symmetric Connectivity by a factor of 2^κ, as shown by an example of n nodes being equidistant on a line.

Acknowledgements

We are grateful to Nickolay Tchervensky for help with the experiments.

References

1. E. Althaus, G. Calinescu, I. Mandoiu, S. Prasad, N. Tchervenski, and A. Zelikovsky, Power Efficient Range Assignment in Ad-hoc Wireless Networks, *Proc. IEEE Wireless Communications and Networking Conference*, 2003.
2. D.M. Blough, M. Leoncini, G. Resta, and P. Santi, On the Symmetric Range Assignment Problem in Wireless Ad Hoc Networks, *Proc. 2nd IFIP International Conference on Theoretical Computer Science*, Montreal, August 2002.
3. T. Calamoneri, and R. Petreschi, An Efficient Orthogonal Grid Drawing Algorithm for Cubic Graphs, *COCOON'95*, Lectures Notes in Computer Science 959, Springer-Verlag, pages 31-40, 1995.
4. G. Calinescu, I. Mandoiu, and A. Zelikovsky, Symmetric Connectivity with Minimum Power Consumption in Radio Networks, *Proc. 2nd IFIP International Conference on Theoretical Computer Science*, Montreal, August 2002.
5. W.T. Chen and N.F. Huang, The Strongly Connecting Problem on Multihop Packet Radio Networks, *IEEE Transactions on Communications*, vol. 37, no. 3, pp. 293-295, Oct. 1989.
6. A. Clementi, P. Crescenzi, P. Penna, G. Rossi and P. Vocca, On the Complexity of Computing Minimum Energy Consumption Broadcast Subgraphs, *18th Annual Symposium on Theoretical Aspects of Computer Science*, LNCS 2010, 2001, pages 121-131.
7. A. Clementi, P. Penna and R. Silvestri, Hardness Results for The Power Range Assignment Problem in Packet Radio Networks, *Proc. 3rd International Workshop Randomization, Approximation and Combinatorial Optimization*, Lecture Notes in Computer Science 1671, pp. 197-208, 1999.
8. A. Clementi, P. Penna and R. Silvestri, The Power Range Assignment Problem in Radio Networks on the Plane, *Proc. 17th Annual Symposium on Theoretical Aspects of Computer Science*, Lecture Notes in Computer Science 1770, pp. 651-660, 2000.

9. A. Clementi, G. Huiban, P. Penna, G. Rossi and Y.C. Verhoeven, Some Recent Theoretical Advances and Open Questions on Energy Consumption in Ad-Hoc Wireless Networks, *3rd Workshop on Approximation and Randomization Algorithms in Communication Networks*, 2002.

10. R. Diestel, *Graph Theory*, 2nd Edition, Graduate Texts in Mathematics, Volume 173, Springer-Verlag, New York, February 2000.

11. J. Edmonds, Edge-disjoint branchings, in *Combinatorial Algorithms*, R. Rustin, Ed., pp. 91-96. Algorithmics Press, New York, 1972.

12. J. Edmonds, Matroid intersection, *Annals of Discrete Mathematics*, No. 4, pp. 185-204, 1979.

13. A. Frank and É. Tardos, An application of submodular flows, *Linear Algebra and its Applications*, vol. 114/115, pp. 329-348, 1989.

14. H. N. Gabow, A matroid approach to finding edge connectivity and packing arborescences, *Proc. 23rd ACM Symposium on Theory of Computing*, pp. 112-122, May 1991.

15. H.N. Gabow, A representation for crossing set families with applications to submodular flow problems, in *Proc. 4th ACM-SIAM Symposium on Discrete Algorithms*, pp. 202-211, Austin, TX, 1993.

16. M.R. Garey, D.S. Johnson and R.E. Tarjan, The Planar Hamiltonian Circuit Problem is NP-complete, *SIAM J. Comput.*, vol. 5, pp. 704-714, 1976.

17. S. Khuller and B. Raghavachari, Improved approximation algorithms for uniform connectivity problems, *Journal of Algorithms* 21, pp. 433-450, 1996.

18. S. Khuller and U. Vishkin, Biconnectivity approximations and graph carvings, *Journal of ACM*, vol. 41, no. 2, pp. 214-235, 1994.

19. L. M. Kirousis, E. Kranakis, D. Krizanc and A. Pelc, Power Consumption in Packet Radio Networks, *Theoretical Computer Science*, vol. 243, no. 1-2, pp. 289-305, 2000. A preliminary version of this papers also appeared in *Proc. 14th Annual Symposium on Theoretical Aspects of Computer Science*, LNCS 1200, pp. 363 - 374, 1997.

20. E. L. Lawler, Matroid intersection algorithms, *Mathematical Programming* 9, pp. 31-56, 1975.

21. E. Lloyd, R. Liu, M. Marathe, R. Ramanathan, and S.S. Ravi, Algorithmic Aspects of Topology Control Problems for Ad hoc Networks, *Proc. 3rd ACM International Symposium on Mobile Ad Hoc Networking and Computing (MobiHoc)*, Lausanne, Switzerland, June 2002.

22. T.S. Rappaport, *Wireless Communications: Principles and Practices*, Prentice Hall, 1996.

23. R. Ramanathan and R. Rosales-Hain, Topology Control of Multihop Wireless Networks Using Transmit Power Adjustment, *IEEE INFOCOM 2000*.

24. J.H. van Lint, R.M. Wilson, *A course in combinatorics*, Cambridge University Press, 1992.

25. P-J. Wan, G. Calinescu, X-Y. Li, and O. Frieder, Minimum Energy Broadcast Routing in Static Ad Hoc Wireless Networks, *IEEE INFOCOM 2001*.

26. R.W. Whitty, Vertex-disjoint paths and edge-disjoint branchings in directed graphs, *J. Graph Theory*, 11(3):349-358, 1987.

27. J.E. Wieselthier, G.D. Nguyen, and A. Ephremides, On the Construction of Energy-Efficient Broadcast and Multicast Trees in Wireless Networks. *IEEE INFOCOM 2000*.

Steiner Systems for Topology-Transparent Access Control in MANETs

Charles J. Colbourn[1,*], Violet R. Syrotiuk[1,**], and Alan C.H. Ling[2]

[1] Computer Science & Engineering, Arizona State University, Tempe, AZ 85287-5406
[2] Computer Science, University of Vermont, Burlington, VT 05405

Abstract. In this paper we examine the combinatorial requirements of topology-transparent transmission schedules for channel access in mobile ad hoc networks. We formulate the problem as a combinatorial question and observe that its solution is a cover-free family. The mathematical properties of certain cover-free families have been studied extensively. Indeed, we show that both existing constructions for topology-transparent schedules (which correspond to orthogonal arrays) give a cover-free family. However, a specific type of cover-free family – called a Steiner system – supports the largest number of nodes for a given frame length. We then explore the minimum and expected throughput for Steiner systems of small strength, first using the acknowledgement scheme proposed earlier and then using a more realistic model of acknowledgements. We contrast these results with the results for comparable orthogonal arrays, indicating some important trade-offs for topology-transparent access control protocols.

1   Introduction

In any network based on a shared broadcast channel, the means by which access to the channel is controlled has a fundamental impact on the overall network performance. While these networks include satellites and local area networks, our interest is in *mobile ad hoc networks* (MANETs). A MANET is a collection of mobile wireless nodes. What distinguishes a MANET from other wireless networks is that it self-organizes without the aid of any centralized control or any fixed infrastructure. Since the radio transmission range of each node is limited, it may be necessary to forward over multiple hops in order for a packet to reach its destination (as such, MANETs have also been called multi-hop and packet-radio networks). This also offers the opportunity for concurrent transmissions when nodes are sufficiently separated. The challenge in medium access control (MAC) protocols for MANETs is to find a satisfactory trade-off between the two objectives of minimizing delay and maximizing throughput.

Of the myriad of access control techniques, our focus is on topology- transparent approaches. Unlike topology-dependent protocols, which recompute access whenever the network topology changes, a *topology-transparent* protocol acts independently of topology change. One class of protocols which may be viewed as

---

* This work was supported in part by ARO grant DAAD 19-01-1-0406.
** This work was supported in part by NSF grant ANI-0105985.

S. Pierre, M. Barbeau, and E. Kranakis (Eds.): ADHOC-NOW 2003, LNCS 2865, pp. 247–258, 2003.
© Springer-Verlag Berlin Heidelberg 2003

topology-transparent is the contention based MAC protocols. Contention based approaches achieve high throughput with a reasonable expected delay but with poor worst-case delay. With increasing interest in multi-media applications, the delay characteristics of contention based MAC protocols do not appear adequate to provide the necessary quality-of-service (QoS) support. While there have been some efforts to make such protocols QoS-aware, in each case the delay guarantee remains probabilistic [1,12,14].

TDMA is an example of a scheduled access control protocol that is trivially topology-transparent. More sophisticated schemes for generating topology-transparent transmission schedules [2,10] depend on two design parameters: N, the number of nodes in the network, and D, the maximum node degree. This creates complex trade-offs between the design parameters and the delay and throughput characteristics of the resulting schedules. For example, while it is often possible to construct schedules that are significantly shorter than TDMA, if the actual node degree exceeds D, the delay guarantee is lost. More exactly, the delay becomes probabilistic rather than deterministic. While the question of what should be done if the protocol fails is important (see [3,16] for some alternatives), we will not address this problem here.

In [16], we observed that existing topology-transparent transmission schedules are instances of orthogonal arrays, and we explored the consequences of this observation on throughput. In this paper we go one step further, looking more carefully at the combinatorial requirements of topology-transparent transmission schedules. This allows us to formulate the problem as a combinatorial question and observe that its solution is a cover-free family. Certain cover-free families have been studied extensively, and rather than derive new mathematical results, we instead show how to use existing results for our application. Our first observation shows that an orthogonal array gives a cover-free family. We then show that a specific type of cover-free family, called a Steiner system, supports the largest number of nodes for a given frame length. We then explore the minimum and expected throughput for Steiner systems of small strength, first using the acknowledgement scheme proposed earlier and then using a more realistic model for acknowledgements. We contrast these results with the results for comparable orthogonal arrays, indicating some important trade-offs for topology-transparent protocols.

The rest of this paper is organized as follows. Section 2 first examines the combinatorial requirements of a topology-transparent transmission schedule, and shows that a cover-free family satisfies the requirements. We also show how cover-free families relate both to orthogonal arrays, and to Steiner systems. In Section 3, we study the selection of parameters of the Steiner system depending on the performance objective of interest. We consider both minimum and expected throughput using an acknowledgment scheme proposed earlier. As well, we introduce a more realistic acknowledgement model and study the resulting frame throughput. We produce our results as a function of neighbourhood size and density, to explore the sensitivity of the actual node degree to the design parameter. Lastly, in Section 4, we summarize and conclude.

2 Cover-Free Families, Orthogonal Arrays, and Steiner Systems

Rather than starting with the existing constructions for topology-transparent transmission schedules, let us instead begin anew by turning the problem of generating a topology-transparent transmission schedule into a combinatorial question. Assume that time is divided into discrete units called *slots* and *frames* are a fixed number n of slots. Suppose that each node $i, 1 \leq i \leq N$, in the network is assigned a transmission schedule $S_i = s_1 s_2 \ldots s_n$ with n slots (i.e., one frame). If $s_j = 1, 1 \leq j \leq n$, then a node may transmit in the slot j, otherwise it is silent (and could receive).

In designing a topology-transparent transmission schedule with design parameters N, the number of nodes in the network and D, the maximum node degree, we are interested in the following combinatorial property. For each node, we want to guarantee that if a node i has at most D neighbours its schedule S_i guarantees a collision-free transmission to each neighbour.

Let us treat each schedule S_i as a subset T_i on $\{1, 2, \ldots, n\}$ by assigning the elements of the subset to correspond to the positions in the schedule, i.e., $j \in T_i$ if $s_j = 1$ in S_i, $j = 1, \ldots, n$ (in essence, S_i is the characteristic vector of the set T_i). Now, the combinatorial problem to ask is for each node i to be given a subset T_i with the property that the union of D or fewer other subsets cannot contain T_i. Expressed mathematically, if $T_j, j = 1, \ldots, D$, are D neighbours of i ($T_j \neq T_i$), then we require that

$$\left(\bigcup_{j=1}^{D} T_j \right) \not\supseteq T_i.$$

This is precisely a D *cover-free family*. These are equivalent to disjunct matrices [6] and to certain superimposed codes [7]; see [5].

Let us first observe that the existing constructions for topology-transparent transmission schedules [2,10] which, as we showed in [16] correspond to an orthogonal array, give a cover-free family.

2.1 An Orthogonal Array Gives a Cover-Free Family

Let V be a set of v symbols, usually denoted by $0, 1, \ldots, v-1$.

Definition 1. *A $k \times v^t$ array A with entries from V is an orthogonal array with v levels and strength t (for some t in the range $0 \leq t \leq k$) if every $t \times v^t$ subarray of A contains each t-tuple based on V exactly once[1] as a column. We denote such an array by $OA(t, k, v)$.*

Table 1 shows an example from [9] of an orthogonal array of strength two with $v = 4$ levels, i.e., $V = \{0, 1, 2, 3\}$. Pick any two rows, say the third and the fourth. Each of the sixteen ordered pairs $(x, y), x, y \in V$ appears the same number of times, once in this case.

---

[1] Here, we assume the index $\lambda = 1$.

Table 1. Orthogonal array $OA(2, 4, 4)$.

```
0 0 0 0 1 1 1 1 2 2 2 2 3 3 3 3
0 1 2 3 0 1 2 3 0 1 2 3 0 1 2 3
0 1 2 3 1 0 3 2 2 3 0 1 3 2 1 0
0 1 3 2 3 2 0 1 2 3 1 0 1 0 2 3
```

In our application, each column gives rise to a transmission schedule. Each column intersects every other in fewer than t positions. For example, the first and the eighth column intersect in no positions, while the first and the second column intersect in a zero in the first position.

The importance of this intersection property is as follows. Select any column. Since any of the other columns can intersect it in at most $t - 1$ positions, any collection of D other columns has the property that our given column differs from all of these D in at least $k - D(t - 1)$ positions. Provided this difference is positive, the column therefore contains at least one symbol appearing in that position, not occurring in any of the D columns in the same position. In our application this means that at least one collision-free slot to each neighbour exists when a node has at most D neighbours. Thus, as long as the number of neighbours is bounded by D, the delay to reach each neighbour is bounded, even when each neighbour is transmitting. Clearly, the orthogonal array gives a D cover-free family.

Many techniques are known for constructing orthogonal arrays, usually classified by the essential ideas that underlie them. There is a classic construction based on Galois fields and finite geometries; both Chlamtac and Faragó [2] and Ju and Li [10] use this construction implicitly though neither observed that they were constructing an orthogonal array. They both employ $OA(t, v, v)$'s when v is a prime power. They therefore restrict attention to the case when $k = v$ (forcing all frame lengths to be v^2 unnecessarily), and indeed by not permitting that $k > v$ they do not obtain the best delay guarantees. The restriction of v to prime powers is also not required, as orthogonal arrays exist for these cases, e.g., $OA(2, 7, 12)$, but k is not as large as v in general.

In the same way that allowing different parameters for orthogonal arrays allows more flexibility in the corresponding schedules, relaxing the parameters further and asking for a cover-free family allows more flexibility yet.

2.2   Steiner Systems

Cover-free families have been studied extensively, most frequently with the objective of maximizing the number of sets in the family. In our application, this corresponds to maximizing the number of nodes, so this is certainly a parameter of interest.

There is a celebrated result of Erdös, Frankl, and Füredi [8] that established bounds on the size of a cover-free family (see also, [13,15] and Theorem 7.3.9 in [6]). Specifically, they established that the extreme value on the size, if achievable, is realized by a Steiner system. Hence in terms of the application, for

Table 2. Steiner system $S(2, 4, 13)$.

```
0 0 0 0 1 1 1 2 2 3 3 4 5
1 2 4 6 2 5 7 3 6 4 7 8 9
3 8 5 a 4 6 b 5 7 6 8 9 a
9 c 7 b a 8 c b 9 c a b c
```

given number of nodes and a given maximum number of neighbours, Steiner systems achieve the shortest frame length of all cover-free families. Thus, they provide not only a solution to our problem, but indeed the best solution in terms of frame length.

Definition 2. *Given three integers t, k, v such that $2 \leq t < k < v$, a Steiner system $S(t, k, v)$ is a v-set V together with a family $\mathcal{B}$ of k-subsets of V (blocks) with the property that every t-subset of V is contained in exactly one block.*

Table 2 shows an example from [4] of a Steiner system on the 13-set $V = \{0, 1, \ldots, 9, a, b, c\}$ together with a family of $\frac{v(v-1)}{k(k-1)} = \frac{13 \cdot 12}{4 \cdot 3} = 13$, 4-subsets of V (the columns). This Steiner system has the property that every 2-subset of V, $\{x, y\}, x, y \in V, x \neq y$ is contained in exactly one column.

While a substantial amount is known about the existence of Steiner systems, in general their existence is not settled [4]. As with orthogonal arrays, there are constructions from finite fields.

Reasons that Steiner systems are of interest for constructing topology- transparent transmission schedules include:

1. Steiner systems admit shorter schedules than orthogonal arrays. This is important since in addition to achieving high throughput, the delay bound is improved. We discuss this issue at length in Section 3.
2. Steiner systems are denser than orthogonal arrays. They can support a larger number of nodes for a given schedule (frame) length.

3   Steiner System Parameter Trade-Offs

The essential difference between the existing constructions of topology- transparent schedules is in the selection of parameters. In [2], the focus is on frame length while in [10] the focus is on throughput.

In [2], an $OA(t, v, v)$ is found for the first $v^t \geq N$, $t \geq 2$, where v is a prime power. In the paper, the interest is in minimizing the frame (or schedule) length in order to minimize delay. The parameters are selected to find a schedule provably shorter than TDMA.

In [10], it is argued that the parameters chosen satisfy the condition on delay but do not maximize minimum throughput. In particular, it is possible to achieve higher minimum throughput at the expense of longer frame length. They select an $OA(t, v, v)$ where $v = 2(t-1)D$ if $\sqrt[t]{N} \leq 2(t-1)D$, and $v = \sqrt[t]{N}$ otherwise.

Intuitively, while Chlamtac and Faragó [2] strive to get *one* free slot per frame, Ju and Li [10] aim to get *many* free slots per frame.

In both studies, however, the figure of merit is minimum throughput measured as number of free slots within a frame divided by frame length. To employ such an analysis, a transmitting node must be able to transmit multiple different packets within a frame. How does it decide to transmit a "new" packet? In this environment, it is expected that collisions occur, and topology-transparency dictates that the collisions cannot be anticipated. Hence an acknowledgement scheme is needed. Both schemes based on orthogonal arrays can transmit in two consecutive slots, and indeed must send different packets in these slots to achieve the minimum throughput in their analyses. Both propose an acknowledgement scheme that involves instantaneous acknowledgment of successful receipt without lengthening the slot. Naturally, this is an optimistic assumption to facilitate the analysis. However, the analysis can be misleading if it leads us to seek many free slots in a frame without an acceptable (realistic) acknowledgment scheme. For purposes of comparison, we consider the throughput measures employed in [2,10,16]. We also adopt a more conservative approach.

We consider a more realistic model for acknowledgements. Rather than a slot by slot acknowledgement, we assume we can piggyback an acknowledgement onto a packet sent from the destination. In the worst case, this might require that the sender wait an entire frame. Hence we define *frame throughput* as the throughput achievable on a per frame basis. This properly incorporates the length of the schedule in the throughput calculation.

In this section we investigate three questions:

1. What is the probability of a successful transmission in a frame?
2. What is the expected throughput?
3. What is the expected frame throughput?

All are functions of the number of active transmitters among the neighbours of a node.

Consider a situation with sender S and receiver R. Let S be a schedule for sender S and $T_1, \ldots, T_{D-1}$ be the subsets that correspond to the schedules of the other active neighbours of R (here, we assume the worst case, when all neighbours are transmitting). Let T_D be the subset corresponding to the schedule for R, and assume that R is also active.

The probability of successful transmission within a frame is just the probability that S has a slot that does not appear in $T_1, \ldots, T_D$. Expected throughput then, is the expected number of such slots. The frame throughput is the expected number of slots over the frame length. This effectively normalizes the expected throughput by frame length allowing easier comparison between Steiner systems.

We derive these measures analytically but present the derivations elsewhere. The most complex derivation is for expected throughput. We did this for schedules that correspond to orthogonal arrays in [16]. This formulation may be used as a basis to derive expected throughput for schedules that correspond to Steiner systems.

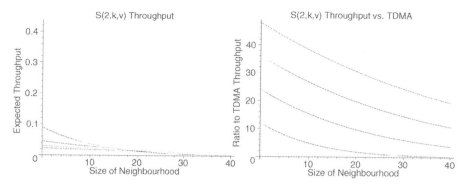

Fig. 1. Expected throughput for (a) $S(2, k, v)$; and (b) versus TDMA, for $k = 3, 6, 9, 12$.

3.1   Numerical Results

The results in this section were obtained using Maple [11], a mathematical software package.

Figure 1(a) plots the expected throughput for $S(2, k, v)$ for $k = 3, 6, 9, 12$ as a function of the number of neighbours. In each of the following cases, $N = v(v - 1)/k(k - 1)$. For $k = 3$, $v = 7, 13, 19, 25$ are considered for $N = 7, 26, 57, 100$ number of nodes, respectively. For $k = 6$, $v = 31, 61, 91, 121$ are considered for $N = 31, 122, 273, 484$ number of nodes, respectively. For $k = 9$, $v = 73, 145, 217, 289$ are considered for $N = 73, 290, 651, 1156$ number of nodes, respectively. Finally, for $k = 12$, $v = 133, 265, 397, 529$ are considered for $N = 133, 530, 1191, 2116$ number of nodes, respectively. In the figure, the y-intercept is given by k/v, and so the curve with the highest y-intercept has the shortest frame length ($k = 3$, $v = 7$). Successive curves with lower y-intercept have successively longer frame length. The shorter the frame, the faster the expected throughput drops to zero. As well, the expected throughput is much more sensitive to changes in neighbourhood size.

In Fig. 1(b), we plot the expected throughput for $S(2, k, v)$ for $k = 3, 6, 9, 12$ over the throughput of TDMA with the same frame length, as a function of the number of neighbours. For example, now the curve with the highest y-intercept is $k = 12$, $v = 529$. This Steiner system supports 2116 nodes, so the expected frame throughput is $\frac{k/v}{1/N} = \frac{12}{529} \cdot 2116 = 48$. In other words, in the best case, this Steiner system has expected throughput that is 48 times that of TDMA with the same frame length. When the ratio of expected throughput to the corresponding TDMA is taken, the curve on the left essentially inverts position on the right. This means that longer frames with more opportunities to transmit are better than shorter frames with fewer opportunities to transmit from the perspective of throughput.

Figure 2(a) plots the more conservative frame throughput for $S(2, k, v)$ for $k = 3, 6, 9, 12$ as a function of neighbourhood size, for the same v's as in the previous figure. Now, the y-intercepts correspond to $1/v$ rather than k/v. Again,

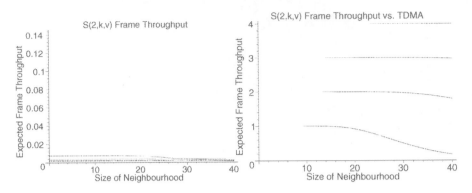

Fig. 2. Frame throughput for (a) $S(2, k, v)$; and (b) versus TDMA, for $k = 3, 6, 9, 12$.

the curves with a shorter frame length have a more pronounced drop than curves with longer frame length. As well, curves with the same k value now show a guarantee (i.e., are horizontal) for up to k neighbours, after which the guarantee degrades.

In Fig. 2(b), we plot the ratio of frame throughput for $S(2, k, v)$ for $k = 3, 6, 9, 12$ over the throughput of TDMA for the same frame length as a function of neighbourhood size, for the same v's as given earlier. Now, we see that the best possible throughput is $\frac{1/v}{1/N} = N/v$ which is $4, 3, 2, 1$ for increasing values of v. Again, the slot guarantee is evident. That is, the curves are horizontal for neighbourhood sizes less than or equal to k and the degrade as the neighbourhood increases. The degradation is slower for the longer frames. The curves whose maximum expected frame throughput equals one correspond to orthogonal arrays $OA(2, v, v)$. Hence it is plainly evident that schedules constructed from Steiner systems are much denser than those constructed from orthogonal arrays, with the potential to yield much higher throughput.

Figure 3(a) plots minimum throughput for $S(2, k, v)$ for $k = 3, 6, 9, 12$ as a function of neighbourhood size for the same values of v as given earlier. Here, the y-intercept is k/v (the same as in Fig. 1), however now the x-intercept is k and is the same for each value of v. This results in the curves dropping to zero much more quickly than in Fig. 1. A curiosity is that the four segments that correspond to the maximum minimum throughput correspond to $S(2, k, v)$ where the smallest frame length v for the given k provides a range of neighbours over which it provides the best minimum throughput. That is, $S(2, 3, 7)$ and $S(2, 12, 133)$ are better over a larger range of neighbours than are $S(2, 6, 31)$ and $S(2, 9, 73)$.

Figure 3(b) plots the ratio of minimum throughput for $S(2, k, v)$ for $k = 3, 6, 9, 12$ over TDMA with the same frame length as a function of neighbourhood size for the same v's. Again we see that the curves invert order when the ratio is considered. Specifically, the curve with the highest y-intercept is $S(2, 12, 529)$ since this is given by $\frac{k/v}{1/N}$ as in Fig. 1. However the x-intercept now corresponds to k as on the figure on the left. Now, the largest v for each k provides the best minimum throughput relative to TDMA.

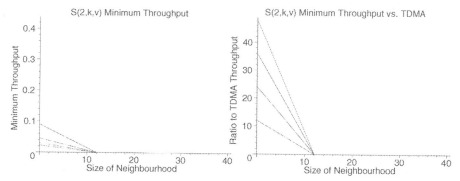

Fig. 3. Minimum throughput for (a) $S(2, k, v)$; and (b) versus TDMA, for $k = 3, 6, 9, 12$.

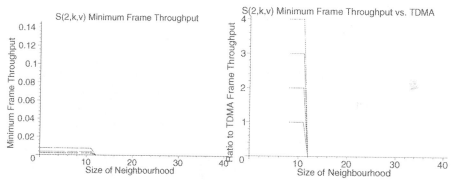

Fig. 4. Minimum frame throughput for (a) $S(2, k, v)$; and (b) versus TDMA, for $k = 3, 6, 9, 12$.

Again, we look at frame throughput, this time the minimum value, in Fig. 4 (for the same k's and v's). Not surprisingly, the minimum frame throughput is lower than when using the more optimistic acknowledgement model. The main difference between this figure and Fig. 2 is the x-intercepts. Here, they correspond to k, clearly showing that with minimum frame throughput, once the neighbourhood exceeds the design parameter, all guarantees are lost immediately. This is also true for the ratio of minimum frame throughput over TDMA with the same frame length (b). This figure also shows that the minimum frame throughput is essentially constant for each k as long as the design parameter is satisfied.

Figure 4 shows us something very important, in addition. Larger Steiner systems give us a minimum frame throughput substantially better than TDMA when the neighbourhood is within the bound. This is in stark constrast with the schemes in [2,10]; they *never* outperform TDMA on minimum frame throughput when orthogonal arrays of strength two are used.

Figure 5 is different from all other figures in that it plots expected throughput versus *density* of the neighbourhood. That is, the x-axis is the percentage of nodes that are neighbours — these are not absolute values, and represent much

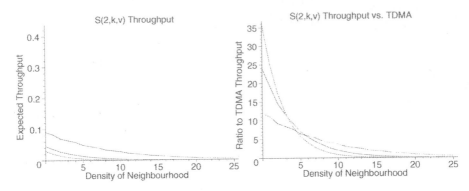

Fig. 5. Throughput versus density for (a) $S(2,k,v)$; and (b) versus TDMA, for $k = 3, 6, 9, 12$.

larger neighbourhood sizes in general. The reason that the curves are jagged is that the closest integer value is taken as the percentage of neighbours, i.e., we do not consider fractional numbers of neighbours. While the figure shows $S(2, k, v)$ for $k = 3, 6, 9, 12$, only the first three values of v for each k are shown since the computations are highly memory and compute intensive. The y-intercepts are the same as in Fig. 1. As a function of neighbourhood density, the expected throughput (a) is more well-behaved than as a function of neighbourhood size. When the ratio of expected throughput to TDMA throughput is considered versus neighbourhood density (b) the curves drop more rapidly as the density increases more rapidly than a linear function.

Finally, Fig. 6 once again plots expected throughput versus neighbourhood size for three Steiner systems that support the same number of nodes, namely $N = 651$ and one orthogonal array that supports a number very close to that (625). Specifically from the top down, the curves correspond to $S(2, 3, 63)$, $S(2, 9, 217)$, $S(2, 26, 651)$ and $OA(2, 26, 25)$. First, we see that the last two curves are essentially indistinguishable from each other. That is, for all intents and purposes, the $S(2, 26, 651)$ and $OA(2, 26, 25)$ give the same performance but the Steiner system supports more nodes. The Steiner system with shorter frame length gives better expected throughput until the neighbourhood is about 20, at which point the curves all cross. Its performance also degrades more rapidly with increasing neighbourhood size.

4   Summary and Conclusions

In this paper, we stepped back and examined anew the combinatorial properties of topology-transparent schedules. The properties were found to correspond precisely to D cover-free families, where D is a design parameter indicating maximum number of neighbours.

Studies of several Steiner systems show the following general trends. Steiner systems admit shorter schedules (frames) than previous cosntructions based or

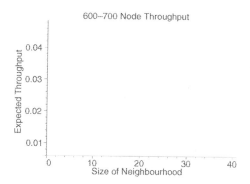

Fig. 6. Expected throughput for Steiner systems for 600-700 nodes.

orthogonal arrays. This is significant for delay sensitive applications such as multi-media. Since Steiner systems are also more dense, they support more nodes for a given frame length and hence achieve higher throughput. While shorter schedules give the best minimum and expected throughput, they also degrade faster as the design parameter D is exceeded. That is, longer schedules are more robust to changes in neighbourhood size. Another general observation is that the Steiner systems that yield longer schedules achieve higher ratios on minimum and expected throughput when compared to TDMA schedules of the same length.

We have characterized the types of solutions topology-transparent transmission schedules require as cover-free families. Using this, along with a more realistic acknowledgement model, we plan to investigate the issue of what to do when the schedule fails due to node mobility causing the design parameter on neighbourhood size to be exceeded. This, together with simulations using mobility models are required to determine how such scheduled topology-transparent protocols compare to contention based protocols.

References

1. M. Benveniste, G. Chesson, M. Hoeben, A. Singla, H. Teunissen, and M. Wentink, Enhanced Distributed Coordination Function (EDCF) proposed draft text, IEEE working document 802.11-01/131r1, March 2001.
2. I. Chlamtac and A. Faragó, "Making Transmission Schedules Immune to Topology Changes in Multi-Hop Packet Radio Networks, *IEEE/ACM Transactions on Networking*, Vol. 2, No. 1, February 1994, pp. 23–29.
3. I. Chlamtac, A. Faragó, and H. Zhang, "Time-Spread Multiple-Access (TSMA) Protocols for Multihop Mobile Radio Networks," *IEEE/ACM Transactions on Networking*, Vol. 5, No. 6, December 1997, pp. 804–812.
4. C.J. Colbourn and J.H. Dinitz (eds.), *The CRC Handbook of Combinatorial Designs*, ©1996 CRC Press, Inc.
5. C.J. Colbourn, J.H. Dinitz, and D.R. Stinson, "Applications of Combinatorial Designs to Communications, Cryptography, and Networking," in *Surveys in Combinatorics, 1999*, J.D. Lamb and D.A. Preece (eds.), London Mathematical Society, Lecture Note Series 267, ©Cambridge University Press, 1999, pp. 37–100.

6. D.-Z. Du and F.K. Hwang, *Combinatorial Group Testing and its Applications*, 2nd edition, ©2000 World Scientific Publishing Co. Pte. Ltd.
7. A. D'yachkov, V. Rykov, and A.M. Rashad, "Superimposed Distance Codes," *Problems Control and Information Theory*, 18 (1989), pp. 237–250.
8. P. Erdös, P. Frankl and Z. Füredi, "Families of Finite Sets in which no Set is Covered by the Union of r Others, *Israel J. Math.* 51 (1985), pp. 79–89.
9. A.S. Hedayat, N.J.A. Sloane, and J. Stufken, *Orthogonal Arrays, Theory and Applications*, ©1999 Springer-Verlag, New York, Inc.
10. J.-H. Ju and V.O.K. Li, "An Optimal Topology-Transparent Scheduling Method in Multihop Packet Radio Networks, *IEEE/ACM Transactions on Networking*, Vol. 6., No. 3, June 1998, pp. 298–306.
11. Maple 8, Waterloo Maple, Inc.
http://www.maplesoft.com/main.html
12. L. Romdhani, Q. Ni, and T. Turletti. "AEDCF: Enhanced Service Differentiation for IEEE 802.11 Wireless Ad-Hoc Networks," *INRIA Research Report*, No. 4544, 2002.
13. M. Ruszinkó, "On the Upper Bound of the Size of the r-cover-free Families," *Journal of Combinatorial Theory, Series A*, 66 (1994), pp. 302–310.
14. J.L. Sobrinho and A.S. Krishnakumar, "Quality-of-Service in Ad Hoc Carrier Sense Multiple Access Wireless Networks," *IEEE Journal on Selected Areas in Communications*, Vol. 17, No. 8, August 1999, pp. 1352–1368.
15. D.R. Stinson, R. Wei and L. Zhu, "Some New Bounds for Cover-Free Families," *Journal of Combinatorial Theory, Series A*, 90 (2000), pp. 224–234.
16. V.R. Syrotiuk, C.J. Colbourn and A.C.H. Ling, *Topology-Transparent Scheduling in MANETs using Orthogonal Arrays*, to appear in *Proceedings of the DIALM-POMC Joint Workshop on Foundations of Mobile Computing*, San Diego, CA, September 19, 2003.

Complexity of Connected Components in Evolving Graphs and the Computation of Multicast Trees in Dynamic Networks[*]

Sandeep Bhadra[1] and Afonso Ferreira[2]

[1] Dept. of Electrical Engineering, Indian Institute of Technology,
Madras, Chennai, India
sandy@ee.iitm.ernet.in
[2] CNRS, I3S & INRIA-Sophia Antipolis, Projet MASCOTTE,
2004 Rt. des Lucioles, BP93, F-06902 Sophia Antipolis, France.
Afonso.Ferreira@inria.fr

Abstract. New technologies and the deployment of mobile and no-
madic services are driving the emergence of complex communications
networks, that have a highly dynamic behavior. This naturally engenders new route-discovery problems under changing conditions over these
networks. Unfortunately, the temporal variations in the topology of dynamic networks are hard to be effectively captured in a classical graph
model. In this paper, we use evolving graphs, which helps capture the
dynamic characteristics of such networks, in order to compute multicast trees with minimum overall transmission time for a class of wireless
mobile dynamic networks. We first show that computing different types
of strongly connected components in evolving digraphs is NP-Complete,
and then propose an algorithm to build all rooted directed minimum
spanning trees in strongly connected dynamic networks.

1   Introduction

Infrastructure-less mobile communication environments, such as mobile ad-hoc
networks and low earth orbiting (LEO) satellite systems, present a paradigm
shift from back-boned networks, such as cellular telephony, in that data is trans-
fered from node to node via peer-to-peer interactions and not over an underlying
backbone of routers. Naturally, this engenders new problems regarding optimal
routing of data under various conditions over these dynamic networks [15].

In this setting, the generalized case of mobile network routing using shortest paths or least cost methods are complicated by the arbitrary movement of
the mobile agents thereby leading to random variations in link costs and connectivity [15]. This variable nature of the topology can be aprehended only by
network updates of the link state between moving nodes, thus creating substantial communication overhead along the link. This naturally motivates studying

---

[*] This work was partially supported by the European RTN project ARACNE, the
European FET project CRESCCO, and the AS CNRS *Dynamo*. It was done while
the first author was visiting the project MASCOTTE, INRIA/CNRS/UNSA.

S. Pierre, M. Barbeau, and E. Kranakis (Eds.): ADHOC-NOW 2003, LNCS 2865, pp. 259–270, 2003.
© Springer-Verlag Berlin Heidelberg 2003

the modeling of such dynamics, and designing algorithms that take it into account [16].

Literature related to route discovery issues in dynamic networks started more than four decades ago, with papers dealing with operations of transport networks (e.g., [5,9,11,10]). Recent work on time-dependent networks deals with flow algorithms in static networks, with edge traversal times that may depend on the number of flow units traversing it at a given moment. If traversal times are discrete, then the approach proposed in [9], namely of expanding the original graph into T layers representing the time steps (also called space-time approach), may work for computing several path-related problems (see [13,14] and references therein). Unfortunately, this approach leads to non-tractable algorithms, since T may be of exponential size.

Predictable dynamics. Note, however, that for the case of LEO satellite systems, Unmanned Aerial Vehicles (UAV), and other mobile networks with pre-destined trajectories of the mobile agents, the network dynamics are somewhat deterministic. Therefore, since the trajectories of the network agents are known in advance, it is possible to exploit this determinism in optimizing routing strategies [6,17,8].

Another setting where the evolution of the network is known was studied in [7]. The authors used the notion of competitive analysis ([1]) on a dynamic setting in order to analyze the quality of a protocol and its on-line choices made, forced by the evolution of the network. At the end of the process, the *history* of the network is formalized as a sequence of graph topologies on which the application can be solved off-line. The *merit* of the protocol is then the ratio of the solution cost found on-line over the optimal off-line cost.

Such networks, where the topology dynamics is known or can be predicted beforehand, are henceforth referred to as *fixed schedule dynamic networks* (FSDN's) (see Figure 1).

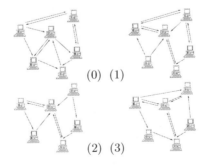

Fig. 1. An FSDN represented as an indexed set of networks. The indices correspond to successive time-steps.

Evolving graphs. Recently, *evolving graphs* [2] have been proposed as a formal abstraction for dynamic networks, and can be suited easily to the case of

FSDN's. Concisely, an evolving graph is an indexed sequence of $\mathcal{T}$ subgraphs of a given graph, where the subgraph at a given index point corresponds to the network connectivity at the time interval indicated by the index number. The time domain is further incorporated into the model by restricting *journeys* (i.e., the equivalent of paths in usual graphs) to *never* move into edges which existed only in past subgraphs (cf. Figure 2 below, and Section 2).

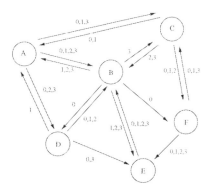

Fig. 2. Evolving digraph corresponding to FSDN in Figure 1. Edges are labeled with corresponding time-steps. Observe that CBF is not a valid journey since BF exists only in the past with respect to CB.

Notice that this model allows for arbitrary changes between two consecutive time steps, with the possible creation and/or deletion of any number of vertices and edges. Evolving graph edges can also be associated with traversal times. In [2], algorithms were proposed for finding *foremost, shortest,* and *fastest* journeys in dynamic mobile networks modeled by evolving graphs. Other path problems in evolving graphs can be found under the *merit* approach [7]. Results proven include finding a sequence of paths that connect a given pair of nodes throughout the system, such that the global routing plus re-routing costs are minimized.

Our work. We focus on the analysis of connectivity properties in FSDN's and the design of algorithms for building directed minimal spanning trees (DMST's) to generate multicast routes in FSDN's. The DMST problem in wireless networks was defined in [12] as finding N minimum weight trees, or arborescences, in a network modeled by a strongly connected digraph with N vertices. A centralized algorithm for finding DMST's in static wireless networks is presented by Chu and Liu [3], and Tarjan [18] provides an efficient implementation of the same. Humblot [12] provides a distributed algorithm for finding DMST's in strongly connected networks. Furthermore, minimum energy multicast trees for wireless networks have also been studied for the static case in [20,19]. In contrast, our approach differs from these in that our algorithm builds DMST's over dynamic mobile networks modeled by evolving digraphs, which can be seen as dynamically changing digraphs.

In this paper, we start by providing, in the next section, basic definitions for various common graph theory terms in the context of evolving digraphs. Following Humblet [12], we define rooted DMST's over strongly connected evolving digraphs. This naturally leads to the question of how to determine if an evolving digraph is strongly connected. In Section 3 we define strongly connected components (SCC's) in evolving digraphs and discover that the unique properties of evolving digraphs yield two types of strongly connected components: standard SCC's and the more loosely defined open strongly connected components (o-SCC's), as it will become clear later. One of our results is that unlike in standard digraphs, finding the strongly connected components in evolving digraphs is not possible in deterministic polynomial time, unless P=NP. In case the evolving digraph is already identified as a strongly connected component, we give in Section 4 an algorithm to compute DMST, which uses a variation of Prim's algorithm [4] for computing minimum spanning trees. For an evolving digraph with maximum outdegree $\mathcal{D}$, our algorithm builds the rooted DMST over a strongly connected component in an evolving digraph in $O(N\mathcal{D}\log\mathcal{T})$ time. Section 5 contains concluding remarks and scope for further research.

2   Graph Theoretic Model

Since we use evolving digraphs as a model for FSDN's throughout this paper, we start with a revision of the basic definitions of terms in the theory of evolving digraphs.

2.1   Evolving Digraphs

Evolving digraphs are defined as follows.

Definition 1 (Evolving Digraphs). *Let a digraph $G(V,E)$ be given, along with an ordered sequence of its subdigraphs, $\mathcal{S}_G = G_0, G_1, \ldots, G_{\mathcal{T}}, \mathcal{T} \in \mathbb{N}$. Then, the system $\mathcal{G} = (G, \mathcal{S}_G)$ is called an evolving digraph.*

We now define some of the main parameters of an evolving digraph. Let $E_{\mathcal{G}} = \bigcup E_i$, and $V_{\mathcal{G}} = \bigcup V_i$. It is clear that $\mathcal{M} = |E_{\mathcal{G}}| \leq |E| = M$ and that $\mathcal{N} = |V_{\mathcal{G}}| \leq |V| = N$. The central notion in evolving graph theory is the restriction imposed upon paths to traverse arcs strictly in non-decreasing order of arc schedule times, implying that there are no paths in $\mathcal{G}$ going to the "past."

Definition 2 (Journeys). *Let P be a path in G_i, under the usual definition. Let $F(P)$ be its first vertex, $L(P)$ be its last vertex, and $|P|$ be its length. We define a journey in $\mathcal{G}$ between two vertices u and v of $V_{\mathcal{G}}$ as a sequence $\mathcal{J}(u,v) = P_{t_1}, P_{t_2}, \ldots, P_{t_k}$, with $t_1 < t_2 < \cdots < t_k$, such that P_{t_i} is a (usually defined) path in G_{t_i} with $F(P_{t_1}) = u, L(P_{t_k}) = v$, and for all $i < k$ it holds that $L(P_{t_i}) = F(P_{t_{i+1}})$.*

Corresponding to each arc in $E_{\mathcal{G}}$ we may define an *arc schedule* as a set of indices indicating the presence of the arc in the respective subdigraphs in $\mathcal{S}_G$

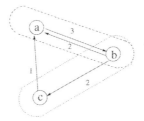

Fig. 3. Open Strongly Connected Components.

Thus, we may alternately define an evolving digraph as a tuple $\mathcal{G} = (V_{\mathcal{G}}, E_{\mathcal{G}})$, where each arc in $E_{\mathcal{G}}$ has an arc schedule defined for it.

Two vertices are said to be *adjacent in $\mathcal{G}$* if and only if they are adjacent in some G_i. The degree of a vertex in $\mathcal{G}$ is defined as its degree in $E_{\mathcal{G}}$.

As usual, a tree in $\mathcal{G}$ could be defined as a connected induced subdigraph of $V_{\mathcal{G}}$ with no circuits in $G(V, E)$. However, such a tree would not be very helpful when studying connectivity issues, since it does not take into account the total order of the subdigraphs in $\mathcal{G}$, and the restrictions it imposes on journeys in $\mathcal{G}$. Therefore, we define a *valid rooted tree in $\mathcal{G}$* as a rooted directed tree in $\mathcal{G}$, where all paths from the root to the leaves are journeys in $\mathcal{G}$.

2.2 Strongly Connected Components and Arborescences

We define an evolving digraph $\mathcal{G}$ to be a *strongly connected* digraph if there exists a journey $\mathcal{J}$ in $\mathcal{G}$ between any two vertices in $V_{\mathcal{G}}$.

Definition 3 (Strongly Connected Component). *Analogous to standard digraphs [4], we define a strongly connected component (SCC) in an evolving digraph as the maximal set of vertices $U_{\mathcal{G}} \subseteq V_{\mathcal{G}}$ such that for any pair $u, v \in U_{\mathcal{G}}$, there exists a journey from u to v and from v to u using only arcs in the Cartesian product $U_{\mathcal{G}} \otimes U_{\mathcal{G}}$.*

Thus, the subdigraph $\mathcal{G}'$ induced by considering vertices in the SCC $U_{\mathcal{G}}$ is a strongly connected digraph. For example, in Figure 3, $\{b, a\}$ forms a SCC since there are journeys from a to b and vice versa which traverse only vertices in the set $\{a, b\}$. In this figure and elsewhere in the paper arcs are labeled with their respective arc schedule times. Note that, unlike standard digraphs, there can be a journey between two vertices in the SCC that traverses vertices outside $U_{\mathcal{G}}$. Thus, it is possible for two vertices $u, v \in U_{\mathcal{G}}$ to establish a journey between them without the constraint that all arcs in the journey must be within $U_{\mathcal{G}} \otimes U_{\mathcal{G}}$. In Figure 3, although there exist journeys from b to c and from c to b, $\{b, c\}$ is not an SCC since the only journey from c to b traverses via a. Indeed the subdigraph induced by $\{b, c\}$ is not strongly connected. So, we offer a looser definition of strong connectivity as follows.

Definition 4. *An open strongly connected component(o-SCC) is the maximal set of vertices $U \subseteq V_{\mathcal{G}}$ such that for any pair $u, v \in U$, there exists a journey from u to v and from v to u.*

A journey between two nodes $u, v \in U$, might need to use nodes $h_i \in V_{\mathcal{G}}, h_i \notin U$ to maintain strong connectivity. The set of such nodes $\{h_i\} = H(u, v)$ are the helping nodes (*h-nodes*) for the vertices u, v.

Consequently, an SCC $U_{\mathcal{G}}$ is an o-SCC with the additional requirement that $H(u, v) = \emptyset \ \forall u, v \in U_{\mathcal{G}}$. Hence the set $\{b, c\}$ in Figure 3 forms a o-SCC with $H(b, c) = \{a\}$ since vertex a is required to form the only journey from b to c, thereby maintaining strong connectivity. Also, since $H(b, c) \neq \emptyset$, $\{b, c\}$ is not an SCC.

For the case of static networks, Humblet [12] defines the concept of rooted spanning trees over strongly connected directed networks. We extend this definition to the case of evolving digraphs as follows. We define a *rooted directed spanning tree* or an *arborescence* over a o-SCC $U_{\mathcal{G}} \in \mathcal{G}$ as a valid rooted directed tree in $\mathcal{G}$ rooted at r which spans all the vertices in $U_{\mathcal{G}}$; thus all the nodes except the root has one and only one incoming arc. Note that the arborescence might need to include h-nodes to reach some vertices in the o-SCC.

3   Complexity of Strongly Connected Components

In this section we will first use the foremost journey algorithm to verify strong connectivity for an FSDN. Then we will prove that the decomposition of a FSDN into (o-) SCC components is NP-Complete.

3.1   The Network Model

A FSDN can be seen as a series of networks $\mathcal{R} = \dots, \mathcal{R}_{t-1}, \mathcal{R}_t, \mathcal{R}_{t+1}, \dots$ over time. We model a FSDN as a dynamic network which has a *presence* matrix $P_E[(u, v), i]$, indicating whether (u, v) is present at time step t_i, for each link (u, v) of $\mathcal{R}$, and another *presence* matrix $P_V[u, i]$, indicating whether u is present at time step t_i, for each node u of $\mathcal{R}$. The network at time t_i is then represented by the subnetwork $\mathcal{R}_{t_i}$ of $\mathcal{R}$, which is obtained by taking the nodes and links of $\mathcal{R}$ for which their corresponding $P[i]$'s indicate they are to be present.

In order to model a fixed-schedule dynamic network by an evolving digraph, it suffices to be given a time window $\mathcal{W}$ of size $\mathcal{T}$, and to work with $\mathcal{G} = (\bigcup \mathcal{R}_i | i \in \mathcal{W}, \text{FSDN}_{|\mathcal{W}})$. Throughout this text, we assume packet based networks – so transmitting one piece of data equals transmitting one packet over an arc. Link transmission time between nodes in the network may allow for the transmission of a packet over several links before a change in the network topology. Correspondingly in the model, considering time between two successive subdigraphs in an evolving digraph as unity, the time taken to cross an arc (u, v) is expressed as a positive delay $w(u, v) \leq 1$. The case where the traversal time is larger than the frequency of topology change would then yield a delay $w(u, v) > 1$. We also implicitly assume conservation of information, i.e. in case a

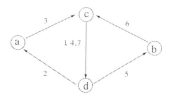

Fig. 4. Overlapping SCC's.

node in the network disappears for any reason, then upon rejoining the network, it will still have all the information that it had received before its disappearance.

3.2 Verification of Strong Connectivity in FSDN's

Given an FSDN network, we must determine if it is strongly connected. It is equivalent to the following proposition over the corresponding evolving digraph.

Proposition 1. *Given an evolving digraph $\mathcal{G}$ with $\mathcal{N}$ nodes and $\mathcal{M}$ links over a sequence of length $\mathcal{T}$, it is possible to determine if it is strongly connected or not in $O(\mathcal{N}\mathcal{M}(log\mathcal{T} + log\mathcal{N}))$ time steps.*

Proof sketch: The *transitive closure* of $\mathcal{G}$ is defined as the digraph $R_{\mathcal{G}} = (V, E_R)$, where $E_R = \{(v_i, v_j) : \exists$ a journey $\mathcal{J}(v_i, v_j)\}$. Hence, $\mathcal{G}$ is strongly connected if the underlying graph of $R_{\mathcal{G}}$ is a complete graph. The verification is executed simply and efficiently by forming the shortest journeys tree for each node in the network using the algorithm proposed in [2]. For $\mathcal{N}$ nodes, the algorithm is repeated $\mathcal{N}$ times, for an overall time of $O(\mathcal{N}\mathcal{M}(log\mathcal{T} + log\mathcal{N}))$. □

3.3 Decomposition into SCC's

Tarjan's algorithm [4], based on the concept of *forefathers* in a depth-first search tree over a digraph, is used to decompose standard digraphs into SCC's. However SCC's in evolving digraphs have the following unique properties, which make it impossible to use Tarjan's algorithm.

Property 1. Two different SCC's can have common vertices.

For example, consider the digraph given in Figure 4, where arcs are labeled with the respective arc schedule times. From the definition of SCC's we see that there are two such components a, c, d and b, c, d which have the common vertices c, d between them.

Property 2. For any two vertices in the SCC (respectively, o-SCC) there may be journeys connecting them which use vertices outside the SCC (respectively, o-SCC).

This stands directly from Property 1. As an example, take in Figure 4 the journey from d to c, which uses vertex a that lies outside the SCC $\{b, c, d\}$.

The main problem calls for decomposing the evolving digraph into all possible SCC's. Consider a subproblem *COMPONENT* defined as follows.

COMPONENT: Given an evolving digraph $\mathcal{G} = (V_{\mathcal{G}}, E_{\mathcal{G}})$ and an integer k, is there a SCC of size k?

We shall subsequently demonstrate that *COMPONENT* is NP-Complete, thereby precluding a polynomial time algorithm for the decomposition problem, unless P=NP.

Theorem 1. COMPONENT *is in NP.*

Proof sketch: Given a subset $V_{\mathcal{G}'}$ of $V_{\mathcal{G}}$ and the integer k, we must have a means of verifying in polynomial time if $V_{\mathcal{G}'}$ is indeed a SCC of size k. First, verify that $|V_{\mathcal{G}'}| = k$. Verifying that the subdigraph $\mathcal{G}'$ induced by $V_{\mathcal{G}'}$ on $\mathcal{G}$ is strongly connected and maximum is possible in polynomial time from Proposition 1. □

We now define a *strong reachability digraph* for an evolving digraph $\mathcal{G}$ as an undirected graph $S_{\mathcal{G}} = (V_{\mathcal{G}}, E_S)$, where $E_S = \{(v_i, v_j)\}$ if and only if $(v_i, v_j) \cup (v_j, v_i) \in R_{\mathcal{G}}$, the transitive closure digraph of $\mathcal{G}$.

To prove the NP-Completeness of *COMPONENT* we reduce the *CLIQUE* problem to *COMPONENT*. *CLIQUE* is formally defined as follows: Given a digraph $G = (V, E)$, and an integer k, is there a clique of size k in G?

Lemma 1. *Finding an SCC in $\mathcal{G}$ is equivalent to finding a maximal clique in $S_{\mathcal{G}}$, the strong connectivity graph of $\mathcal{G}$.*

Proof: Directly from the definitions of strong reachability, SCC and maximal clique, we see that the SCC in $\mathcal{G}$ is equivalent to finding the maximal clique in $S_{\mathcal{G}}$. □

Theorem 2. CLIQUE *can be reduced to* COMPONENT *in polynomial time.*

Proof sketch: Given an undirected graph $G = (V, E)$ and the integer k, we construct an evolving digraph $\mathcal{G} = (V_{\mathcal{G}}, E_{\mathcal{G}})$ as follows (cf. Figure 5):

1. For each node $u_i \in V$ create a node $v_i \in V_{\mathcal{G}}$, a node $h_{ii} \in V_{\mathcal{G}}$, and arcs $(v_i, h_{ii}), (h_{ii}, v_i)$ with arc schedule time 2;
2. For each edge $\{u_i, u_j\} \in E$, do
   (a) create nodes $h_{ij}, h_{ji} \in V_{\mathcal{G}}$,
   (b) create arcs $(v_i, h_{ij}), (h_{ij}, v_i)$, and arcs $(v_j, h_{ji}), (h_{ji}, v_j)$ with arc schedule time 2,
   (c) create arcs $(h_{ij}, v_j), (v_j, h_{ij})$ and arcs $(h_{ji}, v_i), (v_i, h_{ji})$ with arc schedule time 3.
3. Create an SCC connecting all h-nodes. Label these arcs with schedule times 1 and 4.

By construction, $S_{\mathcal{G}}$ contains a clique of size $n' = |\{(h_{ij}, h_{ii}) : 1 \leq i, j \leq |V_{\mathcal{G}}|\}|$ formed of the h-nodes alone. We can then prove that finding an SCC in $\mathcal{G}$ is the same as finding a clique in G, since a clique of size k in G will correspond to a clique of size $n' + k$ in $S_{\mathcal{G}}$, corresponding, in turn, to an SCC of size $n' + k$ in $\mathcal{G}$ (via Lemma 1). □

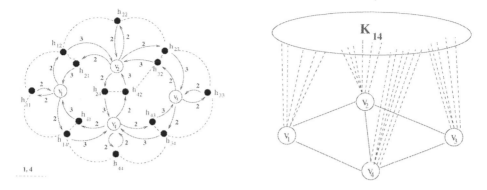

Fig. 5. Construction for Theorem 2.

3.4   Decomposition into o-SCC's

Here we prove the more general result for the case of o-SCC which has a less strict definition than SCC. We define the decision problem as follows.

o-COMPONENT: Given an evolving digraph $\mathcal{G}$ and an integer k, is there a o-SCC of size k?

Although SCC's are a special case of o-SCC's, the NP-Completeness of *COMPONENT* does not directly imply that *o-COMPONENT* is NP-Complete as well. This is because a possible polynomial time algorithm for *o-COMPONENT* need only answer the above decision problem and not identify the o-SCC's of size k, thus making it difficult to verify if at least one o-SCC of size k is an SCC as well (in other words if the set of h-nodes is empty or not for a particular o-SCC of size k). Also, the same digraph $\mathcal{G}$ may contain both an SCC (of indeterminate size) and an o-SCC of size k, so *o-COMPONENT* would always return "yes", ignoring the presence or absence of a SCC of size k, thereby leaving *COMPONENT* unsolved. Conversely, since SCC's are a special case of o-SCC's, proving *o-COMPONENT* to be NP-Complete does not directly imply that *COMPONENT* is NP-Complete as well.

These arguments entail for an independent proof for the NP-Completeness of o-COMPONENT. Fortunately, however, the same widget utilized for the previous reduction can be applied in the current case, yielding the following results.

Theorem 3. o-COMPONENT *is in NP.*

Proof: Same as the proof for Theorem 1.   □

Theorem 4. CLIQUE *can be reduced to* o-COMPONENT *in polynomial time.*

Proof: Given an undirected graph $G = (V, E)$ and the integer $k > 3$, the same arguments used in the proof of Theorem 2 apply here. Indeed, the same widget can be used to reduce *CLIQUE* to o-SCC, since a SCC is a o-SCC where $H = \emptyset$, and in that widget, a max o-SCC is a max SCC, which by Theorem 2 implies the reduction from *CLIQUE*.   □

Theorem 5. o-COMPONENT *is NP-complete.*

Proof: We know that *CLIQUE* is NP-Complete. So from Theorem 3 and Theorem 4, *o-COMPONENT* is NP-Complete.   □

4   Computing the Directed Minimum Spanning Trees

Considering a strongly connected evolving digraph $\mathcal{G}$, the object is to find $\mathcal{N} = |V_{\mathcal{G}}|$ rooted directed minimum spanning trees rooted at each of the nodes $r \in V_{\mathcal{G}}$. Our algorithm is a modification of the Prim-Dijkstra algorithm [4] for finding MST's in undirected standard graphs. The algorithm proceeds by building a fragment which is a subset of the DMST starting from the root r. The property of the fragment $f(r)$ is that it consists of those edges by which information transmitted at the beginning of the time interval from the root r will travel in the shortest time to the vertices included already in the fragment. Having defined a fragment as such, it is easy to see how the algorithm for the DMST proceeds. In the following algorithm we choose from among the set of arcs outgoing from the fragment $f(r)$, the arc with the smallest arc schedule time such that it can form a valid journey starting from the root. A number t_v is associated with each vertex $v \in V_{\mathcal{G}}$ denoting the minimum time required for that vertex to receive the information given that the root r originates the information.

Since each node can transmit information only *after* it has received it, the information cannot pass simultaneously through two edges. Recall that the time required for transmission over one arc is denoted as an arbitrary weight, $w(u, v) < 1$.

Algorithm 1

1. *Start with $f(r) = \emptyset$ and a set V_f containing vertices already considered in fragment $f(r)$.*
2. *$V_f = \{r\}$, $t_r = 1$*
3. *while $V_f \neq V_{\mathcal{G}}$ do*
   (a) *Let Γ_f be the set of all arcs (u_i, v_i) such that $u_i \in V_f$, $v_i \notin V_f$. For each $(u_i, v_i) \in \Gamma_f$, choose the smallest arc schedule time $f_a(u_i, v_i)$, such that $f_a(u_i, v_i) \geq t_{u_i} + w(u_i, v_i)$.*
   (b) *Choose arc (u_j, v_j) where $j = \min_i^{-1}(f_a(u_i, v_i) + w(u_i, v_i))$.*
   (c) *if $f_a(u_j, v_j) = t_{u_j} + w(u_j, v_j)$, then $t_{v_j} \leftarrow f_a(u_j, v_j)$,*
   (d) *else if $f_a(u_j, v_j) - 1 \in \{arc\ schedule\ of\ (u_j, v_j)\}$, then $t_{v_j} \leftarrow t_{u_j} + w(u_j, v_j)$,*
   (e) *else, $t_{v_j} \leftarrow f_a(u_j, v_j) - 1 + w(u_j, v_j)$*
   (f) *add v_j to V_f and (u_j, v_j) to $f(r)$.*

In the above algorithm, an arc schedule time i indicates the presence of the link from time $i - 1$ to i. Note that two cases might arise depending on whether $f_a(u_j, v_j) = t_{u_j} + w(u_j, v_j)$ or $f_a(u_j, v_j) > t_{u_j} + w(u_j, v_j)$. For the first case, the information reaches the node exactly at the time $f_a(u_j, v_j)$. For the other case, i the arc is present both at times $f_a(u_j, v_j) - 1$ and $f_a(u_j, v_j)$, since $w(u_j, v_j) < 1$, the packet will reach v_j in $t_{u_j} + w(u_j, v_j)$. If, however, the arc is not present at time $f_a(u_j, v_j) - 1$, then the transmission process itself starts at the $f_a(u_j, v_j)^{th}$ step (i.e. from time $f_a(u_j, v_j) - 1$ to time $f_a(u_j, v_j)$), thus reaching v_j by time $f_a(u_j, v_j) - 1 + w(u_j, v_j)$.

We remark that a rooted directed tree can also be computed over an o-SCC $V_{G'}$. As a modification for that purpose, V_G must be replaced by $V_{G'}$ and correspondingly, Step 3 of Algorithm 1 should be modified to $V_{G'} \subset V_f$ since the fragment can also contain the h-nodes for the vertices in $V_{G'}$ and the loop can stop once all the vertices are covered.

Algorithm 1 is a greedy algorithm that always chooses the arc that transmits in minimum time. The proof of its correctness is the same as the proof of the Prim-Dijkstra algorithm [4]. If the maximum outdegree of each vertex is $\mathcal{D}$, then each step of increasing the fragment will take $O(\mathcal{N}\mathcal{D}\log\mathcal{T})$ time and the fragment will increase $\mathcal{N}$ times adding up to a total execution time of $O(\mathcal{N}^2\mathcal{D}\log\mathcal{T})$ steps.

5   Conclusion

The two important results in this paper are the intractability of the decomposition into (open) strongly connected components in FSDN's and the construction of DMST's over an already existing strongly connected components.

The first result implies that it is possible to lead a non-strongly connected network towards strong connectedness by adding intermediary agents to serve as hops between two nodes that are out of range from each other. An interesting problem would be to find a way to add such links so as to minimize the number of intermediary (helping) nodes. Another way for further research is to design approximation algorithms for (open) strongly connected components in evolving digraphs.

Acknowledgments

The authors are grateful to Aubin Jarry and Stephane Perennes for very fruitful discussions.

References

1. A. Borodin and R. El-Yaniv. *Online computation and competitive analysis.* Cambridge University Press, 1998.
2. B. Bui-Xuan, A. Ferreira, and A. Jarry. Computing shortest, fastest, and foremost journeys in dynamic networks. *International Journal of Foundations of Computer Science*, 14(2):267–285, April 2003.
3. Y. J. Chu and T. H. Liu. On the shortest arborescence of a directed graph. *Science Sinica*, 14:1396–1400, 1965.

4. T. Cormen, C. Leiserson, and R. Rivest. *Introduction to Algorithms*. The MIT Press, 1990.
5. S.E. Dreyfus. An appraisal of Some Shortest-Path Algorithms. *Operations Research*, 17:269–271, 1969.
6. E. Ekici, I. F. Akyildiz, and M. D. Bender. Datagram routing algorithm for LEO satellite networks. In *IEEE Infocom*, pages 500–508, 2000.
7. A. Faragó and V.R. Syrotiuk. MERIT: A unified framework for routing protocol assessment in mobile ad hoc networks. In *Proc. ACM Mobicom 01*, pages 53–60. ACM, 2001.
8. A. Ferreira, J. Galtier, and P. Penna. Topological design, routing and handover in satellite networks. In I. Stojmenovic, editor, *Handbook of Wireless Networks and Mobile Computing*, pages 473–493. John Wiley and Sons, 2002.
9. L.R. Ford and D.R. Fulkerson. Constructing maximal dynamic flows from static flows. *Operations Research*, 6:419–433, 1958.
10. J. Halpern. Shortest route with time dependent length of edges and limited delay possibilities in nodes. *Zeitschrift für Operations Research*, 21:117–124, 1977.
11. J. Halpern and I. Priess. Shortest path with time constraints on movement and parking. *Networks*, 4:241–253, 1974.
12. P. A. Humblet. A distributed algorithm for minimum weight directed spanning trees. *IEEE Transactions on Communications*, COM-31(6):756–762, 1983.
13. E. Köhler, K. Langkau, and M. Skutella. Time-expanded graphs for flow-dependent transit times. In *proc. ESA'02*, 2002.
14. E. Köhler and M.Skutella. Flows over time with load-dependent transit times. In *Proc. of the 13th Annual ACM-SIAM Symposium on Discrete Algorithms*, pages 174–183, 2002.
15. C. Scheideler. Models and techniques for communication in dynamic networks. In In H. Alt and A. Ferreira, editors, *Proceedings of the 19th International Symposium on Theoretical Aspects of Computer Science*, volume 2285, pages 27–49. Springer-Verlag, March 2002.
16. I. Stojmenovic, editor. *Handbook of Wireless Networks and Mobile Computing*. John Wiley & Sons, February 2002.
17. V. Syrotiuk and C. J. Colbourn. Routing in mobile aerial networks. In *Proceedings of WiOpt'03 – Modeling and Optimization in Mobile, Ad-Hoc and Wireless Networks*, pages 293–302, Sophia Antipolis, March 2003. INRIA.
18. R. E. Tarjan. Finding optimum branchings. *Networks*, pages 25–35, 1977.
19. P.-J. Wan, G. Calinescu, X. Li, and O. Frieder. Minimum-energy broadcast routing in static ad hoc wireless networks. In *Proc. IEEE Infocom*, pages 1162–1171, Anchorage, Alaska, 2001.
20. J. Wieselthier, G. Nguyen, and A. Ephremides. On the construction of energy-efficient broadcast and multicast trees in wireless networks. In *Proc. IEEE Infocom*, pages 585–594, Tel Aviv, 2000.

Mobile Agents for Clustering and Routing in Mobile Ad Hoc Networks

Mieso K. Denko and Qusay H. Mahmoud

Department of Computing and Information Science
University of Guelph
Guelph, Ontario, N1G 2W1, Canada
{denko,qmahmoud}@cis.uoguelph.ca

Abstract. A mobile ad hoc network (MANET) is a dynamic wireless network that can be formed without the need for any pre-existing infrastructure in which each node can act as a router. One of the main challenges in ad hoc networks is the design of robust routing algorithms that adapt to the frequent and randomly changing network topology. Organizing mobile nodes into manageable clusters can reduce routing overhead and provide more scalable solutions. In this paper we propose a mobile agent-based method for clustering and routing in mobile ad hoc networks. All mobile nodes use two agents to perform routing and clustering operations. Using this method, reactive, proactive or hybrid routing schemes can be employed for intra-cluster and inter-cluster routing to improve the performance of routing.

1 Introduction

A mobile ad hoc network (MANET) is a multihop wireless network in which mobile nodes can communicate with each other without the support of any pre-existing infrastructure. In this network environment, each node acts as router and can relay packets to its neighbors. This type of network is characterized by limited bandwidth and battery power, rapidly moving nodes and unpredictable topological changes. In MANETs, one of the main challenges is the design of adaptive and robust routing algorithms.

Routing protocols designed for traditional fixed networks are not suitable for mobile ad hoc networks [13, 15, 20]. As a result, several routing protocols have recently been proposed for MANETs (see for example [3, 4, 5, 8, 13, 15, 20]). These protocols can be classified into three main categories: proactive, reactive and hybrid. Proactive routing protocols update routes periodically or in response to some pre-defined events. Reactive protocols compute routes on demand. Hybrid protocols use features of both reactive and proactive protocols [10]. The main advantage of hybrid protocols is their flexibility in allowing the use of different routing mechanisms within and between clusters. Since the overhead for routing can grow faster than linearly as network size increases [17], routing in a flat architecture faces a scalability problem. Several routing protocols based on clustering architecture have been proposed in recent years (see for example [2, 3, 4, 5, 6, 11, 12 15]).

S. Pierre, M. Barbeau, and E. Kranakis (Eds.): ADHOC-NOW 2003, LNCS 2865, pp. 271–276, 2003.

An important area of research is the application of mobile agents in mobile and wireless networks. A mobile agent is a software entity that can actively migrate among nodes in a heterogeneous network interacting with other agents and service agents [9]. This is ideal for MANETs because mobile agents are capable of supporting asynchronous communication. As a result, several mobile agent-based projects have been proposed. For example, a mobile agent-based routing protocol has been recently proposed in [19]. The protocol uses the combined benefits of traditional ant-based adaptive routing and the Ad-Hoc on Demand Distance Vector (AODV) routing protocol to maintain node connectivity and perform routing.

In this paper, we propose a mobile agent-based method for clustering and routing in MANETs. In this method, mobile agents are used to maintain clustering and routing information at each node in a distributed manner. The information maintained in the routing table is used for intra-cluster and inter-cluster routing. Inter-clustering is performed via the clusterhead.

The rest of this paper is organized as follows. Section 2 describes the benefits of clustering and mobile agents in MANETs. Section 3 presents the proposed agent-based clustering and routing method. Finally, conclusions and future research work are presented in Section 4.

2   Clustering and Mobile Agents in MANETs

Clustering is the partitioning of the network into small manageable groups of nodes. Clustering offers several advantages in mobile ad hoc networks. First, network partitioning improves routing and mobility management [21]. It increases system capacity, reduces signaling and control overhead and minimizes network congestion. This makes the network more scalable and as a result can support a larger network size. Second, clustering stabilizes the network topology and provides a virtual infrastructure for a dynamic network. The clusterhead acts as a base station for its cluster. Third, clustering helps to perform more efficient resource allocation. By assigning different codes to each cluster, MAC resource management can be improved and wireless channels can be used efficiently [1, 7]. It also provides good power management mechanisms. Clusters can be either distinct or overlapping. In the former, each node belongs to only one cluster while in the latter the neighboring clusters can have a common node (gateway or access point) between them. In this paper we consider only distinct clusters.

Several clustering algorithms have been proposed in the literature for grouping nodes into clusters [1, 2, 3, 7, 14, 17]. One of the earliest clustering algorithms is the Linked Cluster Algorithm (LCA) [1] proposed for mobile radio networks. The algorithm uses a distributed control mechanism for neighbor discovery and cluster formation. In this algorithm a node with the lowest ID becomes a clusterhead. In [7], the Lowest-ID (LID) and another distributed clustering algorithm known as Highest-Connectivity (HC) were used for clustering nodes in a multicluster, multihop packet radio network architecture. In the HC algorithm, a node with the highest degree is elected as a clusterhead. A clustering algorithm that combines the LID and HC clustering algorithms was proposed in [2]. The experimental results indicated that the algorithm generates a lower number clusterheads and gateways. In our approach, the mobile agent uses node mobility and link characteristic related parameters for clustering.

A mobile agent is capable of migrating autonomously carrying code, data and state with itself. It can even spawn off child agents anywhere in the network, merge the query results and send back the final result to the source node. Mobile agents can improve bandwidth utilization, reduce communication latency, minimize connection time, and reduce network traffic load.

An attractive application of mobile agents is processing data over unreliable networks, such as MANETs. In such an environment, the low reliability network can be used to transfer agents, rather than a chunk of data, from one node to another. In MANETs, the agents can travel to the nodes of the cluster and collect or process clustering and routing information, without the risk of network disconnection, and then return to its originating node.

In order to deploy mobile agents in MANETs, a suitable mobile agent platform is needed. Conventional mobile agent platforms such as Aglets [16], Concordia [23], and D'Agent [8], to name a few, operate within high-end desktop environments such as Windows and Unix. As a result some research projects have recently been proposed to develop mobile agent platforms for mobile devices. The Lightweight Extensible Agent Platform (LEAP) [18] aims to develop a FIPA-compliant mobile agent platform for mobile devices with services in the area of knowledge and travel management. In our research, we hope to develop our own mobile agent platforms for MANETs.

3   Mobile Agents for Clustering and Routing

In a clustered network, a cluster may be organized into a multilevel hierarchy. A hierarchical clustering architecture can reduce network routing overhead by hiding information about the content of the cluster. Route maintenance procedures and routing table length can be significantly reduced [14]. Such architecture is relatively stable and scalable due to the localized nature of route computation and can be used in MANETs. Most previous works on clustering are based on the design of algorithms that form a 2-hop clustering architecture. These algorithms use a single parameter such as node ID, connectivity, signal strength, mobility, power or some combination of these for cluster formation. The cluster formation process can be slow since the method of gathering the information necessary for cluster formation may not be efficient if clustering with more than 2-hop architecture is desired.

Since mobile agents are autonomous and intelligent entities, they can be used for creating dynamic and adaptive clustering in MANETs [22]. Distributed route computation can be performed at each node. The clustering architecture consists of ordinary nodes, clusterheads and gateways. In our architecture, intra-cluster and inter-cluster routing can be carried out using reactive, proactive or hybrid routing schemes. Since a single criterion is not sufficient for a stable and efficient cluster formation, an aggregate metric that includes node mobility, link quality, available bandwidth, etc. will be maintained and used for cluster formation. This metric will also be used by the clustering agent to assess the quality of the clusterhead periodically.

3.1 Maintaining Routing and Clustering Information

In our proposed architecture, each node has a Routing Mobile Agent (RMA) and a Clustering Static Agent (CSA). These agents operate on top of an agent platform running on top of a Java virtual machine suitable for mobile devices such as the Kilo Virtual Machine (KVM). The CSA maintains clustering information in a clustering table. The clustering table contains IDs of neighbors, the node role (ordinary node, gateway or clusterhead), mobility information, nodal degree and signal strength. The routing mobile agent moves across the network to collect and maintain routing tables while the clustering agent gathers and maintains clustering information.

All inter-cluster routing is performed via the clusterhead. On receiving a packet from a node, the clusterhead forwards the packet to the preferred gateway which in turn forwards it to the adjacent cluster. In the event that the clusterhead does not have this information, it can deploy the mobile agent to get the route to the target destination. Routing and clustering table entries at the clusterheads are updated periodically with up-to-date values based on timestamps.

To localize the agent mobility, the routing agent in a mobile node migrates only within its cluster. The routing mobile agent also updates routing and clustering information during each visit. The agents at the clusterhead also maintain information about other clusterheads and gateways. The clustering agent elects a clusterhead based on the clustering information maintained at each mobile node. Once the clustering architecture stabilizes, each node will have a complete knowledge of its neighbors.

We have devised a migration strategy for routing mobile agents. To find a route, the Routing Mobile Agent (RMA) uses the following algorithm:

```
IF the RMA does not have the route
   then
        The RMA moves to the Clusterhead
        in that Cluster;
        The RMA communicates with the
        Clusterhead;
   IF the Clusterhead has the routing info
      then
          The RMA goes back and updates routes;
   Else
      RMA in clusterhead continues route search
      in other Clusters;
   END IF
END IF
```

3.2 Mobile Nodes and Inter-agent Communication

Each mobile agent will be given a temporary workspace to perform its functions and also to allow multiple mobile agents to co-exist in a host. A hosting node can receive new agents or transport them to other nodes without causing any interference. The agent system hides details about the node and provides access to local resources with the necessary security mechanisms.

Security is a major concern when working in mobile agents and MANETs. There are two types of security concerns: (1) protecting the agent from the agent server, which can be accomplished by devising a security policy that states what agents can and cannot do; and (2) protecting the agent server from the agent, which is almost impossible. While several remedies have been proposed for protecting the agent from the agent server, how can we actually protect an agent from being killed by an agent server? Fortunately, these security issues exist when working with mobile agents competing for a resource. When using mobile agents for clustering and routing, agents are cooperating to deliver a service.

3.3 Performance Metrics

Mobile agents periodically examine the clustering parameters and make cluster size adjustments, perform re-clustering and monitor clusterhead quality. The clustering architecture is evaluated using parameters such as cluster size, clusterhead changes, cluster membership changes, number of clusters, cluster splitting and merging. Each metric is investigated by varying network size, transmission range and node mobility. Our performance metrics for routing are packet delivery ratio, routing overhead and end-to-end delay.

4 Conclusions and Future Work

In this paper we have presented our proposed mobile agent-based method for clustering and routing in mobile ad hoc networks. In this method, each node is equipped with a Clustering Static Agent (CSA) and a Routing Mobile Agent (RMA). The agents are used to collect and maintain routing and clustering information. The method can be used to improve the performance of clustering and routing operations by using agents that support asynchronous, and therefore disconnected operations, and reduces the network traffic in MANETs. Parameters for the performance evaluation of clustering stability and routing performance were identified.

Our future work includes the design and implementation of a mobile agent platform suitable for MANETs. Once we build the platform, we plan to run experiments to help us compare the performance of the proposed method with other methods.

References

1. Baker, D.J., Ephremides, A., Flynn, J.A.: The Design and Simulation of a Mobile Radio Network with Distributed Control. IEEE Journal on Selected Areas in Communications, 2(1): 226-237, January 1984
2. Chen, G., Stojmenovic, I.: Clustering and Routing in Wireless Ad Hoc Networks. Technical Report TR-99-05, Department of Computer Science, SITE, Ottawa, June 1999.
3. Chiang, C.-C.: Routing in Clustered Multi-hop, Mobile Wireless Networks with Fading Channel. In proceedings of IEEE SICON'97, pp.197-211, 1997.
4. Corson, M.S., Ephremides, A.: A Distributed Routing Algorithm for Mobile Wireless Networks. ACM-Baltzer Journal of Wireless Networks, 1(1):61-81, 1995.

5. Das, B., Bharghavan, V.: Routing in Ad Hoc Networks Using Minimum Connected Dominating Sets (MCDS). In proceedings of 1997 IEEE International Conference on Communications (ICC'97), 1997.
6. Denko, M.K., Goddard, W.: Routing Algorithms in Mobile Ad Hoc Networks Using Clustering. In proceedings of the 13th MSc/PhD Annual Conference in Computer Science, University of Stellenbosch, South Africa, pp. 6-18, 1998.
7. Gerla, M., Tsai, J.T.: Multicluster, Mobile, Multimedia Radio network, ACM-Baltzer Journal of Wireless Networks, 1(3): 255-265, 1995.
8. Garcia-Luna-Aceves, J.J., Mosko, M., Perkins, C.E.: A New Approach to On-Demand Loop-Free Routing in Ad Hoc Networks. In proceedings of the 22nd ACM Symposium on Principles of Distributed Computing, Boston, Massachusetts, July 2003.
9. Gray, S.R., Kotz, D., Cybenko, G., Rus, D.: D'Agent - Security in a Multiple-Language, Mobile Agent System. In Vigna, G. (eds.): Mobile Agents and Security, Lecture Notes in Computer Science, vol. 1419. Spring-Verlag, Berlin, Heidelberg, 1998.
10. Haas, Z.J., Perlman, M.: The performance of query control schemes for the zone routing protocol. ACM/IEEE Transactions on Networking, 9(4): 427-438, August 2001.
11. Hou, T.-C., Tsai, T.-W.: An access-based clustering protocol for mobile multihop wireless ad hoc networks. IEEE Journal on Selected Areas in Communications, 19(7):1201-1210, 2001.
12. Jiang, M., Li, J., Tay., Y.C.: Cluster-Based Routing Protocol (CBRP), Internet-Draft, draft-ietf-manet-zone-zrp-02.txt, 1999.
13. Johnson, D., Maltz, D.: Dynamic source routing in ad hoc wireless networks. In Perkins, C.E. (ed.): Ad Hoc Networking. Addison Wesley, pp. 139-172, 2001.
14. Kleinrock, L., Kamoun, F.: Hierarchical Routing for Large Networks: Performance Evaluation and Optimization, Computer Networks, 1(1): 155-174, 1977.
15. Krishna, P., Chatterjee, M., Vaidya, N.H., Pradhan, D.K.: A Cluster-based Approach for Routing in Ad Hoc Networks. In proceedings of the 2nd USENIX Symposium on Mobile and Location Independent Computing, pp. 1-10, 1996.
16. Lange, D. B., Oshima, M.: Programming and Deploying Java Mobile Agents with Aglets. Addison Wesley, 1998.
17. Lauer, G.S.: Packet Radio Routing. In Steenstrup, M. (ed.): Routing in Communication Networks, Prentice Hall, pp. 351-396, 1995.
18. LEAP: http://leap.crm-paris.com.
19. Marwaha, S., Tham, C. K., Srinvasan, D.: Mobile Agent-based Routing Protocol for Mobile Ad Hoc Networks. Symposium of Ad Hoc Networks, IEEE (GLOBCOM 2002), 2002.
20. Perkins, C.E., Bhagwat, P.: Highly Dynamic Destination Sequenced Distance Vector Routing (DSDV) for Mobile Computers. In proceedings of SIGCOMM, pp. 234-244, 1994.
21. Sharony, J.: A Mobile Radio Network Architecture with Dynamically Changing Topology Using Virtual Subnets. ACM/Baltzer Mobile Networks and Applications, 1(1): 75-86, 1996.
22. Sugar, R., Imre S.: Adaptive Clustering Using Mobile Agents in Wireless Ad hoc Networks. Lecture Notes in Computer Science. Spring-Verlag, 2001.
23. Walsh, T., Paciorek, N., Wong, D.: Security and Reliability in Concordia. In proceedings of the 31st Annual Hawaii International Conference on System Science, vol. 7, pp 44-53, 1998.

Routing Update in Ad Hoc Networks

Benjamin Macabéo, Samuel Pierre, and Alejandro Quintero

Mobile Computing and Networking Research Laboratory (LARIM)
Department of Computer Engineering, École Polytechnique de Montréal
C.P. 6079, succ. Centre-Ville, Montréal, Québec, Canada, H3C 3A7
{Benjamin.Macabeo,Samuel.Pierre,Alejandro.Quintero}@polymtl.ca
Tel. (514) 340-3240 ext. 4685, Fax. (514) 340-3240

Abstract. Contrary to cellular networks, ad-hoc networks are a form of mobile networks that function without any fixed infrastructure. This paper proposes a method which improves routing success rates in mobile ad hoc networks. This method is based on the density of the nodes in the neighborhood of a route and on the availability of this neighborhood. The results obtained are encouraging: the data packet loss rate is significantly reduced and the time required to complete a local repair route following a failure decreased significantly.

Index Terms: mobile ad hoc networks, route repair, AODV

1 Introduction

An *ad hoc network* is a mobile wireless network composed of several mobile nodes, likely to communicate together without the required intervention of any centralized management or existing infrastructure. The nodes of these networks must be able to cooperate among themselves to allow communication. The deployment of ad hoc networks is thus largely simplified compared to other forms of mobile networks.

This paper suggests a method which improves the probabilities of success of a local route repair in mobile ad hoc networks (MANETs) by accelerating the process of route reparation after the departure of a node included in the route. Section 2 introduces some background information and related work. Section 3 describes the solution suggested to improve route repairs. Finally, Section 4 presents and analyzes simulation results.

2 Background and Related Work

A routing protocol is a mechanism by which user traffic is directed and transported through a network from a source node to its destination node. It aims to maximize network performance from an application point of view while minimizing the cost imposed on the network in terms of capacity. QoS (Quality of Service) routing protocols search routes with sufficient resources for QoS requirements [3, 7].

S. Pierre, M. Barbeau, and E. Kranakis (Eds.): ADHOC-NOW 2003, LNCS 2865, pp. 277–280, 2003.
© Springer-Verlag Berlin Heidelberg 2003

2.1 Best Effort Protocols

DSDV (Destination Sequenced Distance Vector) is a best effort protocol designed specially for MANETs [4]. It belongs to the class of proactive protocols and uses a version of the distributed Bellman-Ford algorithm which is adapted to ad hoc networks. The routing information associated to each node of the network is recorded in a routing table by each mobile station.

AODV (Ad hoc On Demand Vector) represents an improvement of DSDV [5]. In fact, to synthesize, it takes the advantages of DSDV but limits bandwidth consumption.

2.2 QoS Routing Protocols

In MANET, a route is defined as set of mobile units that contributes to data transmission from source to destination. Quality of service (QoS) consists of a set of characteristics or constraints (bandwidth, hop count, delay, throughput, packet loss rate, etc.) that a connection must guarantee between a source and a destination during the communication to meet the requirements of an application [1].

With the increasing number of applications requiring a certain QoS, the success of mobile ad hoc networks relies heavily on their ability to provide routing protocols that take into account QoS [2, 6].

3 Routing Protocol in Ad Hoc Networks

The routing protocol used here is based on a routing algorithm initiated by the source that takes into account QoS in terms of bandwidth consumption. If a failure occurs during the communication between two nodes, two scenarios can be used to repair the route: a *global route repair* and a *local route repair*. A *global route repair* starts from the source of communication. Although it requires significant time and consumes much bandwidth, this solution is used in most routing protocols. A *local route repair* starts from a node in the neighborhood of the link where the failure occurred. This latter solution offers two advantages: its speed and its low bandwidth consumption.

3.1 Protocol for Route Repair

Our objective aims to ensure the selection of the most easily reparable route among those extracted from the route discovery phase. To achieve this goal, we recommend taking into account the nature of the neighboring nodes composing the network, more particularly the node *density* and their *availability*. The reparation of a route in case of failure can be carried out through local route repair.

We use the availability parameter to establish the ability of a Node A to replace Node B. The availability of a node depends on the nature of the node, the number of packets forwarded by the node as well as their capacity.

We define the density of a node λ as the number of direct neighbors of λ whose available bandwidth is higher than that required by the connection. The density parameter is completely specified by a node and the bandwidth associated with that node.

The discovery phase: We use the route discovery phase as described in the AODV protocol for which we add provisions for the availability and density parameters. These two parameters need to be taken into account in order for our protocol to provide QoS. Thus, in our protocol, the source initiates the routing process upon receiving a connection request. Then, it sends a route request for this connection to all its neighbors. The nodes that receive the message for the first time and that fulfill the QoS requirements propagate the request message towards the destination after the following scenario :

The request message is gradually propagated towards the destination following the aforementioned scenario. Finally, when it arrives at its destination, the destination node initiates a countdown and records all of the incoming request messages.

To *select a route*, we need the parameters contained in each request message that arrived at the destination. It is important to mention that a route containing long sequences of high density nodes will be easier to repair with the local route repair procedure than a route that does not hold that property.

4 Implementation and Results

The modifications to the AODV protocol were implemented using *Opnet Modeler*. The protocol defines four types of packets that can be exchanged between the topology nodes: RERR, REEQ, RREP and DATA.

4.1 Example

Here is a specific example which illustrates the rationale for such modifications to the initial protocol (Figure 1).

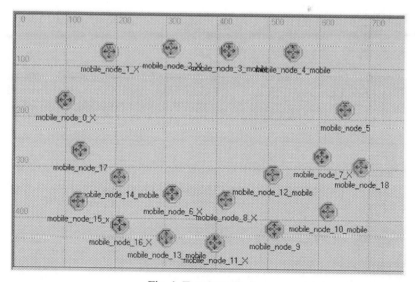

Fig. 1. Topology used

In the configuration presented in Figure 2, Node 0 seeks to establish a communication with Node 5. All of the other nodes behave as routers. Several routes are possible. Among these routes, the route passing through Nodes 1, 2, 3 and 4 is the shortest one from source to destination.

For the sake of clarity, we will only detail the results concerning the departure of Node 12. The data collected for Cases 1 and 2, and in particular the end-to-end (ETE) delays, reveal that no route repair is undertaken following the departure of Node 12 from the network, since this node does not belong to the route used in these cases. On the other hand, we can clearly see that for Case 3, the departure of Node 12 strongly affects the results. Indeed, the ETE delay increases from 0.014 to 0.016 second after the departure of Node 12.

5 Conclusion

The routing method presented in this paper aims to improve QoS management in MANETs by taking into account the *density* of a node, defined as the number of mobile units available in the radio range of the node. Our approach was based on a thorough analysis of the available mechanisms and tools that take into account quality of service in ad hoc networks. We then introduced the concept of density and described how the network could exploit this information to improve the QoS offered.

The described route selection mechanism aims to select the route whose maintenance is the easiest to realize among several routes. The protocol was tested with a given configuration. The results obtained are encouraging: the data packet loss rate is strongly reduced compared to the initial version. In addition, the time required to complete a local route repair following a failure is reduced significantly.

References

1. Chakrabarti S. and Mishra A., "QoS issues in ad hoc wireless networks", *IEEE International Conference on Communications*, 2001, pp. 142–148.
2. Das S., Mukherjee A., Bandyopadhyay S., Paul K., Saha D., "Improving quality-of-service in ad hoc wireless networks with adaptive multi-path routing", IEEE Conference on Global Telecommunications, Vol. 1, 2000, pp. 261 –265.
3. Hongxia S., Hughes, H., "Adaptive QoS routing based on prediction of local performance in ad hoc networks", IEEE Conference on Wireless Communications and Networking, Vol. 2 , 2003, pp. 1191 –1195.
4. Perkins C., Bhagwat P., "Highly dynamic destination-sequenced distance-vector routing for mobile computer", ACM Conference on Communications Architectures, 1994, pp. 234-244.
5. Perkins C., Royer E., "Ad hoc on demand distance vector algorithm", IEEE Workshop on Mobile Computing Systems and Applications WMCSA '99, 1999, pp. 90 –100.
6. Wen-Hwa L., Yu-Chee T., Kuei-Ping S., "A TDMA-based bandwidth reservation protocol for QoS routing in a wireless mobile ad hoc network", IEEE International Conference on Communications, Vol. 5, 2002, pp. 3186–190.
7. Xiaoyan H, K. Xu, M. Gerla, "Scalable Routing Protocols for Mobile Ad Hoc Networks", IEEE Network, Vol. 16, No. 4, 2002, pp. 11-21.

Inter-vehicle Geocast Protocol
Supporting Non-equipped GPS Vehicles*

Abderrahim Benslimane and Abdelmalik Bachir

Laboratoire d'Informatique d'Avignon LIA/CERI
339 chemin des Meinajaries
BP 1228 - 84911 AVIGNON CEDEX 9
{bachir,benslimane}@lia.univ-avignon.fr

Abstract. IVG is a GPS-based Inter-Vehicle Communication protocol used for alarm message dissemination among vehicles in a highway in risk situations. It is based on the principle of wireless ad hoc networks. In this paper, we propose an improvement to IVG towards supporting its interoperability in environments where vehicles "GPS-U" without GPS devices are present. It is also the case, because of obstacles, where certain vehicles have GPS devices but cannot obtain their position via GPS. The proposed solution allows GPS-U vehicle to compute its position with the help of its neighbors that are equipped with GPS devices "GPS-E". Analyses show that the optimal performances of IVG can be reached even when the rate of GPS-U vehicle is 40%.

1 Introduction

Intelligent transportation Systems (ITS) have been investigated for many years in Europe, Japan and North America, with the aim of providing new technologies able to improve safety and efficiency of road transport. Recently, the democratisation of GPS technology and the progress in mobile ad hoc networking have led to the appearance of new inter-vehicle communication protocols [1, 2, 3]. Based on the use of GPS devices, these protocols have been mainly designed for safety driving by the dissemination of urgent information, called alarm messages, in the case of accidents, fogs, etc, among the vehicles. In [1], the proposed solution called RBM Role Based Multicast was designed to overcome fragmentation in the ad hoc network composed by the vehicles and to reduce the number of redundant broadcasts of alarm messages. In [2], two other solutions were proposed, Track Detection (TRADE) and Distance Defer Time (DDT). In TRADE, each vehicle wanting to disseminate an alarm message has to determine positions and driving directions of its neighbors. DDT does not rely on neighbors maintenance, but inserts distance-based defer time slots for each rebroadcast alarm message. When a vehicle executing DDT receives an alarm message, it sets-up a timer in order to determine if it is useful to rebroadcast that message.

---

* This work is supported by CNRS/JemSTIC grant N° SUB/2002/004/DR16.

S. Pierre, M. Barbeau, and E. Kranakis (Eds.): ADHOC-NOW 2003, LNCS 2865, pp. 281–286, 2003.
© Springer-Verlag Berlin Heidelberg 2003

In [3], we proposed IVG, Inter Vehicle Geocast, an inter vehicle message dissemination protocol that improves bandwidth utilization, reduce delays and packet loss since it avoids neighbors maintenance signalling, and overcomes fragmented networks by the use of dynamic relays.

Since all the previous proposed protocols are based on geographical positioning system (i.e. GPS), we analyze in this paper the possibility of the interoperability between GPS-equipped and GPS-unequipped vehicles in IVG, with the aim to give GPS-unequipped vehicles pertinent information about the accident. The solution is based on cooperation between GPS-E vehicles in order to help GPS-U vehicles to get their positions. Although the knowledge of the exact position is not always possible, the GPS-U vehicle can obtain some useful information such as driving direction and distance from the accident.

Several radiolocation systems have been proposed for locating the Mobiles Stations (MS) in cellular systems [4, 5, 6]. To do that, these systems use one or more of the following parameters: signal strength, angle of arrival, time of arrival or their combinations. Recently, a new algorithm Self-Positioning Algorithm (SPA) has been proposed for positioning mobile nodes in wireless ad hoc networks [7] without relying on GPS and not tacking into account inter-vehicle communication. In this paper, we propose another method for GPS-free positioning for IVG [3] taking care on urgent nature of communication. For example, in the case of an accident, vehicles without GPS have to be informed in the right moment. The algorithm should be lightweight and give to the vehicle enough accurate information about the accident. The suggested solution must be temporary while waiting for all the vehicles to be GPS-equipped in the future and the disappearance of GPS-unequipped ones.

The remainder of this paper is organized as follows. In section 2, we give an overview of IVG protocol. In section 3, we present our algorithm of GPS-free positioning for IVG. Section 4 presents a performance evaluation of the proposed algorithm. Finally, we give a conclusion in section 5.

2   IVG Presentation

IVG is mainly designed for effective alarm message dissemination in the ad hoc network of vehicles in a highway. IVG is based on geographical multicast, which consists in determining the multicast group according to the driving direction and the positioning of the vehicles. The multicast is restrained to the so-called risk areas. First, broken vehicle (or accident) begins to broadcast an alarm message to inform the other vehicles of the situation. Since the accident vehicle can just inform its one-hop neighbors, some other vehicles have to rebroadcast the alarm message to inform the vehicles located at more than one hop from the accident. The vehicle that performs the rebroadcast is called relay. Relays in IVG are designated in fully distributed manner. The way with which a node is designated as relay is based on distance defer time algorithm. The node that receives an alarm message does not rebroadcast it immediately but has to wait some time to take a decision about rebroadcast. When the defer time expires, if it does not receive the same alarm message from another node behind it, it deduces that there is no relay node behind it. Thus it has to designate it self as a relay and starts to broadcast the alarm messages in order to inform the vehicles which could be behind it. The defer time of a node (x) receiving a message from another node (s) is inversely proportional to the distance separating them that is to favorite the farthest node to wait less time and to rebroadcast faster. The alarm message must contain some information such as accident position, previous and current positions of the relay from which the message is received. This information is used by the vehicle that received the alarm message in order to determine its location according the accident vehicle [3]. The message is relevant if the vehicle is located in a relevant area and it is received for the first time. When a vehicle receives the same alarm message before its defer timer expires, it concludes that there is another vehicle behind it which is broadcasting the same alarm message. In this situation, the second alarm message is not relevant because the vehicle was already informed about the accident by the first alarm message and it is useless to rebroadcast it because there is a relay behind it that is ensuring the dissemination of this alarm message.

The message dissemination in IVG depends on the rate of vehicles equipped with GPS device in the road. We believe that the success of IVG depends on its performances with GPS-unequipped vehicles. In the next section, we propose a solution that allows the well functioning of IVG even with GPS-unequipped vehicles. The performances of that solution depend on the rate of GPS-unequipped vehicles and on the density of vehicle in the highway.

3 GPS-Unequipped Algorithm

Since each vehicle executing IVG relies on the periodic computation of its driving direction (previous and current positions) some modifications have to be envisaged to make GPS-U vehicles know these positions when the communication with the GPS satellite is not possible. IVG can be executed normally if these positions are accurately known. However, this is not always possible. In some situations, GPS-U vehicles can't obtain their exact previous and current positions. In that case, these vehicles can't participate in the process of alarm message dissemination. However, they can obtain some information about the driving direction and the distance from the accident. This can help the driver to take decisions. For example, if the accident happens in the opposite driving direction according to the accident in a divided highway there will be no need to brake.

In order to obtain and refresh its position, a GPS-U vehicle, say S, periodically broadcasts a PREQ (Position Request) message to its one-hop neighbors. When a GPS-E vehicle receives a PREQ, it creates a PREP (Position Reply) message, includes its current position in that message, and sends it back to S. The knowledge of the exact position of S depends on the number and the positions (not all aligned) of neighbors sending PREP messages. S can compute its exact position if it receives at least three PREP from three different vehicles (Fig. 1).

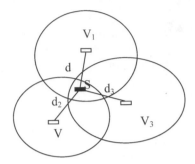

Fig. 1. Location using three non-aligned GPS-E vehicles

When S receives three PREP messages from three different vehicles, say V_1, V_2 and V_3, it uses a radiolocation method (i.e., signal strength) in order to determine the distances d_1, d_2 and d_3 from V_1, V_2 and V_3. In this case the exact position of S can be easily calculated.

The algorithm of IVG can be executed normally if the GPS-U vehicles can compute their positions. In fact, GPS-U vehicle uses PREP messages in order to get its position instead of GPS satellite. However this is not always possible because in some cases, where the number of PREP messages is less than three, the exact position cannot be known. In what follows, we study these cases, when S receives two, one, or zero PREP.

We suppose that S receives answers when it moves from a previous position, S_p, to a current position, S_c. To allow computation of positions and driving directions of vehicles, we distinguish the following situations:
– If S has two neighbors in S_p and three neighbors in S_c, or three neighbors in S_p and two neighbors in S_c, then the exact positions can be known.
– If S has three neighbors in S_p and one neighbor in S_c, or one neighbor in S_p and three neighbors in S_c, then one exact position S_p (Resp. S_c) can be calculated. The second position, called the lacking position, is the intersection of two circles. Hence, if this intersection is in one point, the exact value of the lacking position S_c (Resp. S_p) can be known. Else, the lacking position can be one of the two points of the intersection of the two circles. In some cases, even when the exact values of previous or current positions are not accurately known, the driving direction of vehicle S can be guessed. This is the case where the two possible solutions fall in the same driving direction.

4   Simulations and Analysis

In order to evaluate the performance of the IVG-U algorithm, we model a straight road 10 km long with C lanes in each direction. Each vehicle on the road moves at a constant, randomly chosen velocity. For sake of simplicity, we do not model complex maneuvers like lane changes and overtaking. Furthermore, we uniformly distribute the number of vehicles per kilometer per lane to model the traffic density in the road.

Since the knowledge of the position of a GPS-U vehicle depends on the number of its GPS-E neighbors, we derive a formula giving the mean number of GPS-E neighbors of a GPS-U vehicle: $N(\text{GPS-E}) = \tau \cdot \left(\dfrac{N}{10^3 W} \right) \cdot \left\lceil \overline{H} - 1 \right\rceil$, where $\overline{H}$ is the surface covered by a GPS-U vehicle and τ is the rate of GPS-E. The mean number of vehicles per m^2 is $(N/103W)$, where W is the width of the lane.

Fig. 2 shows the variation of the mean number of GPS-E neighbors of a vehicle according to the variations of the rate of GPS-E vehicles, transmission range and traffic density. We consider four situations according to the density of traffic (N=2, 4, 6 and 8) and four other situation according to the rate of GPS-E (τ = 0.2, 0.4, 0.6 and 0.8). We remark that the mean number of GPS-E vehicles is proportional to the transmission range and the GPS-E vehicles rate. We remark that when τ is greater than 60% that the mean number of GPS-E neighbors is greater that three even with a low transmission range (R=150). This means that all GPS-U vehicles can obtain their positions and IVG performs well.

Two other simulations with τ = 60% and τ = 40% that are not included here, show that with τ around 40% and traffic density is low (N=2) that the mean number of GPS-E neighbors can be less than three when the transmission range is less than 250m. In this situation, the performances are not optimal since not all the GPS-U vehicles can obtain their positions. However, we can envisage that the GPS-U vehicles increase their transmission power to reach ranges more than 250m in order to get more than two GPS-E neighbors, therefore they can compute theirs exact positions.

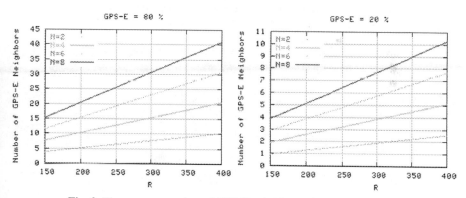

Fig. 2. The average number of GPS-E neighbors with different τ rates

For τ = 40%, curve shows that the number of GPS-E neighbors is always less than three even the transmission range is 400m when the traffic density is low (N=2). In this situation, not all GPS-U vehicles can compute their exact positions. Hence, these vehicles can't be relays in IVG, they are just passive elements.

5 Conclusion

In this paper, we propose an improvement to the basic IVG algorithm towards supporting its interoperability in environments where GPS-U vehicles are present. We show that the performances of IVG are optimal when a GPS-E rate is 60%. We also show that we can improve the performances of our method when GPS-E rate is 40% by the increase of the transmission range.

In some situation where GPS-E rate is less than 20%, the exact positions of such GPS-U vehicles cannot be known even with high transmission power. In that situations, we propose to let these vehicle as passive elements (they don't re-broadcast alarm messages) and we give them some information such as driving direction and distance from the accident. This information can help the driver to take decisions.

We are developing an extension to the ns-2 code of IVG in order to support the presence of GPS-U vehicles. Indeed, we believe that the performances of the proposed method are better than those presented in the mathematical analysis because in the real world some GPS-U vehicles can get their positions and help other GPS-U vehicles. This means that average number of GPS-E vehicles can be higher than the one presented in section 4. Thus the performance of IVG can be optimal even with less than 40% initially GPS-E vehicles.

References

1. L.Briesemeister and G. Hommel, "Overcoming Fragmentation in Mobile Ad Hoc Networks", Journal of Communications and Networks. Vol. 2, N° 3, pp. 182-187, September 2000.
2. M. Sun et al., 'GPS-based Message Broadcast for Adaptive Inter-vehicle Communications", Proc. of IEEE VTC Fall 2000, Boston, MA, 6:2685-2692, September 2000.
3. A. Bachir and A. Benslimane, "A Multicast Protocol in Ad-hoc Networks: Inter-Vehicles Geocast", IEEE VTC-spring 2003, Jeju, Korea, April 2003.
4. James J. Caffery and Gordon L. Stüber, "Overview of Radiolocation in CDMA Cellular Systems", IEEE Communications Magazine pp. 38-45, April 1998.
5. E. K. Wesel, "Wireless Multimedia Communications: Networking Video, Voice and Data", Addition-Wesley, One Jacob Way, Reading Massachusetts 01867 USA, 1998.
6. S. Venkatraman, J. Caffery and H.R. You, "Location Using LOS Range Estimation in NLOS Environments", IEEE VTC Spring, Birmingham, AL, May 2002, pp. 856-860.
7. M.P. Wylie and J. Holtzman, "The non-linear sight problem in mobile location estimation", 5th IEEE International Conference on Universal Personal Communication, 1996.
8. S. Capkun, M. Hamdi and J-P. Hubaux, "GPS-free positioning in mobile ad hoc networks", Hawaii International Conference on System Sciences, 2001.

Cartesian Ad Hoc Routing Protocols*

Larry Hughes, Kafil Shumon, and Ying Zhang

Department of Electrical and Computer Engineering
Dalhousie University
Halifax, Nova Scotia, B3J 2X4, Canada
{larry.hughes,kshumon,yzhang}@dal.ca

Abstract. As ad hoc networks gain in popularity, some of their limitations are becoming apparent, notably power and bandwidth restrictions. Consequently, it is necessary to utilize protocols that reduce power consumption, reduce traffic, and restrict flooding. In this paper, two adaptive, connectionless protocols and their supporting subsystems are described. The protocols, when used with directional antennas, can reduce the number of nodes involved in a transmission, thereby addressing the issue of power consumption and bandwidth utilization.

Keywords: MANET, location awareness, direction awareness.

1 Introduction

A mobile ad hoc network (MANET) is a collection of wireless mobile nodes that are capable of communicating with each other without the use of a network infrastructure or any centralized communication [1]. Like most wireless networks, a MANET is both power and bandwidth sensitive. Communication in a MANET poses special challenges because the network is infrastructureless and topologically dynamic. Energy conservation also plays an important role in the performance of ad hoc networks since most mobile hosts are battery operated. In a relatively dense network with many nodes lying between the source and the destination, these two problems become even more prominent. A number of MANET protocols have been proposed, including on-demand protocols for saving bandwidth, such as DSR [2] and CBRP [3], and for power saving, such as power-aware localized routing [4] and energy conserved routing [5].

Cartesian Ad hoc Routing Protocols (CARPs) are a set of three adaptive, connectionless protocols that address the problems of routing and power consumption in MANETs; they are loosely based on the Cartesian Routing Protocol [6]. Each protocol operates at the physical layer (using directional antennas) and the network layer (through its adaptive protocols); all nodes are location and direction aware. The protocols designed for CARP have three objectives: restrict flooding, reduce power consumption, and reduce traffic. Due to space restrictions only two of the protocols are presented in this paper.

---

* This research is supported by an Atlantic Innovation Fund research grant as part of the Computer Networks and Services Research programme.

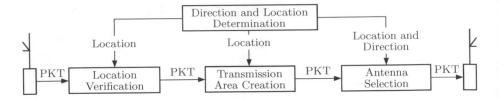

Fig. 1. CARP data flow diagram.

2   CARP

All Cartesian Ad hoc Routing Protocols attempt to restrict transmission to those nodes that lie between the source and the destination. First, a directional antenna is used to create a *bounding box* with a horizontal beamwidth of 90°. Next, the protocol is used to limit the number of forwarding nodes in the bounding box by creating a *transmission area*. The source and destination nodes are at opposite ends of the transmission area; each node within the transmission area is referred to as an *intermediate node*. A *current node* is a node that is forwarding a packet.

Fig. 1 shows the CARP subsystems. When a source node is to transmit a packet, it uses the Transmission Area Creation subsystem to determine the transmission area. Antenna Selection is then employed to select the antenna facing the destination. The Location Verification subsystem of each intermediate node determines whether the node is within the transmission area; if it is, the steps used by the source node are repeated. This process continues until the packet reaches the destination[1].

In addition to its payload, a CARP packet consists of the source address, the destination address, and transmission area information. At a minimum, the transmission area information is the address of the current node (x_c, y_c).

3   Transmission Area with Limiting Angle

If the transmission area has the same shape as the bounding box, unnecessary transmissions may occur especially in dense network. To reduce the number of potential intermediate nodes in the transmission area, the following protocol attempts to restrict the size of the area by employing a *limiting angle*.

The limiting angle, ϕ, defines the shape of the transmission area between the current node, C, and the destination node, D, as shown in Fig. 2. Each intermediate node forms an angle ϕ_i with the current node and the destination node.

---

[1] Since the destination may move during a transmission, a circular *expected zone* is created [7]. Unless otherwise indicated, the expected zone and its related calculations are beyond the scope of this paper.

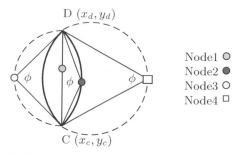

Fig. 2. Transmission area with limiting angle.

Table 1. Nodes in the network.

Value of ϕ	Shape of Transmission Area	Example	Path Length
$\phi = 180°$	line connecting current and destination	Node1	Shortest
$90° < \phi < 180°$	two symmetric minor arcs	Node2	
$\phi = 90°$	circle	Node3	$\downarrow$
$\phi < 90°$	two symmetric major arcs	Node4	Longest

3.1   Transmission Area Creation Subsystem

The value of ϕ is determined by the source. Table 1 shows the relationship between ϕ and the shape of the corresponding transmission area.

As the value of ϕ decreases, the size of the transmission area increases, potentially adding more nodes to the area, increasing the possible route length, and the number of packets. Therefore, there is a trade-off between the robustness of the protocol and the volume of traffic. Different ϕs can be defined based upon the density to determine the shape of the transmission area: the greater the density, the larger the value of ϕ.

Initially, the source node assigns its value of (x_s, y_s) to (x_c, y_c), while the intermediate nodes assign their address, (x_i, y_i), to (x_c, y_c) if they are to forward the packet. The transmission area information for this algorithm includes the value of ϕ_s (the value of ϕ determined by the source).

3.2   Location Verification Subsystem

When a packet arrives at an intermediate node, (x_i, y_i), it contains the limiting angle, ϕ_s, and the addresses of the destination and current nodes, (x_d, y_d) and (x_c, y_c), respectively. From this, the intermediate node can determine its value of ϕ_i as follows:

$$\phi_i = \arctan \frac{(y_c - y_i)(x_d - x_i) - (y_d - y_i)(x_c - x_i)}{(x_c - x_i)(x_d - x_i) + (y_c - y_i)(y_d - y_i)} \tag{1}$$

ϕ_i is then compared with the packet's ϕ_s. If $\phi_i > \phi_s$, the packet will be forwarded; otherwise it is discarded.

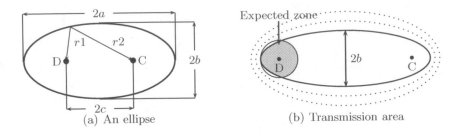

(a) An ellipse   (b) Transmission area

Fig. 3. Fixed path length shapes.

4   Transmission Area with Fixed Path Length

As well as making the transmission area with a limiting angle, the area can also be determined from the path length. Since nodes with a fixed path length form an ellipse, the second CARP algorithm uses an ellipse as the transmission area.

In Fig. 3(a), the current and destination nodes are two foci of an ellipse; the distance between these two nodes is $2c$. The major axis of the ellipse is $2a$ and the minor axis of ellipse is $2b$.

The following equations are of interest:

$$r_1 + r_2 = 2a \tag{2}$$
$$b^2 + c^2 = a^2 \tag{3}$$

All the nodes located on the ellipse boundary have the same path length $2a$, as shown in equation 2, while nodes located inside the ellipse have a shorter path length. These nodes are inside the transmission area.

An expected zone is defined as the overlapping area of a circle (centred at the destination) and the ellipse as shown in Fig. 3(b).

The transmission area information for this algorithm is the current node address (x_c, y_c).

4.1   Transmission Area Creation Subsystem

The parameters a, b and c determine the shape of the ellipse; however since they are correlated as illustrated in equation 3, if any two of them are known, the third can be calculated.

The value of a is related to the radius of the expected zone, r, and the distance between the source and the destination, $2c$. Since the positions of the current and destination nodes are assumed to be fixed at the transmission of the packet, a is determined from the radius of the expected zone, r, which is related to the speed of the destination [7].

An ellipse in a sparse network has a larger value of b than that in a dense network to include more nodes in the area. Fig. 3(b) shows the transmission area with different values of b in networks with different densities.

When an intermediate node forwards the packet, it substitutes the current node co-ordinates with its own to create the transmission area for the next hop

4.2   Location Verification Subsystem

When an intermediate node receives a packet, it calculates the following:

- its distance to the source $r1$ and to the destination $r2$
- distance between source and destination $2c$
- major axis of the ellipse $2a = 2c + 2r$

If $r1 + r2 < 2a$, the node is inside the transmission area and is to forward the packet towards the destination; otherwise it is to discard the packet.

5   Supporting Hardware

Each CARP node must be direction and location aware, in addition, it needs to select the proper antenna(s) for packet transmission.

5.1   Direction and Location Determination Subsystem

The Direction and Location Determination subsystem consists of two distinct units.

A *Direction Unit*, which is responsible for determining magnetic North to make the node direction aware. A magnetoresistive sensor chip can be employed to act like an electronic compass [8]. The compass has a fixed orientation with the antenna subsystem (described below) so that the direction in which each antenna is facing is always known. The sensor gives a deviation angle of $0°$ while facing towards the earth's magnetic North and the angle of deviation increases as the antenna module rotates clockwise and resets after each complete rotation.

The *Location Unit* is responsible for determining the location of the node. Any location detection system, such as GPS [9], can be used to provide the location co-ordinates.

5.2   Antenna Selection Subsystem

This subsystem selects the proper antenna or antennas in the antenna module by taking the destination coordinates from the packet and the local node and direction information provided by the direction and location determination subsystem. The antenna module consists of four directional antennas with each having a horizontal beamwidth of $90°$ and a vertical beamwidth of $180°$.

The appropriate antenna or antennas are then chosen as follows. First, the angle of inclination (θ) is determined with reference to the x-axis between the current and destination nodes using their coordinates. Since a positive inclination with reference to the x-axis is required, $180°$ is added to θ if θ is less than $0°$. Then the angle of inclination is conditioned to determine the direction of the destination node. Next the angle of deviation of the compass is added to θ. Finally, θ is conditioned to be in the range from $0°$ to $360°$.

Once the final θ is calculated the selection of the antenna or antennas can be made easily. When θ is a multiple of $90°$, the two antennas on two sides of the angle are chosen.

6 Concluding Remarks

This paper described two of the Cartesian Ad hoc Routing Protocols. These are adaptive and connectionless routing protocols which:

- restrict any flooding to within the transmission area.
- reduce power consumption of nodes outside the transmission area, since they are not involved in the communication.
- reduce the number of nodes in the communication by dynamically adjusting the transmission area and deploying directional transmission.

In this paper, it has been assumed that the intermediate nodes have a uniform density between the source and destination; however, in a real network environment, this may not be the case. For example, the number of intermediate nodes may appear to be dense, when in reality, there may be a peak around the source or destination only. We are in the process of examining non-uniform network densities with the OPNET modelling tool.

References

1. Ilyas, M., ed., *The Handbook of Ad Hoc Wireless Networks*, CRC Press, Jan. 2003.
2. Johnson, D.B. et al., *The dynamic source routing protocol for mobile ad hoc networks*, IETF Internet Draft. http://www.ietf.org/internet-drafts/draft-ietf-manet-dst-02.txt, 1999. Accessed 12 June 2003.
3. Jiang, M. et al, *The cluster based routing protocol (CBRP) for ad hoc networks*, IETF Internet Draft. http://www.ietf.org/internet-drafts/draft-ietf-manet-cbrp-spec-01.txt, 1999. Accessed 12 June 2003.
4. Stojmenovic, I. and Lin X., *Power-aware localized routing in wireless networks*, IEEE International Parallel and Distributed Symp., 2000.
5. Chang, J-H, and Tassiulas, L., *Energy conserving routing in wireless as hoc networks*, Infocom 2000.
6. Hughes, L. et al, *Cartesian Routing*, Computer Networks, vol. 34, pp. 455 - 466, 2000.
7. Ko, Y.B., and Vaidya, N.H., *Using Location Information In Wireless Ad Hoc Networks*, IEEE 49th Vehicular Technology Conference, pp. 1952-1956, vol. 3, 1999.
8. Stork, T., *Electronic Compass Design Using KMZ51 and KMZ52*, http://www.semiconductors.philips.com/acrobat/applicationnotes/AN00022_COMPASS.pdf. Accessed 13 June 2003.
9. *Instant GPS*, http://www.motorola.com/ies/GPS. Accessed 13 June 2003.

Author Index